FROSTBITE

ALSO BY NICOLA TWILLEY

Until Proven Safe
(cowritten with Geoff Manaugh)

FROSTBITE

How Refrigeration Changed Our Food,

Our Planet, and Ourselves

NICOLA TWILLEY

PENGUIN PRESS

NEW YORK

2024

PENGUIN PRESS
An imprint of Penguin Random House LLC
penguinrandomhouse.com

Some of the material in this book previously appeared, in different form, in the
following publications: *The New York Times Magazine, The New Yorker, Cabinet,*
and *The Container Guide* (eds. Craig Cannon and Tim Hwang)

LIBRARY OF CONGRESS CATALOGING-IN-PUBLICATION DATA
Names: Twilley, Nicola, author.
Title: Frostbite: how refrigeration changed our food,
our planet, and ourselves / Nicola Twilley.
Description: New York: Penguin Press, 2024. |
Includes bibliographical references and index.
Identifiers: LCCN 2023048094 (print) | LCCN 2023048095 (ebook) |
ISBN 9780735223288 (hardcover) | ISBN 9780735223295 (ebook)
Subjects: LCSH: Refrigeration and refrigerating machinery—History. |
Refrigeration and refrigerating machinery—Social aspects—History. |
Cold storage industry—History. | Food supply—Social aspects—History.
Classification: LCC TP492.7 .T95 2024 (print) | LCC TP492.7 (ebook) |
DDC 621.5/609—dc23/eng/20231107
LC record available at https://lccn.loc.gov/2023048094
LC ebook record available at https://lccn.loc.gov/2023048095

Printed in the United States of America
2nd Printing

Designed by Amanda Dewey

For Geoff, for everything

CONTENTS

1 | WELCOME TO THE ARTIFICIAL CRYOSPHERE   1

2 | THE CONQUEST OF COLD   25

I. Stop the Rot   27

II. The Ice Harvest   40

III. A Machine to Produce Cold   55

3 | THE WAY OF ALL FLESH   63

I. Where's the Beef?   65

II. Better Living through Chemistry   82

III. When Muscle Becomes Meat   95

4 | INSIDE THE TIME MACHINE   109

I. Sleeping Beauties   111

II. Oracular Bananas   131

III. Trading Futures   146

5 | A THIRD POLE   161

    I. Meet the Thermo King   163

    II. Reefer Madness   178

    III. Building a New Arctic   191

6 | THE TIP OF THE ICEBERG   215

    I. Cold Case   217

    II. Freshness Guaranteed   230

    III. The Taste of Cold   243

    IV. The Fridge Diet   260

7 | THE END OF COLD   277

    I. The Future of Refrigeration   279

    II. The Future May Not Be Refrigerated   301

Epilogue: Meltdown   313

    Acknowledgments   323
    Selected Sources   329
    Index   373

FROSTBITE

1.

WELCOME TO THE ARTIFICIAL CRYOSPHERE

My first day at Americold's warehouse in Ontario, California, began promptly at 8:00 a.m. Outside, it promised to become the kind of blue-skied, seventy-degree March day that makes greater Los Angeles feel like the paradise it plays on TV. Inside, Anthony Espinoza, the facility's general manager, warned me it was just three degrees in the coolers and between thirty-six and thirty-eight degrees on the dock. "It's minus ten in the freezer," he added, radiating good cheer. "That's the tundra."*

My expedition into the artificial cryosphere—the vast synthetic winter we've built to preserve our food—began here, with a week of shift work in the refrigerated warehouses of Southern California. Americold is one of the largest providers of temperature-controlled warehouse space, not only in the United States but around the world. Globally, the company maintains 1.5 billion cubic feet of cold, storing everything from ground beef destined for school lunch programs to frozen lobsters on their way to upscale restaurant chains like McCormick & Schmick's. In Ontario, most of the 100,000-square-foot warehouse is given over to Danone products: pallet after pallet of Horizon chocolate milk, Land O'Lakes creamer, Silk soy milk, and Greek yogurt, much of which comes from a plant just forty-five minutes away. "They focus on creating food,"

---

*Throughout the book, temperatures are given in Fahrenheit.

explained Espinoza. "We focus on making sure it gets to their customers intact."

Espinoza and his warehouse manager, Kyle Schwedes, had already welcomed two new recruits the day before. "I told them it's very cold, it's very physical, it's very demanding," said Schwedes. He and Espinoza look for a few essential attributes in warehouse candidates. Interpersonal skills and attention to detail are important, but experience driving a stand-up reach forklift is nonnegotiable.

The other essential test is, of course, how the would-be warehouse worker reacts to the frigid temperatures. "I love the cold," said Espinoza. "It preserves me! All our guys have a youthful appearance." (Much to his delight, I had guessed his age as a full decade younger than he actually is.) Still, a lot of people are just not cut out for a career spent inside a gigantic fridge. Later, once I was out on the floor myself, a shift supervisor named Amato took me under his wing. He told me he'd seen dozens of new recruits leave after only a couple of hours in the chiller. "They take off their coats at lunchtime, and poof! They're gone," he said. "It's the rare person that lasts here."

It's the rare person who steps inside a refrigerated warehouse in the first place. Over the course of nearly two decades spent reporting on food for national and international newspapers and magazines, as well as for my own podcast, I've been lucky enough to visit all sorts of unusual, behind-the-scenes locations, from farms and factories to sourdough libraries, experimental orchards, and military labs. Still, until I began writing this book, I had never so much as peeked inside anything larger than a walk-in fridge. You probably have not either.

The refrigerated warehouse is the missing middle in food's journey from farm to table: a black box whose mysterious internal workings allow perishable food to conquer the constraints of both time and space. Even those chefs who are proud to know the life story of each steak they serve, or the foodies who insist on meeting the farmer who raised the meat they eat, would never dream of inquiring as to its storage history—or imagine that beef carcasses have to be electrocuted in order to withstand the rigors of refrigeration without toughening up. Similarly, you may be familiar with the full range of lettuce varietals in the bag of supermarket spring mix sitting in your crisper drawer, but I'd be willing to bet you have no idea that the bag itself is a highly engineered respiratory apparatus, designed in layers of differentially semipermeable films to slow spinach, arugula, and endive metabolism and extend their shelf lives. I certainly didn't.

The cold chain—as the network of warehouses, shipping containers, trucks, display cases, and domestic fridges that keep meat, milk, and more chilled on their journeys from farm to fork is technically known—has become such an essential part of our food system that it is taken for granted. Its extent and operations are opaque even to our elected leaders. During the early days of the COVID-19 pandemic, as supermarket shelves emptied, one industry expert told me that he received a frantic call from a senior official in the British government, asking how many refrigerated trucks and warehouses the country's food supply depended on.

This inadvertent blind spot is a big—and dangerous—mistake. As I've discovered while traveling the world to research this book, it's impossible to make sense of our global food system until you understand the mysterious logic of the all-but-invisible network of thermal control that underpins it. What we eat, what it tastes like, where it's grown, and how it affects both our health and that of our planet: these things shape our daily lives as well as our continued existence as a species, and they've been entirely transformed by manufactured cold.

In 2012, the Royal Society—the UK's national academy of science—declared refrigeration the most important invention in the history of food and drink. Judged in terms of its impact on a range of criteria, including productivity and health, refrigeration was deemed more significant than the knife, the oven, the plow, and even the millennia of selective breeding that gave us the livestock, fruits, and vegetables we recognize today. It is also a much more recent development: our ancestors learned to control fire before modern humans even evolved, but our ability to command cold at will dates back little more than 150 years. Mechanical cooling—refrigeration produced by human artifice, as opposed to the natural chill offered by weather-dependent snow and ice—wasn't achieved until the mid-1700s, it wasn't commercialized until the late 1800s, and it wasn't domesticated until the 1920s.

Today, a century later, nearly three-quarters of everything on the average American plate is processed, packaged, shipped, stored, and/or sold under refrigeration. The United States already boasts an estimated 5.5 billion cubic feet of refrigerated space—a third polar region of sorts. This is an almost unimaginably large volume: the tallest mountain on Earth, Everest, occupies only roughly two-thirds that amount of space from base to peak.

As the developing world begins to build its own American-style cold chains, the expansion of this manufactured Arctic has accelerated. According to the most recent statistics from the Global Cold Chain Alliance, the world's chilled and frozen warehouse space increased by nearly 20 percent between 2018 and 2020—a leap that still left most of the planet's citizens provisioned using less than a sixth of the cold-storage capacity required to feed the average American. (The standard domestic fridge, at between twenty and twenty-five cubic feet, is just a tiny closet in the distributed, McMansion-sized pantry it takes to preserve our perishables.) While ecologists and explorers concern themselves with the shrinking natural cryosphere—Earth's frozen poles and permafrost—this alternate, entirely artificial cryosphere is expanding virtually unnoticed, all around us.

My fascination with the cold chain began about fifteen years ago, when the farm-to-table movement was picking up steam in the media. While my fellow food journalists were writing about feedlots and fast food, or locavores and edible schoolyards, I got stuck on the conjunction. What about the *to*? What happened between the farms and the tables?

Almost immediately, I realized that my humble home fridge was merely the tip of the iceberg. Considered as a connected whole, the cold chain seemed to me as worthy of awe as the Pyramids of Giza—a continuous monument of engineered winter that has remade our entire relationship with food, for better and for worse. Industry insiders clearly had a grip on its geography and mechanics; a handful of historians had traced particular elements of its evolution; scattered scientists were engaged in analyzing its various effects on everything from the flavor of tomatoes to the contents of our gut microbiome; and, more recently, policymakers have begun to worry about its environmental impact. Yet, to my surprise, no one had tried to tie those threads together into a coherent narrative— a story that could help us understand the extraordinary scope and implications of the refrigeration revolution.

As I pored over my copy of the International Association of Refrigerated Warehouses' annual directory of member facilities, I couldn't help but wonder: Where are the Shackleton and Scott of the artificial cryosphere? Why is no one embarking on bold expeditions into its farthest-flung corners, braving its icy wastes, mapping its unexplored contours, meeting its inhabitants, and chronicling its customs? Then I realized that perhaps I should put on some thermal underwear and do it myself. This book is the result.

In the course of my adventures, I have been inside places most of us will never see and met people most of us have never heard of—despite the fact that they stand between us and hunger. As we continue together on the following pages, we'll visit the landmarks of the artificial cryosphere, from the vast caves where Kraft stores America's national cheese reserve to the Arctic vault where the future of farming is safeguarded by refrigeration. We'll tease out cooling's unwritten history, rummaging through untouched archives and tracking down forgotten pioneers. We'll

get to know forklift operators, fridge designers, a frozen-dumpling billionaire, and the world's only refrigerator dating expert. Most important,
we'll arrive at an understanding of the true stakes of our refrigerated food
system.

Of course, humanity's mastery of cold has been turned to many other
fascinating uses over the past century, from data centers to medicine, air-
conditioning to ice rinks. In this book, I have limited my inquiries to
food alone: the means by which refrigeration has created the most radical change in how we live. As the Royal Society said, explaining the reasoning behind its decision, refrigeration has been a blessing, "responsible
for bringing a more varied, interesting, nutritious, and more affordable diet
to an ever increasing number of people."

But as the developing world undergoes the transformation that the
United States underwent during the twentieth century, it is time to make
a full accounting of the cold chain's costs, as well as its benefits. Refrigeration has changed our height, our health, and our family dynamics;
it has reshaped our kitchens, ports, and cities; and it has reconfigured
global economics and politics. It spawned Tupperware and the TV dinner, it served as midwife to the shopping trolley and the hoodie, and it
sounded the death knell for several species. Most urgent, mechanical
cooling makes a growing and significant contribution to global warming, based on the power required to run it as well as the super-greenhouse
gases that circulate within many cooling systems. With unfortunate
irony, the spread of the artificial cryosphere turns out to be one of the
leading culprits in the disappearance of its natural counterpart.

To its earliest pioneers, control of cold endowed humankind with
godlike powers over the otherwise immutable forces of decay and loss,
unlocking limitless abundance by removing the constraints of distance
and the cycles of seasonality. Today, our dietary dependence on refrigeration is almost complete—and human control over nature has never
seemed less sustainable. Cooking may have made us human but, to mis-
quote Paul Theroux's utopian protagonist in *The Mosquito Coast*, is ice

really civilization? What would happen to our dinner plates, our cities, and our environment if we cast off its frosty fetters?

First of all, though, what actually goes on inside the (usually white) boxes that house the artificial cryosphere?

Before I was allowed anywhere near Americold's refrigerated warehouse floor, I had two hours of safety training to complete. Warehouse work is already one of the most dangerous jobs in the United States, and many of those risks can be traced back to the forklift. These little cubes on wheels look like oversize bumper cars with two silver prongs attached, but they are surprisingly tricky to operate. Tweaking the angle of the fork so that the truck doesn't tip over when reaching for a heavy pallet load depends on experience and intuition. Steering is done using two levers, both of which are incredibly sensitive; on one of them, the controls are also inverted, so that a left turn will take you to the right. "If you want some horror, watch YouTube forklift accidents," said Anthony Espinoza. "If you crash into the racking hard enough to knock it over, you get a domino effect and the entire roof will come down."

In addition to the standard forklift-driving and pallet-unloading accidents, the cold-storage environment presents dozens of additional risks. In a frozen warehouse, the floor glitters with ice crystals, leading to slips and falls. The ammonia used in the refrigeration system is deadly. A few years earlier, Espinoza told me, he'd experienced a chemical leak when a pipe was accidentally ruptured by an out-of-control forklift. Within three minutes, the entire dock was filled with a white cloud. "When you see that, you're seeing death," he said. "Ammonia wants moisture—it wants your eyeballs and your crevices."

The biggest challenge, however, comes from the very same quality that makes refrigeration so powerful: cold's ability to slow everything down. The microbes and enzymes that would normally be spoiling the yogurt and curdling the milk become sluggish in the chilled air, but so do the humans charged with loading and unloading those dairy products. Even computers cease to function in the deep freeze, so companies like Honeywell produce a special range of barcode sensors and laptops equipped with internal heaters and screen defrosters. At minus twenty and below, tape doesn't stick properly, rubber becomes brittle, cardboard is stiffer—and all those minor obstacles seem more like insurmountable challenges to a cold-slowed brain.

A medical mnemonic describes the effects of excessive cold on the human body as the "umbles": the underdressed or overexposed individual starts to grumble, mumble, fumble, and stumble. "Cold stupid" is mountaineering slang for the way that thought processes congeal after spending too long at a low temperature. As early as 1895, the cold-storage industry's first trade journal, *Ice and Refrigeration*, pointed out that "extreme cold, as is well known, exerts a benumbing influence upon the mental faculties." By way of example, the author referred to an account of the retreat of Napoleon and his troops from Moscow, during which a doctor noted that, at five degrees, "many of the soldiers were found to have forgotten the names of the most ordinary things about them." For context, the average frozen food warehouse is held between five and twenty degrees below zero, although specialist facilities for storage of particularly delicate foods such as tuna can go as low as minus eighty; the South Pole averages minus seventy-four during its chilliest months; while the mean temperature at the summit of Mount Everest in winter is a comparatively balmy minus thirty-one degrees.

As polar explorers and mountaineers know all too well, long before amnesia sets in—let alone hypothermia or frostbite—human performance slips when the mercury drops. The energy needed for fast and focused exertion is siphoned off to help maintain body heat. Espinoza told me that engineers at Americold HQ carefully calculate work plans to take that lowered productivity into account. The persistent background discomfort of being cold—numb fingers and toes, runny noses, and teary eyes—is distracting. Cold also inhibits peripheral vision, reaction times, and coordination, a phenomenon researchers blame in part on slowed neural connectivity but mostly on the fact that, when the brain is focused on bodily discomfort, it can't really concentrate on anything else. Compared side by side, employees in a refrigerated warehouse will move and think more slowly, and are more likely to punch the wrong buttons on their forklift trucks and touch screens, than their counterparts in a dry-goods facility.

This effect seems to hold true throughout the natural world: speed, agility, and mental acuity are directly correlated, via metabolic rates, with body temperature. Colder almost always means slower and dumber. One recent study showed that warm-blooded marine predators such as seals and whales tend to cluster in the coolest parts of the ocean, not because they find the chill congenial but rather because, under those conditions, their piscine prey is "slow, stupid, and cold"—and thus easier to catch.

Even when one is not being hunted by killer whales, cold-induced lapses can be deadly, as depicted in what might be the refrigerated warehouse's most high-profile literary appearance. A chapter of Tom Wolfe's 1998 novel, *A Man in Full*, is devoted to the struggles of the young Conrad Hensley as he stoically heaves beef shanks, frozen into eighty-pound bricks, during a night shift in Croker Global Foods' freezer unit, in Richmond, California. In his first few months on the job, Hensley has already seen colleagues felled. One "had wrenched his back doing practically nothing, and now he couldn't walk," he recalls before heading in for his shift. "Last week one of the Okies, Junior Frye, had had his ankle crushed by a pallet sliding across a patch of ice." That night, after a few hours in the "frigid gray dusk" of what Hensley has dubbed "the Suicidal Freezer Unit," a colleague turns his jack too fast, spins out of control on the frosty floor, and nearly kills them both.

In one of the training videos, an Americold associate named Jason Carter related his own cautionary tale—a "fifteen-second mess-up" that led to seven and a half hours in the trauma unit. It was the Saturday of

Labor Day weekend, and he had loaded one pallet too many on his fork-lift. "I caught a bar, hit the back of my head, and knocked into my computer screen," he explained. "I broke thirty-seven different bones and both eye sockets—my kids were real tore up by it."

Espinoza told me that both safety and morale had improved dramatically in recent years. A sign near the entrance proclaimed the facility had gone 1,045 days without incident. "Still, it's a tough environment," he said. "It's not all blue skies and roses."

Once I was deemed ready to start my first shift at Americold, I was introduced to my buddy, an old-timer named Jimmy Ambrosi. Together we went through one final safety check and some warm-up stretches. Then I followed him into room 1, fumbling my way through the curtain of heavy-gauge vinyl strips that separate the docking zone from the coolers. "Welcome to Disneyland," Ambrosi yelled as I gazed up into a series of narrow slot canyons whose walls were made of Stonyfield Farm, Dannon Light + Fit, COCOYO nondairy, and Oikos Greek yogurts. "Just under seven thousand pallet positions and we're ninety-five percent full. I'm going to guess there's a hundred million dollars' worth of product in here right now," he calculated. At the supermarket, it's normal to encounter more yogurt than I would eat in a year. This was more yogurt than I could eat in a lifetime: thousands upon thousands of cartons, all packed into chunky cardboard cubes on wooden pallets, each cube swathed in plastic wrap and stacked on steel racking, reaching three stories up to the ceiling's girder skeleton and receding into the distance as far as the eye could see.

Lighting uses energy and emits heat, so a perpetual blue-gray gloom prevailed inside the windowless cooler and freezer rooms. Pools of deeper blue light traveled along the icy concrete floor, projected from the LED spotlights mounted on each forklift truck, in order to forewarn others of its emergence from the canyon depths. Each time a forklift reversed, a chorus of beeps pierced the unending roar of enormous fans. Everything seemed dimmed and muffled—even the air felt dense.*

It also smelled funny: a distinctive, slightly metallic base note that I grew to recognize as the underlying smell of the artificial cryosphere. Everyone who works in cold storage knows it, even if they struggle to describe it. "There's no way to explain it to someone who hasn't been in there for an eight-hour shift," one industry veteran told me. "It just smells kind of weird." "It's a pleasant odor—or at least it isn't nasty," said Adam Feiges, who grew up in his family's cold-storage business before selling it to Americold. "It's cardboard, wood, foam insulation, oil, and what I always just think of as the smell of cold."

At Americold, new hires are assigned to stocking—moving freshly delivered perishables into the cooler or freezer—and picking, or getting them out again, for their first ninety days, so that's where I began. The racks store pallets three deep and six high, and product has to go out in the date order it came in. The Ontario facility typically receives 120 truckloads a day, and because it runs more or less at capacity, putting new pallets away often means shuffling old pallets around first, as if at a valet car park. Ambrosi's headset told us what to pick up and where it should go, and we used a scanner gun to log each pallet's final destination.

"The system we have isn't that smart," said Ambrosi. "It doesn't tell you what order to pick the pallet for maximum efficiency or to balance the weight load and compression." After a while, he promised me, I'd get the hang of it. "You have to marry up your tiers and heights," he explained, referring to the vertical and horizontal layout of the boxes we were stacking on each pallet. "It's all individual judgment, but there's different patterns we can use."

To make this daily game of warehouse Jenga even more challenging, certain products can't go next to each other. In the cooler, you have to

---

*There is a scientific explanation for this: in the cold, air becomes thicker, meaning that both sound and light travel more slowly through it.

make sure foods containing allergens—soy, nuts, dairy, wheat—aren't touching; in the freezer, that's okay. Organic products shouldn't sit underneath conventional ones; raw foods mustn't be stacked above cooked. "You have to think about odor," added Espinoza. "Onions and seafood can be quite potent." Pizza sauce and pepperoni are also disturbingly pungent: just a few hours spent picking Schwan's Big Daddy's Pepperoni and Freschetta Supreme sausage frozen pizzas was enough to make my woolen beanie and furry coat collar stink for days. Like natural fibers, bread and cheese have a tendency to absorb the odors to which they're exposed, as does ice cream, which can't even be stored in the same room as the pizzas.

"Ice cream's a whole different level of complexity," said Espinoza. "It's mostly air, so you can't stack it because it will compress."* Making sure food at the bottom of the pallet doesn't get squashed by the weight of the food above is another requirement for successful stocking, as is ensuring that the finished cube is evenly weighted, so as not to risk tipping the forklift.

Determined to earn my cryospheric credentials, I was thrilled when I took just 24 minutes and 40 seconds of my HQ-allotted 29:32 to complete my first picking task. But after less than half an hour in the freezer, the chill had crept in. My fingers and toes were numb; my nose wouldn't stop running. "New guys will put tissue up their noses," said Amato. "You see guys who want to work, and they just can't handle the cold." Men with facial hair grew miniature icicles in their mustaches and beards; the one guy wearing glasses had to stop and wipe off the condensation from his breath every few minutes. Everyone gets sick in their first

---

*In nonpremium brands, a pint of ice cream is, on average, 50 percent air. This leads to all sorts of logistical complications. National brands of ice cream have to use different formulations for different regions to take into account the thinner air at higher elevations. "You can't truck it from Washington to Georgia," Espinoza told me. "The Rockies," he explained, shaking his head.

few months on the job, I was told; colds, coughs, and "freezer flu" are a year-round phenomenon in the cold-storage industry.

Although the idea that *working* in the cold would lead to *catching* cold makes intuitive sense, scientists have only just discovered why. Until recently, the general uptick in respiratory illness in the winter was blamed on the fact that people spend more time indoors together, swapping viruses, in cooler weather. That's likely a factor, but cold is also directly, not just indirectly, responsible for making us sick, thanks to a previously unknown immune mechanism: cells in our nostrils that are capable of detecting incoming microbes and releasing a swarm of tiny little antiviral bubbles to surround and neutralize them. According to the Boston-based team behind the breakthrough, at forty degrees, nostril cells release significantly fewer and less potent defensive bubbles than they do at seventy-five degrees, making it easier for viruses to stage a successful infection.

This recent discovery aside, much remains to be learned about the health impact of working in the cold or the ways in which humans adapt to life in the cryosphere, either natural or artificial. Over the years, researchers studying Arctic explorers, lumberjacks, miners, and refrigerated warehouse employees have consistently found that exposure to cold results in a rapid pulse rate as well as higher blood pressure, as the heart works harder to pump cold-thickened blood around cold-constricted blood vessels.

Cold stress can literally be measured in the blood of forklift truck operators: scientists found that plasma levels of the fight-or-flight hormone noradrenaline were significantly higher after a shift in the freezer, as opposed to the same amount of time in an ambient warehouse. Muscles contract and tendons tighten in the cold, making them more prone to strains and tears; inhaling cold air can trigger bronchial spasms, inducing asthma or even a chronic pulmonary condition known as Eskimo lung. There's even some evidence that low temperatures wreak psychological harm by increasing feelings of loneliness, rejection, and exclusion.

After fifteen years in the Siberian gulag, including "solitary confinement in ice," Russian writer Varlam Shalamov concluded that "the main means for depraving the soul is the cold."

On the other hand, a growing body of research is focused on the ways in which exposure to cold temperatures improves insulin sensitivity and blood-sugar control. Although references to the health benefits of cold-water immersion date back to ancient Egypt, over the past decade, ice baths and so-called wild swimming have become the focus of both medical research and social media hype as something of a cure-all. Denis Blondin, a Quebecois scientist and former personal trainer whose research is focused on the therapeutic impacts of cold exposure, told me that when he sits people in a specially designed cooling suit for three hours, they experience the metabolic equivalent of a medium exercise training session. "It's the combination of muscle contractions from shivering and the stimulus that you're getting from the cold that really improve your ability to handle glucose in your blood," he said. "There are some changes to your lipid profiles, too, but there's not enough data on that yet."

Still, despite the growing body of evidence for cold's benefits, Blondin cautioned me against the conclusion that working a shift in an Americold freezer would automatically make me healthier. Participants in Blondin's studies spend their three hours of low-temperature exposure sitting still, absorbing cold's benefits without the need to pump their syrupy blood around narrowed vasculature or strain their cold-contracted muscles. Amato, Jimmy, and the rest of my new colleagues spend their eight hours in near-constant motion. According to Blondin, it's likely that both things can be true: "You can be at higher risk if you work in a cold environment *and* the cold can be beneficial."

Even the question of cold habituation is complicated. Dozens of studies seem to show that, after repeated exposure to freezing temperatures, Quebecois postmen and Japanese pearl divers alike have trained their bodies to work better in the cold. Other researchers have argued that the fact that they shiver less and feel warmer isn't due to some kind of physiological adaptation but rather that frequent exposure to the pain of cold has muted its warning signal in their brains, in the same way you might eventually tune out the itch of a mosquito bite.

What I can say is that, as the days went on, I did not get used to the cold at all. In fact, the discomfort seemed cumulative: the longer I spent inside an Americold warehouse, the more I wanted to get out. Carlos, a round-faced forklift operator, told me he found it easier to just not warm up. "Even in winter, I blast the AC in the car on the drive home," he said. "For the guys that come in on the bus, it's hell." Amato, who initially claimed that working in the cold was "like a giant ice pack—all your aches and pains go away," confessed that he still found having to spend an hour or two in the freezer pretty miserable. "Don't think about it," he advised me. "You have to focus on what you're doing or else you'll mis-pick or you'll go to the wrong location."

At the end of my first shift, torn between joy at my imminent return to ambient temperature and a deep-seated desire for approval, I asked Amato if he would hire me. "I would, and here's why," he said. "Your nose isn't very red."

A part from a buddy system and break-time group exercises in the forklift truck battery-charging room, which is the toastiest spot in the entire facility, the main way Americold tried to protect me against the cold's discomfort and dangers was by equipping me with specially de-signed clothing. I lumbered through the warehouse clad in heavy-duty boots, padded bib overalls, a hat, gloves, and an enormous jacket. The gear I was wearing carried the Americold logo, but it was all made by a company named RefrigiWear, which invented the concept of a cold storage–specific wardrobe and still dominates this category today. The jacket—a stiff, squarish black nylon anorak with a faux-fur collar, a sil-ver reflective strip around the midsection, and a built-in elasticated waistline—was potentially the least flattering item of clothing I'd ever owned. I couldn't have cared less, because, in print on the label, it promised to keep me comfortable at temperatures as low as minus fifty degrees.

"This is how we figure that out," Shawn Deaton told me, his southern drawl almost drowned out by the deafening roar of stainless-steel fans blowing cold air into a small glass box. I had traveled to the Thermal Protection Lab at North Carolina State University to visit one of the handful of facilities in the United States capable of measuring exactly how much warmth my RefrigiWear ensemble could offer. Deaton manages operations across the university's textiles complex, including its biological- and chemical-protection labs. As a literal warm-up, he had already introduced me to PyroMan: a flame-resistant, polyester resin and fiberglass humanoid model hanging, gimp-like, from heavy chains in a room-sized oven. A technician charged the gas lines in order to demonstrate, and I watched, open-mouthed behind fire-resistant glass, as a ring of eight flamethrowers ignited and, in seconds, enveloped PyroMan in a two-thousand-degree fireball.

Much of Deaton and his colleagues' research is focused on extreme heat. They've used PyroMan to test the protection offered by new foam insulation in firefighter uniforms and to study how best to balance burn prevention with portability in firefighting shelter design. But the North Carolina facility is also home to a couple of refrigerated rooms, and Deaton led me into one: a glass-walled, steel-floored cubicle a little longer and narrower than a regular parking spot. In the test chamber with us, suspended from a sturdy metal frame, was a life-size gray manikin wearing a black woolen hat, a navy-blue turtleneck, dad jeans, Velcro sneakers, and a pair of mittens.

"Here's one of our thermal sweating manikins," said Deaton. "This one has a breathing mechanism, so we call him Darth Vader. The other one is Anakin the Manikin, and of course we have Hand Solo for gloves." Vader looked oddly terrified—his deep, round eyes seemed to bore into me, until I returned his gaze directly and realized that his eye sockets were literally sockets. "He's a little freaky-looking," Deaton agreed. "Those are the hookups for the power and the controls." Both models are instrumented with 122 sensors that record "skin" temperature all over their bodies, as well as at their core. Deaton lifted Vader's turtleneck to reveal a flat silver body dotted with little pores, out of which "sweat" (distilled water, piped in through ports on his cheeks) can either leak slowly or flow profusely, as necessary.

We stepped outside the cubicle to view Darth Vader's vitals on a screen. Arriving at a temperature rating for a jacket is relatively simple: the chamber is refrigerated while the manikin is programmed to remain at ninety-five degrees, despite constantly losing heat to the cooler air all around it. Depending on the test, the manikin can also be programmed to simulate walking, with a sort of stiff-legged march. The sensors allow Deaton to see what parts of the manikin's body are losing the most heat, which helps garment designers fix problem areas, but they also let him calculate just how much energy is needed to keep its skin surface at ninety-five degrees. The difference between how hard it is for Vader to stay warm nude and while wearing a RefrigiWear jacket is the insulation rating of the garment.

Despite the technology's promise of precision, Deaton warned me that I shouldn't put too much faith in the resulting number. For one thing, both Darth Vader and Anakin are technically unisex, which in this case means a castrated male. Deaton and his colleagues are working to create an internationally recognized female manikin standard. "It hadn't been seen as necessary until recently," he explained. For another, each of us has our own metabolic rate, as well as varying degrees and forms of internal padding, surface area, and body hair, and even a particular ratio of slow- to fast-twitch muscle fibers, all of which add up to make an individual more or less cold tolerant. "If it says minus twenty, that doesn't necessarily mean you're going to be comfortable at minus twenty," Deaton said. "At the end of the day, comfort's tough. Any of those extremities—

fingers, toes—when they get cold, they get cold, and then forget it, you're cold. The rest doesn't matter."

"It's almost always fingers," said Ryan Silberman, co-CEO of Refrigi-Wear. "It's hard to keep fingers warm." Before Silberman joined the company, he worked at a cold-storage warehouse that stored frozen hops for Anheuser-Busch. "I remember when I came to RefrigiWear, I was so excited," he told me. "I thought: *It doesn't matter what happens, I will be solving problems at seventy-two degrees from this point forward.*"

I caught up with Silberman at the airport, and while he waited in line for a preflight hazelnut latte, he told me that the company was founded in 1954 by Mortimer Malden and Myron Breakstone, scions of a cottage cheese empire and great-uncle and grandfather of Silberman's co-CEO. "Back then, people just wore their regular winter jackets inside," said Silberman.

The hoodie—iconic wardrobe staple of skateboarders, hip-hop stars, and normcore tech-industry bros—is one of the very earliest examples of purpose-built protective wear for refrigerated-warehouse workers. Champion, which was originally known as the Knickerbocker Knitting Mills of Rochester, New York, started out making woolen underwear favored by cold-storage employees, before developing a line of sturdy sweatshirts for student athletes. In the 1930s, they introduced a hooded version of their sweatshirt, which was gratefully adopted by refrigerated-warehouse workers and football players alike. "We still sell a ton of hoodies," said Silberman. "We've had a shift over the years, where some of the safety managers don't want hoods because it limits field of vision, but I'm always shocked by how many hoodies we sell."

Thanks to his family's dairy business, Myron Breakstone was familiar with the limitations of long underwear, hooded sweatshirts, and wool coats and caps when it came to keeping warm while working in the cold. He and his cousin Mort partnered with DuPont's newly formed Textile Fibers Department to create the first innovation in insulated clothing in decades, using all-new petrochemical products: a wind- and water-resistant nylon shell layered on top of synthetic polyester hollow-fiber insulation. The very first jacket they introduced—the Iron-Tuff Minus 50—is basically the same model I was given at Americold two-thirds of a century later.

RefrigiWear products are also worn by Iditarod dogsled racers, New Mexican molybdenum miners, and the men who built the Trans-Alaska Pipeline, but the company's core market remains refrigerated-warehouse workers. "We're just trying to protect people that have to go to work in an environment that allows America to eat at scale," said Silberman.

As I rotated through Americold's Southern California facilities over the course of a week, I started to figure out the language and rhythms of refrigerated storage. I began to anticipate the arrival of the day's first deliveries (often cut flowers) and the last (Dodger Stadium hot dogs, made from pigs who had arrived at the Smithfield slaughterhouse in Vernon, a few miles away, just that morning). In the cooler, food often came in one day and went out the next; in the freezer, pallets might sit for a year or even two. Wednesday and Thursday were the slowest days. Friday was "rock-and-roll time," my new colleagues told me, because logistics companies take advantage of the weekend to get product to the East Coast for Monday morning.

Although the warehouses remain stuck in eternal, unchanging winter year-round, there are still seasonal cues. My refrigeration immersion took place in March—busy season for fish and lamb, thanks to the lingering cultural influence of Lenten fasts and springtime celebrations. "We build seafood up in the fourth quarter and ship it out from January through April," Amato told me. While I was working at Compton, we received 3,500 pounds of boxed lamb flown in on Qantas just to meet Easter demand. "Sometimes I think, *Who eats all this pie?*" Amato said, gesturing at racks full of frozen key lime, s'mores, and cookies and cream desserts. "But it's Easter; people need pies." Meanwhile, the first turkeys were starting to come in from Montana to be stockpiled for Thanksgiving; frozen pizzas and TV dinners accumulate over the summer in anticipation of back-to-school and football season in the fall.

Beneath these annual rhythms, the aseasonal fulfillment of American desires continued unabated. For hour after hour, we received, stocked, restocked, and picked: frozen guava juice in barrels, destined for a Dr. Smoothie bottling plant; cans of refrigerated peanut butter paste, imported from Argentina to fill M&M's and Clif Bars; pallets stacked with rolls of X-ray film for local hospitals; and thousands and thousands of freshly baked King's Hawaiian buns, trucked in hot from Torrance—a thermal disruption for which Americold charges extra. "We need to bring it down slowly to keep the moisture in the product. Bread will crystallize if it's cooled too fast," explained Espinoza. "People think it's so fresh and soft," said Carlos. "I tell them it's been in here for months, and they don't believe me."

Americold's Compton facility also offers a range of "protein processing and packaging services," all carried out in a white-walled room where the temperature hovers just below freezing. In a brief break from pallet picking, Cesar, who told me he had been a butcher back home in Peru, walked me through how he slices, grinds, brushes off the bone chips, and vacuum-packages meat. A huge stainless-steel drum, called a tumbler, was for marinating; an equally gigantic tub fed the grinder. "I love this machine," Cesar said, showing me a squat metal box with a conical hopper on top. "It can do sausage, it can do meatballs," he said, showing me how chunks of meat get sucked down from the hopper into a whir of silver, screw-shaped blades. Within seconds, the gloppy splats of flesh being slapped against the machine's metal innards were replaced by a rhythmic grinding sound; then, with an audible *whoosh*, the meat was extruded out in a long, fat cylinder.

These value-added services mean additional revenue for Americold. "We like to make the product work while it's in storage," explained Espinoza. Cesar told me it also helps make his job more interesting. Certainly, as we spent the rest of the afternoon in the freezer picking boxes labeled "Carl's Jr. Patties" to send to Riverside and receiving boxes of Kobe beef, airfreighted to LAX from Tokyo, I couldn't help but feel like a cog of cold-stiffened flesh in a perpetual protein-shuffling machine. Box after box of Asian white shrimp and imitation crab meat was stacked forty feet high to the ceiling. One pallet held gallons of beef blood in milk cartons; the label on another set of boxes said that they contained bull pizzles, hearts, and livers. "We call that 'misc.,'" said Cesar. "This is trim, this is all going into burgers." There were entire lamb carcasses from New Zealand, wrapped in canvas and nestled together nose to tail on the wooden pallet, as if sharing a bunk bed; there were huge plastic bag–lined cardboard boxes full of ground meat destined for school lunch programs.

At break time, I rubbed my frozen fingers together, and we talked about basketball and Los Angeles traffic and why the automated voice in the Americold headsets was so weird. Carlos told me this wasn't a dream job, for sure, but he liked the security, the sick pay, and the 401(k) match. "If you can go to school, go to school," he said. "But if you haven't got a lot of talent, this is okay." Frank, who had swapped his company-issued beanie for a Raiders hat, chimed in. "I never imagined a place like this could exist before I worked here," he said. "I go to the store now and I think, *I probably picked this*. What we do—I feel it's invisible."

"Before, I used to see this stuff and I didn't think about how it got there," agreed Carlos. "You see the guys in the store filling the shelves, but how did it get to them? No one ever thinks about that. But if it wasn't for me, that pepperoni pizza wouldn't be there."

2.

THE CONQUEST OF COLD

I. Stop the Rot

"If it makes you feel any better, I've got two engineering degrees and I studied thermodynamics from one of the premier scientists in the field," said Kipp Bradford. "And I could do a ton of great math, but I didn't have a clue how a refrigerator worked."

Bradford is an engineer and cofounder of an HVAC start-up. Although his interest lies in inventing the future of cooling, after I confessed my own ignorance, he invited me to spend an April morning with him in his garage in Pawtucket, Rhode Island, building an old-school fridge.

Flush with excitement at the prospect of being able to create cold from scratch, I bragged about my plans to my taxi driver, who was suitably impressed. "If I knew how to make a fridge, I'd move back to Haiti and open a business," he said before giving me his card so that we could go into partnership together. "I'll do your imports for you, because I know how to get things to Haiti," he said. "And you'll make the fridges."

"Why not?" said Bradford when I told him about my prospective new enterprise. He poured me a coffee before we got started, then gestured at a workbench covered in shiny copper tubing, clamps, and corrugated aluminum. "It's absurdly easy."

This nonchalance would have come as a shock to generations of scientists, including many of the big names—Leonardo da Vinci, Francis Bacon, Galileo Galilei, Robert Boyle, Isaac Newton—who all tried, and failed, to establish where cold comes from. Bacon died from a chill caught while trying to freeze a chicken using snow; Boyle complained that he had "never handled any part of Natural Philosophy, that was so Troublesome, and full of Hardships, as this has proved." In the 1600s, cold was a mystery whose effects were little studied and whose source was unknown. Some believed it came from the far north, others that its origins lay in wind, water, or deep underground, and still others speculated that it was a property of invisible "frigorifick atoms" or "cold corpuscles."

René Descartes, who compared certainty about the very existence of cold unfavorably to certainty about that of God, came closest to our modern scientific understanding in his *Meditations*, published in 1641. "If it is true that cold is merely the absence of heat," he argued, "then an idea that represents cold to me as something real and positive will not inappropriately be called false."

Two centuries later, when scientists formulated the first two laws of thermodynamics, they realized that Descartes was right—about cold, if not necessarily God. There is no such thing as cold, in that it's not a thing or a force or a property that exists and is measurable in its own right. Cold is, as Descartes speculated, the absence of heat; cooling is thus the sensation of loss as heat is transferred elsewhere. Building a fridge doesn't involve the creation of cold, as I'd imagined—it's simply a matter of finding a way to move heat from inside a container to outside.

"We're going to cut, bend, join, evacuate, and then charge," said Bradford, clearing space on the workbench in front of us to lay out the internal organs of our future fridge. Above us hung a couple of bike frames; behind stood snow shovels, pruning shears, and rakes. Curious sparrows hopped in and out through the open garage door, cheeping furiously. Our fridge-building exercise was, I quickly realized, more akin to putting together a chicken pie from a rotisserie hen, store-bought puff pastry, and a bouillon cube than to making the whole thing from scratch. Bradford had ordered all the components online—we simply had to connect them together correctly.

"There are four parts that matter," explained Bradford, introducing them to me in turn. The compressor was a black plastic–sheathed cylinder the size of a can of beer, manufactured by Samsung. The condenser, a book-sized concertina of rippled aluminum, came from Thailand. The evaporator was an off-white rectangle about as big as a legal-size sheet of paper that looked sort of like a flattened, abstract-art version of a circuit board. The final player in the fridge quartet was so small I didn't spot it at first—a little coil of thin copper wire called a capillary tube.

After cutting a couple of lengths of copper pipe and bending them to connect the parts into a loop—compressor to evaporator to capillary tube to condenser, and round to the compressor again—we put on goggles. "Now for the fun part," Bradford announced as he wielded the blue flame of his oxyacetylene torch like a paintbrush to seal all the joints. We pressurized the circuit to make sure there were no leaks by filling it full of nitrogen from a tank Bradford kept in the corner, next to his leaf blower. Then he hooked up a vacuum pump, which looked like a heavy-duty flashlight and sounded like a swarm of metallic bees. Once that had pulled all the air out of the system, we opened up a small, aqua-blue tank of liquid 1,1,1,2-Tetrafluoroethane, better known in the industry as R-134a, and, with a *whoosh*, charged the entire loop with refrigerant.

"This is exactly what's in your home fridge," Bradford said. "I could go to your house, rip out your fridge's guts, and replace it with this, and you wouldn't notice any difference." The machines that cool refrigerated warehouses operate on the same principle, albeit supersized and using different components and refrigerant chemicals.

"I'm going to turn on the power to the compressor," said Bradford. "This is where the magic happens—we're going to start feeling cold any second now." With a low purr, the compressor began pumping refrigerant around the circuit, and in less than a minute, the drops of coffee I'd spilled earlier froze into icy brown slush. While Bradford tweaked and fine-tuned everything, I laid my hands on the evaporator and watched as my fingertips turned white. It felt like the closest to sorcery I'd ever come.

This system is called vapor-compression refrigeration, and although it's not the only means of making cold in use today, it is by far the most common. The evaporator—the off-white circuit board—is the part that sits inside the back wall of your fridge, pulling heat out of the box and all the food in it. The compressor is typically housed in its own built-in cubby at the back of the appliance, plugged into a power outlet. The condenser is always on the outside, either mounted on the back as a dusty set of coils, woven back and forth over a grid like a metal loom, or, in newer models, hidden behind a kickplate at the bottom.

The cycle relies on three key principles, all of which were vaguely familiar from high school science class: the thermodynamic truth that heat always moves from warmer to cooler; the law of physics that says the molecules in a liquid absorb heat when they boil, changing state to become a gas; and, finally, the fact that you can raise or lower the temperature at which a liquid evaporates by raising or lowering the pressure it is under. (This is the same phenomenon that causes a kettle to boil at a lower temperature up in the mountains, where the pressure of the atmosphere is lower.)

The cooling part of the process is surprisingly simple. At the point a mostly liquid refrigerant enters the evaporator, it's at very low pressure—so low that, as it loops around inside the back wall of the fridge, it can't help but begin to boil. As it boils, its molecules suck heat from their surroundings in order to move faster and expand into a gas. That thermal gradient produces the desirable side effect of cooling the contents of the fridge.

So far, so good—but to keep the fridge cool, that refrigerant needs to perform the same trick again and again, which means it needs to become a liquid again. This is where the rest of the system comes in. Although R-134a is a gas when it gets to the end of the evaporator, it's a *cold* gas—colder than room temperature. In an explanation that was completely counterintuitive yet also made perfect sense, Bradford explained that R-134a could not dump all the heat it picked up from inside the fridge and cool down into a liquid again *unless we warmed it up*. Only once our R-134a was hotter than the garage we were standing in would that heat flow out from the refrigerant and into the air around us.

The compressor essentially performs the labor of carrying the refrigerant uphill, so that it can coast back down the thermal gradient again, absorbing heat as it goes. Plugged into the wall, the little black cylinder draws in energy in the form of electricity and uses it to pump a piston back and forth inside the R-134a-filled canister. This raises both the temperature and the pressure of the refrigerant: the speed at which its molecules are traveling and the rate at which they collide. Once the compressor has done its work, the R-134a will be a dense cloud of superheated vapor.

From there, the gas flows into the corrugated aluminum condenser—the external coils on your fridge. All that surface area allows the hot refrigerant to dump its energy into the cooler air of the room, so that by the end of the loop, its molecules are sluggish enough to condense into a still-quite-warm liquid. (This is also why the back of your fridge or the floor just in front of it feels toasty.) "By the time it comes out of the compressor, it's at maybe a hundred and forty degrees," confirmed Bradford. "By the time it gets down to the end of the condenser, it's probably at ninety degrees."

Then it hits a choke point: the expansion valve. This long and extremely skinny tube restricts the amount of refrigerant flowing through, creating an area of lower pressure on the other side. The refrigerant makes it back to the evaporator as a low-pressure liquid—cool enough, and with a low enough boiling point, to suck in all the heat inside the fridge as it turns into a gas once again.

As long as your circuit doesn't leak and the compressor remains connected to a source of power, you can move heat from inside the box to outside it for eternity—or at least until the moving parts of the compressor wear out. "The compressors are rated for fifty years," Bradford reassured me. "Your fridge should run, maintenance free, for longer than you own your house."

A refrigerator is an underappreciated engineering marvel, I realized—a reliable, relatively simple box that, without fuss or fanfare, harnesses the powers of nature to supernatural effect, performing the daily miracle of delaying matter's inevitable decomposition and death.

In December 2011, hundreds of thousands of Britons tuned in to watch what happened to food stored inside a very different box. As part of a BBC initiative called After Life: The Strange Science of Decay, a team of scientists, engineers, and technicians built a replica kitchen, filled it with food, enclosed it inside a glass cube, and placed it on a soundstage at the Edinburgh Zoo for two months. The resulting "rot box"—complete with time-lapse cameras trained on its microbial, mold, and maggot inhabitants—was the refrigerator's diametric opposite, a visceral illustration of what happens without cold.

As biologist George McGavin explained in the subsequent documentary, "on the inside is all the food you'd expect as if a family were just about to have a party": a dish of chili con carne, some cooked rice, a fruit bowl and veg box, raw fish and chicken set out on a baking tray, and some plastic-wrapped burgers. "Also on the inside," said McGavin, a round-cheeked biologist with a light Scottish burr, "are the bacteria and fungal spores that are going to start the process of decay." Rubbing his hands together, he added, "I can't wait to see what happens!" Watching food rot turns out to be strangely mesmerizing, and over the course of the next eight weeks, the British public gazed in fascinated horror, hands over mouths and noses wrinkled in disgust, as the rot box lived up to its name.

Although the melons, sweet corn, lettuce, and strawberries in the kitchen seemed perfectly fresh at the beginning of the experiment, in reality, the minute a fruit or vegetable is harvested, it starts to deteriorate. In the fruit bowl, the peaches and apples looked pretty good, but their texture, flavor, and nutrient content were already starting to suffer. Cut off from the roots and leaves that supplied them with food and water, the fruits had already resorted to self-cannibalism, consuming themselves from the inside in a desperate attempt to keep their cell metabolism going. The corn and peas burned through half their stock of sugar in the first few hours in the box, while the celery stalks and the head of lettuce became limp as they drained their internal reservoirs of water.

Within a day, billions of bacteria started taking advantage of the plant and animal cells' weakened state to launch their own attack. The sandwiches sagged and the fish's shiny scales dulled, becoming first tacky, then slimy. The chicken, in particular, looked puffy and bloated, its skin mottled purple and yellow. The room was humid, hot, and, McGavin reported, already beginning to stink.

By the end of the first week, the molds had taken over. Fungi typically grow more slowly than bacteria, but they are more adaptable and resilient, strengths that lead to their eventual triumph. Inside their rustic, slatted box, the vegetables had utterly succumbed: ravaged by decay, they had collapsed upon themselves and were swathed in a blanket of fluffy mold. Close up, each individual snow-white filament was sprinkled with thistle-like spore heads, glittering like crystals.

The bread had disappeared, rewoven into a green-gray tapestry of penicillium. The chili con carne was buried beneath a thick crust of mold, and the burgers were dotted with white fur, their plastic wrapping stretched taut over a buildup of gas. The chicken had lost its shape entirely, slumped formlessly over the edge of the baking tray, leaking cloudy fluid. As he entered the box, McGavin noted that the room's initial ran-cid and ammoniac stink had been replaced by a yeasty, sickly sweet pall—but not before those early, stronger-smelling gases produced by the bacteria had served as a beacon to attract flies to the meat and fish, in order to lay their eggs.

It took another seven days for those eggs to hatch, at which point the box's maggot population exploded into a gelatinous blob of white grubs, sucking and undulating their way through the burgers and chili. The chicken was seething with them too, gradually dissolving into a froth of yellowy mucus. The fish had long since deliquesced into brown goo, with only its bones and cartilage remaining. It was almost impossible not to gag, even shielded from the smell by the double glazing of the rot box's glass walls and my own TV screen.

Armed with UV lights and microscopes, and accompanied by a small army of experts, McGavin did his best to narrate the slow-moving horror movie that is food spoilage. At the end of two months, the box's contents were unrecognizable and McGavin seemed thrilled. "Little is left of the fresh food we began with, and what remains will continue its inexorable journey," he intoned. "What we have witnessed in the box is a process of renewal that we are all part of." Broken down into the basic building blocks of life, he explained, the atoms formerly known as chicken and strawberries would be recycled into radishes, then humans, a process he and his scientific colleagues made visible by chemically tagging mole-cules of nitrogen and then following them as they were extricated from one organism, only to be taken up and incorporated into another.

McGavin's point is that if rot were stopped—if bacteria, fungi, and insects didn't decompose plants and animals in order to release essential nutrients—Earth would rapidly become uninhabitable. In a thought experiment aimed at imagining a world without microbes, scientists painted a grim picture of biomass piling up into "vast reservoirs" of waste, disrupting the "biogeochemical recycling upon which all life ultimately depends." In conclusion, they wrote, without rot, "we predict complete societal collapse only within a year or so, linked to catastrophic failure of the food supply chain."

F ood spoilage is clearly a matter of perspective: biodegradation is terrific, except where our dinner is concerned. Perishable foods such as fish, meat, milk, and produce are simply more vulnerable to rot, and succumb to it more quickly, than other organic matter. Packed with nourishment and water, they are as perfectly equipped to promote growth in microbes as in humans.

From the very first time our hunter-gatherer ancestors killed an animal too big to eat in one go on a hot summer day, humans have known that fresh food has a limited life span, although not necessarily why. While a majority of twenty-first-century Britons are sure enough of their next meal to be able to treat decomposing food as entertainment, for much of human history—and still today, in many parts of the world— most people were and are uncomfortably familiar with hunger. Food lost to decay would have represented an unbearable waste.

Microbes and people have thus engaged in a millennia-long form of interspecies warfare: a sustained competition for food, in which the bacteria and fungi attempt to secure their dinner by excreting objectionable or toxic chemicals, and humans have, in response and over time, developed an equally impressive arsenal of antimicrobial weapons, or preservation techniques, as they are more commonly known.

Before the 1860s, when French scientist Louis Pasteur discovered the souring of milk was caused by microorganisms that could be destroyed using heat, these kinds of techniques were developed empirically, without any real understanding of why they prevented rot. Their invention was motivated by the pressing need to ward off future want: "preserving," as Italian author Girolamo Sineri has written, "is anxiety in its purest form."

Using the sun and wind to dry food is thought to be the oldest method of preservation. Archaeologists working in the Middle East have found evidence of desiccated meat, its life-enabling water evaporated away, from as early as 12,000 BCE. This transmogrification must have seemed like something of a miracle at the time. After just a week spent air-drying in the sun, strips of meat would remain edible for up to two years, as opposed to less than two days unpreserved. The development of salting and sugaring—chemical means to achieve the same end of reduced water availability—wasn't far behind. By 3000 BCE, the Sumerians were packing salt around fish for storage in amphorae, and the ancient Greeks regularly preserved fruits in honey. In regions without ready access to salt, and before industrialization made sugar affordable, people also resorted to making food extremely acidic or alkaline using vinegar or lye, boiling it to kill microbes, or smothering it under an oxygen-depriving layer of fat.

Many techniques take a multipronged approach, using both chemical and physical methods in combination for additional preservative power. Smoking meat or fish dries it out but also deposits cell-killing chemicals on the food's surface. Cheese—memorably characterized by author Clifton Fadiman as "milk's leap toward immortality"—relies on lactic acid bacteria to sour the milk, followed by the addition of salt and a mixture of enzymes called rennet to help it coagulate into curds, in order to arrive at a long-lasting and, fortuitously, more portable end product with greatly reduced water levels.

In other cases, the battle became even more sophisticated: rather than relying solely on brute force, humans recruited fungal allies in an attempt to harness and control rot as opposed to halt it altogether. In Asia, for example, fish, cabbage, and soybeans were mixed with salt as well as moldy rice to encourage the growth of beneficial microbes whose busy fermentation both repels their noxious cousins and produces deliciously tangy, funky flavors—the precursors of today's soy sauce, sushi, and kimchi. This partial decomposition can be seen as a kind of negotiated preservation truce.

All of these intense, punchy flavors—salty, funky, smoky, sour, and sweet—are the signature of prerefrigeration cuisine. While our hunter-gatherer ancestors consumed an extremely varied diet that archaeologists estimate included dozens of different kinds of fresh foods, the introduction of agriculture a little more than ten thousand years ago radically streamlined that menu. Wheat, barley, rice, and corn fed early farming societies reasonably reliably and with painful monotony. Salt beef, smoked salmon, pickled peppers, fermented cabbage, and candied fruit would have been valued as desperately needed flavor bombs as well as essential food reserves—an edible insurance policy that made harvest gluts bankable against leaner seasons.

Indeed, many of the world's most delicious foods are the result of humanity's millennia-long war against rot: from stinky cheese, smoked salmon, and salami to miso, marmalade, and membrillo. Even the gelatinous pleasures of Scandinavian lutefisk or Chinese century eggs have their devotees. Most of these preserved foods are also incredibly long-lasting. What they are not, however, is the same as fresh: the chemical and physical transformations required to vanquish microbes inevitably also destroy the food's original flavor, texture, and appearance. Raspberry jam may well be the perfect topping for a scone, but it is not likely to be mistaken for fresh raspberries; a sauerkraut-free hot dog is, in the opinion of many, incomplete, but you could be forgiven for not inferring the relish's relationship to cabbage.

Still, these preservation methods mostly worked, they often tasted good, and many of them even had the side benefit of making food more portable: a chunk of cheese or stick of salami can be brought along on a day's journey much more easily than a jar of milk or a pig. Of course, transporting food over long distances was not a particularly pressing concern for most of human history. People have almost always lived in extremely close proximity to the plants and animals that formed the basis of their diet. Even in 1800, after Britain's Agricultural and Industrial Revolutions were well underway, only 3 percent of the world's population

lived in cities. For a majority-rural population, food preservation func-
tioned less as a distribution mechanism and more as a static, seasonal
hedge, blunting the agricultural year's cycles of feasting and scarcity
rather than removing them altogether.

Sailors and soldiers were the exception: the bulk of their diet was
formed of stored food, and they relied on hardtack or biscuit, salt pork,
brined and smoked cheese, and dried peas and beans, supplemented by
whatever fresh food they could catch or forage from the land or sea.
Feeding armies on the move posed such a serious problem that it inspired the next major breakthrough in antispoilage technology: canning.
At the end of the eighteenth century, the French introduced universal
conscription to raise the largest army the world had yet seen: more than
a million men, under the command of a young, ambitious Napoleon
Bonaparte. The local countryside buckled under the pressure of feeding
this plundering horde, and in 1795 the French government issued a challenge: a generous cash prize of twelve thousand francs for a new method
of preserving food.

Nicolas Appert, a talented chef with no formal education, wondered
whether the method he used to put up sugared fruit in glass jars might
be applied to the problem of conserving soup, vegetables, beef stew, and
beans. "A dynamic and jovial little man," according to French historian
Maguelonne Toussaint-Samat, Appert began his experiments by funnel-
ing peas and boiled beef into old champagne bottles, corking them, and
sitting them in hot-water baths for varying lengths of time. As curiosity
became obsession, Appert sold his Parisian confectionery business and
retired to a small town just outside the city, where he spent the better
part of a decade perfecting his method.

In 1803, Appert delivered the first batch of preserved food to the
French navy for field-testing. The contents of his bottles received rave
reviews: the beef was pronounced "very edible," while the beans and
green peas had "all the freshness and flavor of freshly picked vegetables."
Appert was awarded the prize and promptly used the money to finance

more experiments. Rather than patent his technique, he published a book of detailed instructions so that anyone could master "l'art de conserver." Perhaps unsurprisingly, he died a pauper. Despite being formally recognized as "a benefactor of humanity" by the French government, even his wife eventually left him, and he ended up buried in a mass grave.

Appert's glass bottles brought summertime peaches and corn to February dinner tables and allowed landlocked eaters to enjoy their first taste of unsalted seafood—though it took the invention of the tin can, on the other side of the Channel, to make canned food truly portable, and the evolution of the can opener from primitive lever to rotating wheel, in the 1920s, to make it user-friendly. But long boiling times meant that the can's contents were also often tasteless, mushy, or rubbery. Tinned tomatoes might be a better substitute for fresh than their sun-dried counterparts, or than vinegar-preserved ketchup, but they're never going to fool anybody that they are straight off the vine. And tomatoes are the rare example of a food whose nutritional profile is improved by canning—in most cases, heating reduces the vitamin content of vegetables.

By the start of the twentieth century, humans had developed a variety of effective and more-or-less delicious strategies to stop the rot. But preserved food was labor-intensive. It frequently required the assistance of costly materials, such as sugar or aluminum, and in the case of cheese and other cultured and fermented foods, it demanded nonnegotiable stretches of time. During the preservation process, food often shrank, the lost water weight making it more expensive by volume. Most important, it didn't look or taste anything like fresh. As a result, most people still ate seasonally and locally most of the time. While that is held up as ideal in contemporary culinary philosophy, in a prerefrigeration food system, it entailed hardship for everyone but the very rich. In the late winter and early spring, the average diet could be extremely monotonous and often deficient in important micronutrients. Farmers had to dispose of all of their harvest at once, bringing prices down, and processing the resulting glut demanded hours of labor, frequently from the unpaid women of the house. Meanwhile, a fresh peach would have been a rare treat available once a year—or not at all, for the urban poor.

The introduction of refrigeration—first natural, then mechanical—changed all of that, overturning millennia of dietary history. In some ways, it brought us full circle: for the first time since the introduction of agriculture, humans began to be able to eat a little more like our Paleolithic ancestors again, choosing from a vast array of hundreds of different kinds of foods, all fresh and preservative-free. Ultimately, however, the systematic use of cold opened the door to an entirely new chapter in human nutrition, one in which we overcame not just rot but seasonality and geography as well.

II. The Ice Harvest

There are only a handful of places left in the United States that still practice the forgotten art of harvesting winter. One of them is the Thompson Ice House in South Bristol, Maine—a small wooden shack by the side of State Route 129. In 1826, Asa Thompson dammed a brook on his property to create a shallow, one-acre pond, built an icehouse on its shore, and got into the frozen-water business. Five generations of the family kept the annual harvest going until 1985, half a century after the rest of the New England ice industry had evaporated. Ken Lincoln, who grew up nearby and began working on the last commercial harvests aged just nine, helped reopen the icehouse as a living museum in 1990.

By the time I arrived on a bright but bitter February morning, determined to participate in the annual ice cutting, several older men in worn overalls and plaid had already etched an oblong grid on the snow-dusted surface of the lake using an ice plow, leaving ghostly outlines of white rectangles, each just under two feet wide and three feet long, stretching into the distance. Lincoln was busy coaxing a century-old, gas-powered, sled-mounted circular saw into action, in order to carve out the channel leading to the icehouse ramp.

Inside the timber-shingled icehouse, a couple of layers of ice blocks already glowed blue white. They had been harvested a week earlier at the bidding of a Discovery Channel film crew. Lincoln told me that the ice-cutting scene in the prologue to Disney's *Frozen* had led to a surge of interest in the process. Still, the few dozen folks clustered around the coffee and doughnuts set up on a folding table in the parking lot were mostly from neighboring communities, though some visitors had driven up from Portland and beyond. Many told me they were regulars, returning each year with thermos flasks full of chili, woolen blankets, and deck chairs from which to watch the fun.

Without any formal announcement or fanfare, the day's harvest got underway, using a collection of found and donated antique tools. Working in pairs, the men cut along parallel scored lines using heavy, two-handled ice saws. When the rectangular raft floated free, they grabbed long-handled chisel-like tools called breaker bars, which, when dropped on the scored lines, broke the blocks up crosswise. Finally, they steered each individual cake into the icehouse channel, using a pike pole to gently nudge and drag the slow-moving white cuboids through a narrow band of black water.

After the initial row had been harvested as a demonstration, it was open season on the tools. The gathering crowd of locals and visitors was free to try their hand at sawing, breaking, and steering the ice. I can report that sawing was the quickest way to warm up—dragging the unwieldy iron blades through twelve inches of ice was quite a workout. Steering was also surprisingly tricky: the cakes weigh more than three hundred pounds each and moved through the water with the slow solidity of a container ship. Snapping individual blocks off the longer raft using the breaker bar was, however, the most rewarding experience—one correctly positioned thrust and, with a gentle crack, a newly carved block emerged in all its crystalline perfection.

The business of getting the harvested ice into the house was a little less amateur friendly. A pair of guys in fluorescent fishing waterproofs dunked the ice cakes and maneuvered them onto an underwater sled attached to a pulley system. The horsepower was provided by a man in a battered pickup truck, rather than the more historically accurate Clydesdales. (A couple of horses were present but were otherwise employed giving sleigh rides to local kids.) Once the dunkers gave the signal, the pickup driver reversed, and a single glistening, translucent, oversize cube was hauled on its sled up the ramp to the top of a rickety wooden scaffold, where it paused for a brief moment, scattering rainbows in all directions like a giant prism, before picking up speed on its slide down the other side.

Inside the icehouse, the younger generation performed the slightly more hair-raising work of positioning the cakes. As the block paused for a moment at the top of its ascent, someone would shout "Ice inbound!" and, like a combination of curling and Tetris played using pike poles, the wranglers would crouch, ready to guide the heavy block, sliding fast on its glistening undercarriage, into the next open slot. Sometimes it worked, and the block accelerated into its assigned parking space with a satisfying *thunk*. Mostly it did not, and the speeding ice cake careened off course, crashing into walls, chipping chunks off other blocks in a sparkling spray, and occasionally, the old-timers told me, breaking ankles. The more tightly the ice is packed, the longer it lasts, so broken bits were used to patch any gaps. The jagged chunks glittered like milky-blue quartz, throwing shards of light across the warm wooden walls.

The ice we harvested that crisp February day could, carefully packed, last through the summer. These days, however, the Thompson Lake ice is mostly used up in one epic burst: a July ice cream social. Volunteers—many of whom, like me, had participated in the harvest—returned to the lake; we chipped up a couple of blocks and mixed them with salt to freeze ice cream in old-fashioned, hand-cranked machines. In the critter-dense humidity of an East Coast summer, bathing in the clear turquoise radiance of the remaining ice cakes while waiting for the ice cream to set was already refreshing—a melting monolith of cold. I enjoyed a scoop each of creamy strawberry and mint chocolate chip on the shores of the almost unrecognizable lake, its dark waters now abuzz with dragonflies and decorated with lily pads.

By the end of the day, there were just a few shrunken blocks left in the icehouse, their glistening sides shaggy with the straw that had blanketed them as insulation. Afterward, any remaining ice is sold, a dollar a cake on the honor system, to sport fishermen. With fewer air bubbles, natural ice melts more slowly than the man-made kind, and a good block will apparently last up to a week out on the water. In an odd way, this usage—an ice cream party first, followed, almost incidentally, by food storage—reflects refrigeration's curious pathway to adoption, from pleasurable treat to practical necessity.

In the late 1960s, Sylvia Beamon visited one of her rural hometown's most intriguing attractions. Royston lies a few miles south of Cambridge, England, and is home to an artificial cave whose walls are covered in mysterious carvings—circles within circles, shields and swords, a man holding a skull and a candle, a horse, and an Earth goddess–like figure, as well as a couple of crucifixion scenes.

No record of Royston Cave's age or purpose has been found to this day, but earlier researchers speculated that it might have been an ancient pagan temple, a Roman sepulchre, a medieval private chapel, or a hermit's hole. Intrigued, Beamon—a mother of four with no formal qualifications in the field—embarked on her own investigation, eventually concluding that the cave likely served as an overnight butter storage site for the Knights Templar, who were known to sell their wares in the nearby market in the late 1100s and early 1200s. (This remains the prevailing theory today, based on the striking similarity of many of the designs to those known to have been carved by members of the mysterious order on the walls of a French dungeon.)

Her interest in all things subterranean piqued, Beamon applied for a place to study archaeology at Cambridge University. A week before she began her studies, she presented her research on Royston Cave to the newly founded Société Française d'Etude des Souterrains.* "Then came the questions, and the word *glacière* came up several times," Beamon recalled. Could Royston Cave be an icehouse? Beamon, who at that point had never heard of such a thing, was captivated by the concept. In her final year at Cambridge, she focused her dissertation research on a pit in a natural cave on the island of Jersey that she had become convinced was used by Neanderthals as a protorefrigerator to store mammoth meat about 140,000 years ago—a theory that, if correct, would make it one of the earliest known examples of deliberate food storage by humans. To flesh out her hypothesis, she wanted to cite evidence that meat would last longer packed with ice in a pit—but, to her astonishment, she discovered that no one seemed to have conducted such a study.† Icehouses, she was told, were not considered a worthy subject for research.‡ She would have to do it herself.

One winter morning in 1980, Beamon woke up to find a fresh layer of snow blanketing her backyard and blocking the roads around Royston. Thrilled, she quickly enlisted the help of one of her children and a neighboring student, neither of whom could get to work or school. Then she scanned her fridge. "I went for a cooked lamb chop, a lidded jam jar

---

*Inspired by this French group devoted to the study of human activities underground, Beamon founded a sister organization in the United Kingdom named Subterranea Britannica. My account of her Paleolithic preservation experiment is drawn from the pages of the society's fascinating triannual magazine.

†Francis Bacon's doomed chicken experiment was the lone example Beamon could find in the annals of science.

‡The literature on perishable food provisioning in ancient history remains thin to this day, confirmed Monica Smith, an archaeologist and anthropologist who studies early cities and their household activities. "Archaeologically, we're mostly concerned with grain," she admitted. "And all these other things—art and that kind of stuff."

containing whole pasteurized milk, two frozen fish fingers and two unfrozen fish fingers," she recalled. (*Fish finger* is the British term for fish stick.)

Together they filled a three-foot-diameter depression in her backyard with a foot of snow, "packing it as hard as possible into solid ice with the back of our spades." In went the lamb chop and a jam jar full of milk, and on went another foot of hard-packed snow. The fish fingers formed the next layer in this icy protein sandwich, and the whole thing was sealed up with yet another foot of snow. "We tidied up and smoothed the sides down leaving no holes or gaps," Beamon remembered. Then she went back inside, had a cup of tea to warm up, and recruited her mother, who lived nearby, to take daily thermometer readings of the pit's internal temperature.

Within a few hours, the Beamon family dog had dug up the two unfrozen fish fingers. By day five, as the snow started to melt, Beamon noticed that the other two had become exposed. "Birds were pecking and eating the breaded outside, but the fish was actually still frozen," she recorded. It took three weeks for all the ice to disappear, at which point Beamon drank the milk and nibbled on the chop. The milk fat had separated, but "it tasted alright," as did the chop. "I gave the rest to the cat who thought Christmas had come and ate it all!" she reported. "Cats are notoriously fussy so I knew there was nothing wrong with it."

Although neither jam jars nor breaded fish sticks were available 140,000 years ago, Beamon's thrown-together experiment was one of the first documented attempts to gather data on how long food buried in snow remains edible. To this day, much experimental archaeological research starts from this kind of simple, ad hoc trial, to establish proof of concept.* Over the years, as Beamon repeated her experiment with

---

*A few years ago, researchers found new evidence that between 420,000 and 200,000 years ago, ancient hominins likely deliberately stored deer bones in a cave in what is now Israel in order to eat the marrow later. The team of archaeologists behind the finding performed their own Beamon-style trial in an attempt to gauge the feasibility of their hypothesis. One of them told *The New York Times* that, after nine weeks, the cave-aged marrow tasted "not bad"—"like a bland sausage, without salt, and a little stale."

increasing levels of sophistication and period authenticity, she established that, packed correctly, a snow pit would remain cool for a full five weeks, keeping the food stored in it good for almost as long. "If early man had placed milk or cheeses within covered pots there was no reason why such food would not have become frozen and preserved for three weeks or more," she concluded.

Beamon's research helped demonstrate that humans have likely known about and used cold's spoilage-suspension powers for millennia—probably as long as we have been salting meat and sugaring fruit. Putting a fish on ice buys time: the microbes that coat its surface are rendered sluggish, eating, excreting, moving, and reproducing at a snail's pace. Putting grapes on ice not only delays the growth of the fungi that live on their skin but also has the added advantage of slowing down the fruit cells' own respiration, and thus the speed with which they burn through their internal water and sugar reserves. If enough heat is removed to reach the freezing point of water, cold's preservative power is multiplied: sharp-edged ice crystals form, reducing the amount of water available to bacteria and fungi while damaging their cell walls.

The problem is that, while we have been able to add heat to food at will ever since our Stone Age ancestors first gained control of fire, we have not been able to reliably remove it for much more than a century. We have been dependent, instead, on the vagaries of naturally occurring cold—the chill of an underground cave or cellar, breeze-borne evaporative cooling, and the fleeting, counterseasonal frost of stored snow and ice.

Paleolithic pits aside, the earliest written evidence for cold storage dates back almost four thousand years. Built by imperial decree and located on the banks of the Euphrates River in present-day Syria, these icehouses were described on cuneiform tablets as heavily guarded structures nearly twenty feet deep and forty feet long, lined with tamarisk boughs. Records from the time also show that ice was in such demand that, just three days after one shipment reached the city, likely from mountains to the north, it had almost completely sold out.

An ancient Chinese book of poems, the *Shih-ching*, contains the next reference, from a little more than three thousand years ago. "D'iong, D'iong! Chop ice 'neath the second moon, store it 'neath third," runs Ezra Pound's translation of a verse from the Songs of Pin. Based on the writings of Athenaeus, Plutarch, and others, it seems as though ancient Greek and Roman elites would regularly purchase snow brought down from the mountains to chill their wine and keep their shrimp fresh in summer. Still, it was an expensive luxury, which helps explain the aggrieved tone of a letter written by Roman author and administrator Pliny the Younger to a friend following his failure to show up at a dinner party: "I had prepared, you must know, a lettuce apiece, three snails, two eggs, and a barley cake, with some sweet wine and snow; the snow most certainly I shall charge to your account, as a rarity that will not keep."

Fast-forward to the 1600s, and the Italian peninsula "was entering a new Ice Age," according to food writer Elizabeth David. A few decades earlier, Neapolitan polymath Giambattista della Porta had discovered that adding salt to ice lowered its freezing temperature, meaning that custards could be turned into ice cream and wine into slushies. From Sicily to Florence, David wrote, "everyone with aspirations to comfortable and elegant living . . . had a private snow store in the grounds of their town mansion and country villa." British visitors wrote home that "a Scarcity of snow would raise a Mutiny at Naples, as much as a dearth of Corn or Provisions in another country." Cold's desirability made heists inevitable. David quotes a lengthy and sternly worded missive from a Medici cardinal to the Roman chief of police on the topic of diverted snow shipments and icehouse security.

A century later, the British had caught up, and no fashionable estate was without a subterranean icehouse—often designed by the era's big-name architects, including John Soane and Nicholas Hawksmoor—in which frozen water harvested in the winter could be stored until summer. As Beamon, who went on to spend more than a decade surveying the UK to create a gazetteer of these otherwise overlooked structures, put it: "It is difficult to comprehend today the sheer numbers of icehouses that existed"—"not tens or hundreds, but thousands." For the most part, she found, local authorities and planning departments were no longer aware of the existence of these subterranean structures, conjuring up an evocative image of the British Isles as a sort of ice-pocked Swiss cheese, riddled with forgotten voids. To this day, ice wells are regularly rediscovered by developers in London as they attempt to sink pilings or excavate basements into unexpectedly hollow earth. In rural areas, according to Bruce Walker, an expert in Scottish vernacular architecture, the entrances to subterranean icehouses, which were typically built at a distance from any other structure and seemed to open directly into the earth, are likely responsible for many of Britain's legends of secret tunnels and hobbit-like creatures.

But despite this vast infrastructure built to harness it, natural refrigeration remained refractory. Even in northern climates, a warm winter would leave ice in short supply. Icehouses were built to a variety of designs, and some didn't work well—they lacked adequate ventilation to remove the heat generated as ice melted or sufficient drainage to carry off the resulting meltwater. None other than George Washington, first president of the United States, had problems keeping ice in a specially built cellar on his Mount Vernon estate. His diary entry for Sunday, June 5, 1785, records the disappointment: "Opened the well in my cellar in which I had laid up a store of ice but there was not the smallest particle remaining."

Such failures were more of an irritation than a genuine catastrophe. For the most part, the ice that was harvested and loaded into these structures at vast effort and expense was destined for decadence. It was employed almost exactly as ancient Romans had used snow more than a millennium earlier: to chill wine, delicate fruits, and creamy desserts, for the hedonistic thrill of consuming frosty refreshments amid summer's heat. In the early 1800s, Jane Austen wrote to her sister, "For Elegance & Ease & Luxury . . . I shall eat Ice & drink French wine, & be above vulgar economy."

This focus on the most delightful—some might say frivolous—applications of cooling didn't come about because the elites of Renaissance Italy, Georgian or Regency-era Britain, and colonial America were unaware of cold's food-preserving potential. It just didn't seem either desirable or practical to use such a precious and limited resource on a task as simultaneously mundane and massive as chilling, say, the entire meat supply chain. What's more, in a world without huge distances between producers and consumers, or the expectation that all social classes could enjoy the same food all year round, the benefits of doing so weren't nearly as self-evident as the pleasures of wine slushies and ice cream.

In her survey, Beamon documented very few examples of icehouse-based food storage—and those she found were mostly limited to delicate fruits. At one stately home in the West Midlands, pears and peaches were suspended in wooden trays from the roof of the icehouse. "Gamekeepers were known to sharpen a stake from the coppice and to spear the juicy William pears from the access lid," she writes. Fishermen were also early adopters. From the 1780s, ice was collected from Scottish lochs in winter and stored to keep salmon fresh during its six-day journey to London in the spring and summer; later, trawlers would take it out to sea, but, as Beamon notes, "the supplies of ice were scarcely adequate, and had to be used sparingly and only for the most valuable species."

Ultimately, natural ice was just too expensive, too unreliable, and too ephemeral to rely on for large-scale food preservation. Until 1805, that is, when a short, slight high school dropout named Frederic Tudor launched a new industry: the international frozen-water trade.

In the mid-1800s, America's abundance of freshwater lakes and bitterly cold winters was seen as a valuable natural resource, equivalent to Saudi oil. Its commercial exploitation set the stage for cold to be not just democratized but industrialized.

When he began harvesting and shipping ice in 1805, Frederic Tudor never imagined that he was setting the stage for a domestic dependence on cold. He had grown up, like many wealthy New Englanders, enjoying ice cream and chilled drinks made possible by a small icehouse on his family's summer estate. Blessed with indulgent parents and a generous allowance, Tudor decided, aged thirteen, that further education was a waste of time. He quit an apprenticeship at a Boston store soon after and hung out on the family's estate instead, hunting, fishing, and dreaming up quixotic get-rich-quick schemes.

At first, the plan to sell ice seemed like another of the teenager's hare-brained ideas. After he accompanied his tubercular brother John Henry on a convalescent trip to Havana, during which the two New Englanders suffered mightily in the tropical heat, it occurred to him that no Cuban would be able to resist the lure of a frosty beverage, if only the ice could somehow be brought to him. "A man who has drank his drinks cold at the same expense for one week can never be presented with them warm again," he explained, outlining his plan to create ice addicts out of Havana drinkers by supplying the stuff for free to the city's bartenders for a limited time—then, once their customers were hooked, charging retail.

Despite the fact that he had no idea whether, or how, ice could be preserved on a long voyage to the tropics, Tudor was supremely confident of success. He wrote to a wealthy Boston politician from whom he hoped to secure funding that he and business partners would shortly be in possession of "fortunes larger than we shall know what to do with." The politician declined the investment opportunity, as did almost everyone but Tudor's wealthy new brother-in-law, who recalled later, in his autobiography, that "the idea was considered so utterly absurd by the sober-minded merchants as to be the vagary of a disordered brain."

The list of obstacles Tudor had failed to anticipate was lengthy. He had to buy his own boat after he couldn't find a single shipowner willing to take a cargo that would likely melt en route, ruining everything else in the hold before it drained into the sea, leaving the vessel short on ballast and hard to handle. He had not foreseen that the freezing temperatures necessary for ice to form would also leave Boston's harbor icebound, meaning that he needed to build huge icehouses to store his cargo until it could be shipped. And he hadn't imagined that Caribbean islanders might have no idea what to do with ice, let alone a place to store it. He estimated that he lost between $3,000 and $4,000 on his first voyage to Martinique in 1806—up to $100,000 in today's money. Worse still, the very next year, his father came home from an extended trip to Europe having somehow mislaid the family fortune.

Tudor blamed everyone and everything but himself: the weather; the ignorant New Englanders who declined to fund him and the equally un-enlightened tropical islanders who failed to recognize ice's benefits; corrupt authorities and faithless friends and family; even President Thomas Jefferson for a temporary ban on shipments to Cuba. "Have not all my undertakings in the eventful Ice business been attended by a villainous train of events against which no calculation could be made?" he complained in his journal. "They have cured me of superfluous gaiety. They have made my head gray; but they have not driven me to despair." Despite this bluster, one string of diary entries consists simply of the word ANXIETY printed in larger and larger block caps; later, motivational notes reminded him that he was still young and had plenty of time to find an alternative career. "You may yet get back into the old road," he advised himself. "Sell out in the best way you can and become a regular man." Less than a decade into his venture, he had been arrested three times for debt and jailed twice.

Nonetheless, Tudor carried on building the world's first refrigeration business from the ground up. His foreman, Nathaniel Wyeth, developed all the tools I used at Thompson Lake—the ice plow, breaker bars, and pike poles—and optimized the design of the icehouse itself, an architectural technology that was perhaps Tudor's biggest contribution to the industrialization of frozen water. Unlike the underground stone- or brick-lined icehouses catalogued by Sylvia Beamon, Tudor's icehouses were entirely aboveground and built of wood, with a layer of sawdust between double walls as insulation. Once the temperature rose above freezing, there was no way of preventing harvested ice from melting—the best that Tudor could hope for was to slow its disappearance. His revolutionary shingle-and-sawdust design, with its steeply pitched roof to dissipate the rising heat given off by a mass of slowly melting ice and a subterranean drain to siphon off meltwater, worked rather well, reducing shrinkage to less than 10 percent.

While Wyeth handled the technology, Tudor focused on business development. He gave ice cream–making demonstrations to confectioners, he offered coffee shop owners a water-cooling jug of his own design, and he came up with ice-block subscription models—customers could sign up for one or two deliveries a day, on a monthly plan. He even designed and built some of the earliest domestic iceboxes, which he called "Little Ice Houses," so that customers could store their daily allowance of ice at home.

Meanwhile, despite his self-pitying journal entries, Tudor had to admit that the nascent ice industry enjoyed some unique advantages. Ships departing New England ports were generally light on their outbound voyages, and frequently resorted to carrying stones as ballast, which they simply tossed overboard at their destination in order to return with foreign cargo. Once they were convinced that most of Tudor's ice wouldn't melt in transit, they gladly carried it at low rates: even a discounted cargo made more economic sense than a pile of rocks. Before the ice trade, the sawdust from Maine's timber mills had been similarly worthless—indeed, it often built up in rivers and caused flooding—meaning that Tudor could acquire vital insulation materials at a knockdown price. Perhaps most fortuitously, as Gavin Weightman, author of *The Frozen Water Trade*, explains, "since it could be classified as neither mining nor farming," the ice trade was not subject to any taxes.*

---

*A similar category confusion arose in the UK when the first shipment of ice arrived in London in 1822. As *The Times* reported, "the commodity being foreign, it was clear it should be entered at the Custom-house of London but whether under the head of *produce* or of

---

Ice-harvesting season also coincided with a quiet period for farm-hands, who could thus be persuaded to undertake what Sylvia Beamon describes as a "dreaded, drenching, dangerous job." Tudor is said to have plied his workers with whiskey to steel them against ice's rigors. "Its demands are peremptory, and if not instantly obeyed, it weeps itself away," a Massachusetts man complained at the time. "It is wet and heavy, sharp and cutting, and without grit or grain enough to keep quiet, it is ever uneasy." "Hot grog is the only game to keep one's toes from falling off here," confirmed another.

Gradually, the frozen-water trade took off. In the 1820s, there was already enough of a market to convince Asa Thompson, in Maine, to get into the business. By 1837, Boston got through several thousand tons of ice per year, and trade more or less doubled every three years for the next decade. New England ice found customers up and down the East Coast and as far away as London, Peru, and Calcutta. Tudor, now known as the Boston Ice King, was just one ice merchant among many—albeit an extremely rich one, hailed in the media as "a great public benefactor."

Visitors to the United States remarked upon the country's ice habit, often with envy. When a middle-aged Englishwoman, Sarah Mytton Maury, reminisced about her 1840s visit to the United States to visit her sister, she wrote, "Of all the luxuries in America I most enjoyed the ice—its use was then rare and expensive in England." Jugs of ice water cooled her bedroom on hot nights, friends received her with a glass of iced lemonade or a sherry cobbler with "huge crystals floating about," and dinner parties on sweaty August evenings always culminated in ice cream. The mint julep, popularized as a mixed drink in the eighteenth-century American South, became, with the addition of ice, the refreshing cocktail we know today. As one hostess reminded Maury, "Whenever you hear America abused, remember the ice."

Tudor died a wealthy man in 1864, and the industry he founded

---

manufacture, was a very puzzling question. After much dispute, it was proposed to cut the knot, by entering the commodity as *foreign* fabric."

continued to grow. By 1879, when the first national report on the frozen-water trade was issued, an estimated eight million tons were harvested annually—although, thanks to melt, only about five million tons reached consumers. Among them were the poor: the rise of street vendors selling small chunks of ice or "penny licks" of ice cream made the delights of cold accessible to all for the first time. More important, the ready availability of cheap cold allowed refrigeration to demonstrate its value in the food industry.

Alongside the adoption of the domestic icebox, fishermen began to take ice with them to preserve their catches, allowing them to stay at sea for longer and harvest deepwater stocks. Farmers realized ice could save them from having to travel to market in the middle of the night just to prevent their butter from melting before it could be sold. Cakes of natural ice were the refrigerant in the breweries set up by a wave of German immigrants in the 1850s, enabling them to make lager beer all year round; they cooled the country's first cold-storage warehouse, built to store poultry at New York City's Fulton Market in 1865; and they were packed alongside pioneering rail shipments of slaughtered meat in the 1870s. An industry that began by stimulating a decadent appetite for chilled drinks and ice cream had, inadvertently, ended up proving cold's commercial value for preserving food.

Tudor never seemed to worry that the natural-ice industry would be threatened by mechanical refrigeration. While he was building his first icehouse in Havana in 1807, he assured authorities there that the entire concept of manufacturing ice was "absurd." His confidence was reasonable: at the beginning of the nineteenth century, scientists had only a shaky grasp on thermodynamics, which made it difficult to explain the formation of natural cold, let alone re-create it artificially.

Yet the success of Tudor's frozen-water industry motivated an entire generation of would-be synthetic-cold manufacturers. "By slowing down time's destructive work, *le froid* increases the power and resources of Man," said one of them, a Frenchman named Charles Tellier. But what if man also had the power to *control* cold, producing it in quantity and at will?

III. *A Machine to Produce Cold*

"The sun was shining its brightest from a blue sky at 1:30 o'clock in the afternoon, and the pleasure-seeking people stood under the shadow of the storage warehouse," read the front-page story in *The Piqua Daily Call* on Tuesday, July 11, 1893. The warehouse in question had been dubbed the "Greatest Refrigerator on Earth" and was "one of the most notable landmarks" at the 1893 Chicago World's Fair. An article in the *Clearwater Echo*, published shortly before the fair opened, heralded the facility as a thrilling addition to the exposition in its own right. "To the average citizen the process of ice-making is a profound mystery," the paper's Chicago correspondent wrote. "Here he will be afforded an excellent chance to study the thing in detail."

Inside were four engines, operated by three immense boilers, powering a machine capable of turning out forty thousand tons of ice a day. The building's cold-storage vaults held perishable food for the fair's vendors, so that, as the *Echo*'s correspondent noted, "there will be no necessity of drinking warm water or eating melted butter in the fair grounds." The entire structure was clad in staff (an artificial stone made out of plaster of Paris and cement), spray-painted white, and ornamented to resemble a Venetian palazzo, complete with cupola-topped towers in the corners.* To add to its grandeur, the seventy-foot-tall engine-room smokestack, which also served to vent ammonia fumes from the ice-making machine, was disguised as a campanile, complete with another

---

*All the fair's temporary structures were clad in this gleaming white plaster, leading to the exposition's nickname: the White City. Katharine Lee Bates, the lesbian feminist professor whose poem became the lyrics to "America the Beautiful," with its reference to its "alabaster cities" that "gleam, undimmed by human tears," visited the fair just days before the fire, on July 1, 1893. She published her poem two years later, after the tragedy of July 11.

cupola. For scale, grandeur, and sheer wonder, it easily rivaled the fair's other attractions, which included a series of gardens, lagoons, and canals landscaped by Frederick Law Olmsted of Central Park fame, the world's first Ferris wheel, the world's first moving walkway, and the first commercial movie theater, at which Eadweard Muybridge showed his studies of animal locomotion.

But the reason the Greatest Refrigerator on Earth made the front pages of almost every newspaper in America on July 11, 1893, was horrifying rather than wondrous. THE COLD STORAGE WAREHOUSE GOES UP LIKE TINDER and TERRIFIED THOUSANDS SAW BRAVE MEN LEAP OR FALL TO DEATH, read the headlines. "Thousands of people, bent on pleasure, saw death in its most appalling and heartrending form at the World's Fair to-day," wrote *The New York Times*.

The fire began in the smokestack early in the afternoon of Monday, July 10. The World's Fair company of firemen gallantly rushed in and, without ladders, scaled the tower—only to realize that the flames had spread along the chimney's length, cutting them off from the ground. The fair's attendees, numbering in the tens of thousands, had, by this time, begun to gather around the warehouse, watching with increasing dread.

One man had the presence of mind to slide down the hose. "When he appeared on the roof, terribly burned but alive, a great shout went up," the *Times*' account continued. But the flames quickly swallowed the hose, and the rest of the firemen realized they "must jump or die where they were."

Two or three at a time, they took the "terrible leap," falling seventy feet to land on the tar-and-gravel roof, "only to be so solidly imbedded in the sticky, yielding composition that they could not have extricated themselves had they been granted the strength to do so." The screams of those being burned alive were audible to all.

The crowd panicked. "For nearly an hour it seemed as if the exposition buildings were doomed," explained *The Union County Journal* of Marysville, Ohio. "The whole crowd at the fair was gathered at the scene, and became almost riotous with terror." From their national pavilions, the French marine corps and the Spanish military guard joined forces with the cowboys from Buffalo Bill's Wild West Show to try to move people back from the blazing building, for fear that the ice factory's ammonia tanks might explode.

Within two hours, the sparkling white palazzo of cold was reduced to a smoking ruin. Over the next few days, workers struggled to dig the bodies out from beneath a tangle of melted cooling pipes. In all, more than a dozen people were confirmed dead, with many more severely injured or missing. The structure's blackened frame subsequently became one of the fair's main, if morbid, attractions.

The conflagration's cause was never ascertained, but Chief Swenie of the Chicago Fire Department blamed the ammonia in the ice-making machinery. "That cold storage building would never have been built if the city authorities had possessed a voice in the matter," declared Mayor Carter Harrison.

At the time, in the 1890s, mechanical refrigeration was still a novelty. Ice-making machines were expensive and enormous but also temperamental: they were prone to leaks, fires, and explosions. The cold-storage building at the Chicago World's Fair wasn't insured, because it was considered too risky. There were fewer than a thousand such machines in the entire country; just fifteen years earlier, in 1875, there were only a few dozen prototypes, each unique.

Although the era of machine-made cold had just begun, its underlying mechanism was developed more than a century earlier, in 1755, when a Scottish doctor by the name of William Cullen became the first person to freeze water without the use of natural ice. Cullen taught medicine at the universities of both Glasgow and Edinburgh, where he was renowned for giving lively and enthusiastic lectures that attracted large numbers of students. His cold-related investigations seem to have been a hobby, prompted by an observation made by one of his pupils that, when a mercury thermometer that had been immersed in wine was removed, the temperature fell by two or three degrees. (Benjamin Franklin, future US founding father, had noticed this phenomenon a few years later, on the other side of the pond, and speculated that alcohol-soaked bandages might help doctors draw the heat from burns more effectively or, indeed, freeze "a man to death on a warm summer's day.")

Cullen took this insight and drew on his scientific predecessors' earlier experiments involving vacuum pumps and the evaporation of volatile fluids. After years of tinkering—early attempts used vinegar, brandy, mint, and even chili oil as a refrigerant—he succeeded in creating the first device capable of turning water into ice on demand. His system consisted of a pair of glass vessels placed in a vacuum chamber so that the liquid in one—"nitrous ether," a mixture of alcohol and nitrate salts that boiled at an extremely low temperature—evaporated at high speed. As the liquid ether turned into a gas, it removed energy from the air around it, finally making it cold enough to freeze the water in the other vessel.

The basic principle behind today's refrigerators had been demonstrated. "Such a means of producing cold and to so great a degree, has not, so far as I know, been observed before, and it seems to deserve being further examined," he wrote in an essay summarizing his findings. His suggestion seems to have found few takers. To paraphrase poet Robert Browning, humanity's reach had exceeded its grasp: the ability to create cold on demand had preceded the ability to imagine what such a power might enable.

It wasn't until Frederic Tudor industrialized the production, distribution, and storage of natural ice in the early decades of the nineteenth century that entrepreneurial inventors began to see the benefits of taking control of cold out of nature's hands. In 1834, an American engineer in London, Jacob Perkins, designed a system that used Cullen's basic setup but added the condenser and compressor necessary to recycle the ether in an endless loop in order to produce continuous cold. He hired a millwright to build a small-scale demonstration model—but, although it worked, he dropped the project to focus on the more lucrative business of developing high-pressure steam engines for the burgeoning British railway network.

Finally, in the 1850s, came a flowering of cryospheric creativity—and commercialization. Independently, a Florida physician and a Connecticut engineer both developed working prototypes, but it was an Australian who succeeded in selling the world's first ice-making machine. The first shipment of natural ice from Boston had arrived in frost-free Australia in 1839, but the enormous distance led to delays, excessive melt, and high prices—all of which made it easier for a man-made substitute to compete. Journalist James Harrison, the son of a Scottish salmon fisher, had settled in Geelong, just outside Melbourne, where he founded the city's first morning newspaper. In order to print in the heat, he used to wipe the type with ether—as it evaporated, it cooled the metal and prevented smudging. Harrison took that insight and recruited a blacksmith friend to help him build a steam engine–powered, ether-based, vapor-compression ice-making machine in a Skunk Works–style cave near the Barwon River.

After at least two serious explosions, one of which left Harrison hospitalized, and a return trip to the UK to consult with steam engine manufacturers in order to refine his design, he was able to sell the resulting contraption to both Truman's brewery in London and the Glasgow & Thunder brewery in Bendigo, Australia. As J. E. Siebel, the author of the refrigeration industry's first textbook, published in 1895, put it, it was the brewers who "were ready to pay the lesson money which always must be paid when a great scientific discovery is to be translated into practical usefulness." Just as the desire for beer is thought to have motivated early hunter-gatherers to take up farming, breweries provided the all-important early investments in mechanical refrigeration: two technologies that remade the world, both fueled by the human desire for intoxication.

War, as is often the case, also provided an impetus. In the 1860s, the American Civil War cut the Southern states off from the shipments of lake and river ice upon which they had become dependent, and several inventors seized the chance to build prototype ice machines as replacements. San Antonio, Texas, had three by 1867, while New Orleans smuggled in two from France through the Union blockade. These early units produced such limited quantities that the ice was mostly reserved for hospitals and the wounded, rather than for cocktails and Popsicles. Nevertheless, they still represented mechanical refrigeration's first real toehold in the United States. Meanwhile, in India, where Tudor's natural ice was significantly more expensive due to its long journey, the first successful steam-powered ice machine was up and running by 1878.

At first, mechanical refrigeration was mostly used for making ice to cool food and drinks, as opposed to cooling spaces in which to store food and drinks. That conceptual leap was also pioneered by brewers. During the nineteenth century, millions of German immigrants arrived in the United States. They brought with them a taste for lager beer, which, unlike English-style ale, requires consistent low temperatures to ferment. Without ice, lagering caves in Saint Louis, Milwaukee, and even New York City became too hot in the summer to brew all year round. Meanwhile, American beer consumption skyrocketed following the Civil War—it went from fewer than four gallons a year per person to twenty-one during the five decades following 1865.

In the late 1800s, the brewing industry was thus, unsurprisingly, among the largest consumers of natural ice in America. But in the face of recurring "ice famines" caused by unseasonably warm winters, as well as growing competition for ice from consumers in America's booming East Coast cities, mechanical refrigeration increasingly seemed like a cost-effective option for the beer industry. Besides, ice wasn't ideal: as it melted, the temperature fluctuated, causing fermentation problems, and damp cellars often grew mold. Brewers gradually realized that, rather than use refrigeration machinery to make ice, they could cut out the middleman and chill their wort by flowing it around pipes filled with evaporating refrigerant. Soon afterward, it occurred to them that those same pipes, if mounted on the ceiling, could cool the entire cellar.

S. Liebmann's Sons in Brooklyn was the first American brewery to install a refrigeration machine, with better-known names such as Busch and Pabst following close behind. In early marketing materials published by the De La Vergne company, a commercial refrigeration manufacturer founded by a former brewer, an anonymous customer extolled the virtues of machine-made cold, noting that now his "cellars and fermenting rooms are kept as cool as I desire, while the air in them is dry and fresh, a marked contrast with the condition when ice is used and which only a brewer can appreciate."

During these first decades, refrigeration was in its experimental phase, with each new machine incorporating different attempts to solve the technology's manifold safety, efficiency, and engineering problems. In Grasse, France, a Trappist monk who wanted to cool the wine produced at his monastery invented the first hermetically sealed compressor, which went a long way toward fixing the leakiness that plagued early machines. Others came up with better valves, refinements in condenser design, and new refrigerant liquids. Widespread electrification, which began to roll out across America, Britain, and Germany from the 1880s, also helped: as steam-powered machinery was replaced by electric pumps and motors, refrigerators gradually shrank to more manageable dimensions and expense.

Still, as late as 1907, New York City, already a modern metropolis filled with automobiles and skyscrapers, relied on natural ice, harvested from lakes upstate and brought down the Hudson River on barges. It wasn't until the 1930s that mechanical refrigeration conclusively triumphed over the natural-ice trade. In surprisingly large part, its eventual success was due to pollution. As American cities and factories grew, in an era that predated environmental regulation, more and more untreated waste was dumped into nearby lakes and rivers. And as the germ theory of disease became accepted medical wisdom and people realized that bacteria did not die in frozen water, drinks cooled with chunks of natural ice began to be associated with health scares, from typhoid to diarrhea.

Once the natural-ice trade lost to its synthetic rival, it vanished almost without a trace, replaced by the artificial cryosphere of cold-storage warehouses and domestic refrigerators for which it had laid the foundation. Most of the subterranean icehouses of the UK were abandoned and forgotten by all but the bats who frequently occupy them in winter as a hibernacle. In the United States, where icehouses like the one at Thompson Lake in Maine used to be found on the edge of any sizable pond or river in the northern states, the disappearance was even more complete. The sawdust-stuffed wooden structures have a tendency either to burn or to fall down—after several summers of housing ice as it slowly melted, they almost all developed a pronounced sag, leaning ever further southward. Even the one I helped fill at Thompson Lake is a reconstruction.

3.

THE WAY OF
ALL FLESH

I. Where's the Beef?

Gustavus Swift achieved fame and fortune by refrigerating—and, in so doing, revolutionizing—the American meat supply. "He was not to change the world's maps, nor make military history," his eldest son, Louis, wrote in his biography of his father, *The Yankee of the Yards*, co-authored with journalist Arthur Van Vlissingen Jr. "Instead, he was the human instrument by which destiny transformed the world's sources and supplies of an essential class of foodstuffs." According to his son, Swift wasn't driven by grand ambition or a desire to benefit mankind. Instead, his motivation sprang from a much less lofty place: a mania for saving money.

By way of illustration, Louis explained that one of the places Swift most liked to visit was Bubbly Creek, the notoriously fetid open sewer that conveyed blood and entrails from the Union Stock Yards into the Chicago River. Upton Sinclair's novel *The Jungle*, a fictionalized account of working in the city's meat-packing plants, offers an explanation of how the sewer came by its moniker: "The grease and chemicals that are poured into it undergo all sorts of strange transformations, which are the cause of its name; it is constantly in motion, as if huge fish were feeding in it, or great leviathans disporting themselves in its depths. Bubbles of carbonic acid gas will rise to the surface and burst and make rings two or three feet wide. Here and there the grease and filth have caked solid, and the creek looks like a bed of lava; chickens walk about on it, feeding, and many times an unwary stranger has started to stroll across and vanished temporarily."

Swift knew exactly where his company's outfall emptied into Bubbly Creek. At the risk of ruining his boots, he went there on a regular basis, in order to see whether he could spot any fat in the effluent flowing from his packing plant. If he did, someone was going to hear about it. "Father did not believe in sparing overmuch the feelings of the man who needed correction or guidance or reproof," wrote Louis. "He felt that if a man needed talking to, the talk had better be strong and to the point."

Swift's disappointment was not evidence of an advanced environmental conscience. Rather, he abhorred waste, and fat in Bubbly Creek was fat that wasn't being monetized as margarine. It was being wasted—"and to my father any waste was too much!"

Louis credited his father's frugality to his ancestors, "who for two hundred and fifty years had fought a none too equal battle with the miserly sands of Cape Cod." Born a younger son in a family of twelve in the tiny village of Sagamore, Massachusetts, just before the major depression of the 1840s, Swift "did not have much of a chance." But he had a gimlet eye for waste, his elder brothers were butchers, and he came of age just as natural ice began its adoption in the food industry.

A t the start of the nineteenth century, when Frederic Tudor first began shipping and selling ice, more than nine out of every ten Americans still lived in the countryside and ate what they could grow or gather and preserve. Because food availability was still seasonal and local, there was no average American diet in 1800, and poor, indigenous, or enslaved people were often unable to access enough to eat. Still, researchers agree that most pre–Civil War Americans sustained themselves quite adequately on a diet of corn, wheat, oats, and meat—lots of meat, more meat than we eat today, whether it was squirrel, beaver, salted and pickled pork, or freshly slaughtered beef in the autumn. Animal fats and dairy, newly laid eggs in the springtime, fish where available, and freshly harvested and dried, pickled, or sugared fruits and vegetables made up the balance.

Of course, many of those foods were not produced, or readily available, in an urban environment—but in the 1800s American cities were few and relatively tiny. In 1839, when Gustavus Swift was born, New York City held fewer than half a million people.

That began to change—rapidly—when Swift was still a teenager. Between 1850 and 1860, the population of New York City nearly doubled, reaching more than a million; the population of Philadelphia more than quadrupled. Through the second half of the century, hundreds and hundreds of thousands of both freshly arrived immigrants and rural Americans moved to the nation's cities, looking for work in newfangled factories.

As long as cities had existed, feeding them had been a tricky business. London, which became the largest city the world had ever seen in the 1830s, was the first to struggle in modern times, and thus, unsurprisingly, the first to study the matter scientifically. Although its population jumped by nearly a million people between 1841 and 1861, reaching nearly three million, postharvest supply chains remained necessarily short for perishable goods such as meat and milk. "Eighty miles was the farthest distance from which carcasses ever came," British physician Andrew Wynter wrote in 1854.

Some livestock was raised in the city. Throughout the 1800s, Londoners could consume eggs from local chickens, as well as milk from cows that spent most of their lives in the dark, housed in basement dairies under the Strand and only occasionally hauled up and sent out of town for a quick respite graze aboveground, in the weak sunlight of northern pastures. In 1856, when British writer George Dodd published an in-depth examination of where the city's food came from, he noted that at least three thousand pigs were kept in "a group of wretched tenements" in Kensington, where "some of the pigs lived in the houses and even under the beds." Manhattan's porcine inhabitants were similarly plentiful. During his first visit to America in 1842, Charles Dickens ran into "two portly sows" and "a select party of half-a-dozen gentlemen hogs" strolling down Broadway, while in 1848, *The New York Times* described the "Pigtowns" of Central Park as full of Irish immigrants living in "shanties in which . . . little ones of Celtic and swinish origin lie miscellaneously, with billy-goats here and there interspersed."

Most meat eaten in cities walked itself to market, often over enormous distances. Much of the lamb consumed in ancient Rome had traveled hundreds of miles on its own four legs; in George Dodd's day, cattle regularly embarked on a three-week death march from Scotland in the autumn, in order to be slaughtered in central London's Smithfield Market. Colonial Americans established the same kind of meat supply chains in the New World, where even turkeys trotted to town. In his 1912 book, *The Wayside Inns on the Lancaster Roadside between Philadelphia and Lancaster*, local historian Julius F. Sachse described "one of the curious sights, common in the fall of the year, half a century ago": "flocks or droves of fowls, generally turkeys, but now and then also geese, being driven towards the city." Even nudged by "shooers" armed with long poles, these turkey trains moved at about a mile an hour, a speed that decreased as the day wore on, "for as it commenced to grow dark, the birds were determined to go to roost, and then the fun commenced."

These supply chains were not exactly optimized, which meant that, although the rural majority ate plenty of meat, it wasn't common fare in cities, except among the elite. Preindustrial meat was also likely quite tough by today's taste, and generally slaughtered in less than hygienic conditions—but it was usually extremely fresh. The issue, particularly in the larger cities, was that it was expensive, and that there was never enough.

The 1800s were also the moment when Western governments began to count everything—and all their censuses and surveys seemed to reveal evidence of a "meat famine." In Britain, the population increased fivefold between 1800 and 1914, while the meat supply remained static. In an 1868 address to the Society for the Encouragement of Arts, Manufactures and Commerce (now better known as the Royal Society of Arts), chemist Wentworth Lascelles Scott reported that he had run the numbers, and not only was the United Kingdom confronted with a "terrible deficiency" of animal food, but neither homegrown meat production nor European herds could feasibly fill the gap. "As we cannot hope, then, to manufacture meat, where from, and how shall we obtain it?" he asked. This was, it was generally acknowledged, "the great food-question" of the era.

This obsession with maximizing meat intake was prompted by recent findings from the relatively new discipline of organic chemistry. In the 1830s, European chemists had isolated and named protein. Based on an experiment in which dogs fed a diet consisting of just carbohydrates and fat died, one of the field's leading lights, Justus von Liebig, mistakenly concluded that protein was the only truly nutritive element in food. Protein both built muscle and provided the energy to flex it, he asserted, while carbohydrates existed purely to help respiration function smoothly. Factory owners, generals, and governments—indeed, anyone who wanted to squeeze the most productivity out of their minions—began to concern themselves with the adequacy of the workers' meat and dairy intake. "Manufacturing demanded energetic, flesh-fed men, but meat supplies and their prices were on a most unsatisfactory basis," concluded an analysis of the situation authored by an engineer and a meat industry expert.

Frustratingly, on the thinly populated Great Plains and Texas rangelands in the United States, the endless pampas of South America, and the verdant hillsides of New Zealand, there were cattle and sheep aplenty. Imagine the torments of the meat-starved Londoner when Australians slaughtered their herds of sheep for fleece and tallow and left the meat to rot for lack of sufficient local mouths to feed. Argentinians had the gall to complain of their burdensome surplus, reporting that their livestock multiplied "in such numbers that, were it not for the dogs that devour the calves and other tender animals, they would devastate the country." In at least one instance, "so great was the congestion" that a flock of sheep was driven over a cliff, just to get rid of it.

"We see evidence of the presence in other lands of vast stores of animal food, a fraction of which, were it only here, would strengthen our people, diminish our poor-rates, and almost make the East of London happy," lamented Scott. "The entire question now concentrates into one sentence—how is meat, or any similar product, to be prevented from undergoing that curious change we call putrefaction?" Experts in chemistry—the recently rationalized offspring of the occult alchemical quest for an elixir vitae, a substance that would indefinitely prolong life—were called upon to recommit themselves to the pursuit of immortality, at least in beef.

Just a couple of years prior to Scott's call to arms, the Society had set up a special committee of fellows to consider this issue. Many chemists and other enterprising minds had already turned their attention to the challenge: Scott counted several hundred patents for novel methods of food preservation issued in Great Britain alone. At the Great Exhibition of 1851—the first of its kind, held in the Crystal Palace in London— hundreds of examples of these preserved foods were on display, many of which were listed in an intriguingly titled pamphlet: "Substances used as food, as exemplified in the Great Exhibition."

American contributions to the protein-preservation challenge included Gail Borden's "meat-biscuits" (a "dry, inodorous, flat, little cake" that rehydrated into a nutritious soup, "somewhat of the consistency of sago") as well as Charles Alden's vermicelli-like dried fish bits. Justus von Liebig's concentrated Extract of Meat—the protein panic–era predecessor of today's bouillon cube—promised much but turned out not to contain meat's vital nutrients: "like the play of *Hamlet* without the character of Hamlet," as one physician put it. There was powdered beef, beef compressed under hydraulic pressure, mutton fumigated with sulfur gas, and mutton coated in creosote, all of which, the pamphlet admitted, "had lost much of the freshness and flavour peculiar to newly-killed meat." For a while, Scott's fellow fellows thought charqui—sundried, salted meat traditionally eaten by the Quechua people, and the ancestor of today's jerky—was the solution they'd been looking for, and "jerked beef banquets" were held in London in the 1860s. It did not catch on. A contemporary report notes that, in terms of taste, texture, and aesthetic appeal, charqui was easily mistaken for "a roll of rubber roofing felt."

Of all the many methods of preserving meat—coatings, antiseptic injections, fumigation, compression, desiccation, and more—the one that routinely failed to excite enthusiasm was cold. Wentworth Lascelles Scott voiced the opinion of many when he declared that, while "keeping meat, &c., in ice or freezing mixtures, the construction of ice-safes, &c." certainly prevented putrefaction, these methods were "all necessarily of very limited application": expensive, unreliable, and, worst of all, liable to "impair" meat's nutritional value. Ice was fine for certain uses—for cooling drinks, making ice cream, and helping a fisherman preserve his catch—but to the experts of the day, it seemed to have no bearing on the problem of urban-scale meat preservation.

Then came the proliferation of steamships and railways, knitting distant parts of the planet together. Suddenly, those vast reserves of previously off-limits but livestock-friendly land were substantially closer, close enough that they could finally be mined for meat. Once the railways reached Chicago in the 1850s, live cattle could be shipped to the East Coast. From the Pennsylvania Railroad terminus in New Jersey, barges ferried unhappy midwestern steers across the Hudson to Manhattan slaughterhouses.

The price of a steak in New York City dropped immediately. More cattle were coming to town; western beef cost less than the locally raised kind because land and corn were cheaper in the West than in the more densely populated East; and beef that used fossil fuels to travel rather than its own energy reserves lost less weight en route, leaving more of itself available to sell. But there were still significant downsides. Killing even more animals in ever more crowded urban centers left nearby streets "all asmear with filth and fat and blood and foam," as a horrified Pip put it in Charles Dickens's *Great Expectations*. The sounds were harrowing, the smell perhaps worse. Even butchers admitted to occasionally losing their lunch in summertime. People who lived in cities wanted cheap, fresh meat, but not the sensory horrors that came with such scaled-up urban slaughter.

Gustavus Swift had started working for his brother, a local butcher, at the age of fourteen. In those days, butchers bought live cattle at the local market, slaughtering them and dressing the meat themselves. Many of the animals the brothers purchased had been brought in from the West by rail, and young Swift couldn't help but rue "the waste of buying cattle which had passed through the hands of too many middlemen." Worse, only a little over half of each steer was meat, which meant that a lot of bones, guts, and gristle were being shipped a thousand miles only to be thrown away or sold for pennies. In his thirties, he moved to Chicago, "to buy the cattle nearer to their source, to eliminate these extra charges."

It wasn't just the freight fees paid to transport the inedible portion of a steer that chafed at Swift. The cost of feeding the cattle in transit and the losses incurred when animals became bruised or died on the long journey also rankled. Perhaps most irritatingly of all, shipping live cattle for individual local butchers to slaughter meant missing out on the revenue from by-products. Intestines that could have been sold as perfectly good sausage casing, shinbones that might have become knife handles, blood and fat that should rightfully have been monetized as fertilizer and margarine—all instead represented a gigantic and unbearable waste.*

The obvious solution was to kill the cattle out West and ship just the beef. The challenge was that the travel time from Chicago was a week or more, which meant that, except in midwinter, the meat would be rotten by the time it reached East Coast cities.† Add another ten to twelve days for the steamship voyage to the United Kingdom, and the putrefaction problem was real. But by the time Swift moved to Chicago in 1875, Frederic Tudor's company had been successfully transporting ice around the globe in boats for decades. Why shouldn't ice also travel by railroad, helping keep dressed beef cool? Despite the chemists' misgivings, Swift decided to give cold a chance.

"'That crazy man Swift,' the wiseacres called him," wrote Louis. "It was one of those things which everyone knew couldn't be done." One of Swift's oldest business partners back east, James Hathaway, broke with him over it. "Hathaway knew, as did everyone else, a thousand reasons why nobody could sell Chicago-dressed beef in the East, and why the East would continue to eat meat from cattle shipped alive for slaughter at the point of consumption." Even Swift's own family doubted him: "'Stave's Wild West scheme' it came to be known among the Cape Cod relatives."

Like Tudor before him, Swift didn't listen to the doubters. (Indeed, Louis makes it clear that Swift never admitted he might be wrong, ending conversations in which it was clear he was in error by saying, "Let's talk about something else.") And like Tudor's before him, his road to success was "long and wearying" and filled with more difficulties than he could have imagined. "His technical troubles with the cars and with

---

*Louis Swift claims that his father was the likely source of the common saying that Chicago packers used every part of the pig but the squeal. It "probably had its inception in a remark my father once made, when the by-products utilization was complete, that 'Now we use all of the hog except his grunt.'"
†In midwinter it might easily have frozen, thawed, and frozen again by the time it reached New York City, with predictably deleterious effects on its eating quality.

getting the meat chilled properly before hanging it in the cars very nearly broke him," according to Louis.

By the 1860s, a handful of inventors had patented designs for railcars that were cooled by natural ice. Swift tried them all but found that none worked to his liking. In some, the carcasses would touch the ice and become freezer burned, others left the beef soaked in meltwater and rotten, and still others stayed so warm the meat spoiled. Ventilation proved particularly vexing: if the cold air didn't circulate throughout the car, its preservative powers were uneven. For a while, Swift was optimistic about an iced railcar with a fan attached to its axle. It worked perfectly, until the train stopped moving. In short, Louis concluded, "we lost a good deal of meat." Car lots of beef would arrive on the East Coast "good only for dumping into the bay at Fall River."

Dissatisfied, Swift hired an engineer, and together they worked out a new design that combined the best features of existing patents with some improvements of their own. Getting the meat chilled properly before loading it into the cars presented another obstacle. The sheer quantity of cattle that needed to be slaughtered to fill a dozen railcars overwhelmed Swift's existing ice-cooled chilling rooms. "It was like bringing in so many loads of hot bricks," Louis recalled. "They raised the temperature effectively and frequently held it high through the night until the next day's fresh kill came in to reinforce them. If this kept up for two or three days, perhaps every carcass in the cooler was still warm to the touch"— and bound to deteriorate long before it got to market.

Although he was losing more money than he could afford, Swift seemed undaunted. "'The trouble is, we don't quite know how to do it right,' he would admit with not a trace of discouragement," Louis wrote. "'We'll get it, though. We'll learn.'" After years of effort and long days spent watching the thermometer, Swift solved the carcass-chilling problem only to confront yet another challenge: the railway companies wouldn't build his refrigerator cars, and they wouldn't take his business even if he built them himself. They already owned cars designed to ship live animals, which brought them double the tonnage. Unsurprisingly, Swift's scheme, which promised to halve their revenue and undercut their existing infrastructural investment, didn't appeal.

Swift persuaded a Michigan firm to build his cars and an upstart Canadian railway company whose line ran along the border, with spurs down to Chicago and Boston, to haul them. He bought up the rights to ice from lakes across Wisconsin and southern Ontario and built a chain of icing stations alongside the tracks to top up the railcars every couple of hundred miles.

It was a massive investment—and it worked. By 1880, the Swift refrigerator car and the Swift method of slaughtering and packing beef in Chicago were an indisputable success. "The savings through dressing beef in Chicago instead of shipping live cattle east constituted so large a sum per head that Swift beef [. . .] could be sold below the market and still leave a handsome margin," gloated Louis. "From the time he had his refrigerator car lines going, he was the largest slaughterer of beef."

Where Swift led, his competitors followed, and the amount of beef being shipped from Chicago skyrocketed over the next decade, growing from little more than a rounding error in the late 1870s to surpass the total tonnage of live animals—half of which was inedible hoof, hide, and bone—by the late 1880s. The dead-meat trade was here to stay.

A t the same moment that Swift had begun his early experiments in ice-cooled train transportation, a handful of even more ambitious individuals were attempting to ship meat from the New World to the Old under mechanical refrigeration. As we've seen, this technology was very much in its infancy in the 1870s; it was considered too untried and risky to insure as late as 1893, when the Chicago World's Fair took place. Given these all-too-evident dangers, as well as the unwieldy dimensions and erratic performance of early machinery, moving meat between continents under mechanical refrigeration seemed even less likely to succeed than shipping it cross-country using ice. To its pioneers, however, the quest to keep meat cold as it voyaged overseas was more of a calling than a business enterprise. "I feel, as I have always felt, that there is no work on the world's carpet greater than this in which I have been engaged," declared Thomas Mort, an Australian wool broker, of his early meat-freezing efforts. "Where the food is, the people are not; and where the people are, the food is not. It is, however . . . within the power of man to adjust these things."

New York livestock dealer Timothy C. Eastman had managed to successfully ship a cargo of beef from the United States to London using natural ice in 1875, presenting Queen Victoria with a two-week-old double sirloin that she pronounced "very good." As Swift had demonstrated, for the duration of a five- or six-day train journey in the United States, the cooling provided by ice, topped up at regular intervals, sufficed to keep meat edible. But for a ship crossing the Atlantic, ice was less than ideal. With no opportunity to replenish supplies during the voyage, a full quarter of the boat's cargo hold had to be sacrificed to ice. For Argentinians and Australians, the distance between their huge herds of cattle and sheep and the meat-deprived cities of Europe was simply too great to be bridged by natural cold.

Charles Tellier, the French engineer whom we last encountered exalting cold's power to slow time, had determined at an early age that his efforts would be directed exclusively toward engineering problems whose solution would, as he saw it, improve the human condition.* It was Baron Haussmann, then engaged in demolishing the city's medieval neighborhoods, who first suggested that Tellier should consider cold—making refrigerated transportation one of the lesser-known side effects of

---

*Among his early initiatives was a scheme to divert Parisian sewage from the Seine and process it into fertilizer and biofuel. Recently, construction commenced on the city's first neighborhood-scale urine reclamation scheme designed to do just that, but in 1855 Tellier's peecycling proposal was not met with enthusiasm.

Haussmannization, alongside modern sewage systems and iconic Parisian boulevards.

Tellier quickly realized that this new technology of refrigeration, if only it could be made seaworthy, held the potential to transform scarcity into abundance at a global scale, feeding the French while providing New World farmers with much-needed income. He threw himself into his new venture. On one particularly eventful day in the lab, he was splashed with acid, badly burned, and nearly blinded when one of his early experiments exploded. Soon he had invented his own improved refrigerating machine—one that, as his patent application noted, was designed to work even at sea, where the "motions of the ship" would cause other designs to leak and fail.

Nonetheless, at its first trial, in 1868, his refrigerator also failed. It stopped working twenty-three days into the voyage between France and Uruguay and couldn't be repaired; the meat had to be eaten on board. Tellier refined his design and spent eight years trying to secure the funding to try again. The SS *Frigorifique* set sail from Rouen, France, in September 1876, carrying three Tellier refrigerating engines and several hundred pounds of beef. It arrived triumphant, its cargo still edible, in Buenos Aires on Christmas Day.* (Tellier's goal was to import South American meat to France, not vice versa, but as his machines had to be manufactured in France, their trials necessarily took place in the other direction of travel.)

Despite this successful proof of concept, Tellier struggled to find investors willing to fund further shipments. "I heard no encouragement from the government or the capitalists generally, and there things remained," he wrote. "I turned my attention to other problems." He designed a solar-powered water pump for use in Africa and embarked on "the study of replacing coal" with energy derived from air. "Before I die, I hope to be able to discover the process by which it can be

---

*Although the meat had not rotted, the deputy president of the Argentine Rural Society reported that its appearance was marred by "dark spots" and that "the flavor of the most part of it was rather unpleasant," for which he blamed the use of turf as ballast.

accomplished," he told the *Cold Storage and Ice Trade Journal*. Sadly, he did not—and having failed to find commercial success with his inventions, Tellier ended up "in dire straits," according to a cabled news item in *The New York Times*. He perished, the *Times* continued, "in terrible agony," suffering from both hunger and cold.

During the decades Tellier spent trying to ship meat from South America to France, a handful of inventors, engineers, and entrepreneurs had also been trying—and failing—to ship meat from Australia to Britain. Although Tellier managed it first, the Australians eventually triumphed, financially as well as practically, but not before fires, leaks, and engineering difficulties had taken their toll.*

Among the disappointed was wool broker Thomas Mort, whose enthusiasm and optimism for the frozen-meat trade had once known no bounds, but who ended up describing it as a "diabolical idea." At the grand opening of his freezing plant in Sydney, he admitted to the assembled dignitaries, "I can tell you that not once but a thousand times have I wished . . . Mr. Nicolle [his engineer] and myself had never been born." Shortly thereafter, the cooling pipes burst before his trial shipment even left harbor, ruining the entire cargo. "This failure was a terrible blow to Mr. Mort, and hastened his death," according to contemporary reports.

Fortunately for the meat-hungry British, success was close at hand. Henry Bell, a Glasgow butcher who dealt in T. C. Eastman's ice-chilled beef, had firsthand experience of both the demand for imported meat and the downsides of ice-based cooling. He hired J. J. Coleman, a Scottish engineer, who devised a refrigerating machine so compact that ships could afford to carry a backup, just in case. The Bell-Coleman refrigerator led to a series of firsts: the first New York–London shipment of beef chilled by mechanical refrigeration, in 1879; the first successful shipment

---

*As a French doctor in Buenos Aires noted, with a certain bitterness, "as has often happened in the history of industries, it has been the French who have made the discoveries, and the English who have turned them to account to their profit."

of frozen Australian beef and mutton to London, in 1880; and finally, the first shipment of mutton all the way from New Zealand, in 1882.

British butchers were dubious but, after examining the "fine big sheep," pronounced the mutton "as perfect as meat could be." The *Daily Telegraph* declared that the beef was "in such good condition that neither by its appearance in the butchers' shops, nor by any peculiarity of flavour when cooked for the table, could it be distinguished from freshly killed English meat." Rejoicing ensued. The time had finally arrived when, as the unlucky Mort had prophesied, "the various portions of the earth will each give forth their products for the use of each and of all; that the over-abundance of one country will make up for the deficiency of another." Mort had already died by this point, but no matter: cold had triumphed. "Climate, seasons, plenty, scarcity, distance, will all shake hands, and out of the commingling will come enough for all."

The rise of the dead-meat trade ushered in a new era in both production and consumption. Perhaps most predictably, it triggered a beef boom: the cattle population of the United States more than doubled, from 15 million in 1870 to 35 million in 1900.

To kill that many steer, Swift and his competitors developed the "disassembly line," in which each carcass was kept in continuous motion on an overhead rail. Individual workers performed the same job on one dead body after another: skinning, gutting, splitting backbones, removing organs, and draining blood. Henry Ford claimed to have come up with the idea for his Model T assembly line, an innovation that would, in turn, revolutionize manufacturing, after seeing "an overhead trolley that the Chicago packers used in dressing beef."

Meat-packing jobs became low skilled, low paid, repetitive, and exploitative; butchery, as a skilled profession, all but died out. Up and down the East Coast and throughout the UK, butchers who used to slaughter cattle for their customers started selling dead meat instead. "Of course, the 'butcher' is not now a butcher, but a meat retailer," industry analysts James Critchell and Joseph Raymond wrote in 1912. "The frozen meat trade has brought about the change."

Refrigeration changed where cattle were raised, as well as where they were killed. Previously, farmers in New England, upstate New York, or the mid-Atlantic were able to make up for the fact that their land and feed prices were higher than those of their Western competitors by avoiding railway freight fees. With the closing of local slaughterhouses, they, too, increasingly got out of the livestock business. Ranching in remote, rural Texas and the Great Plains became newly profitable, contributing to the ongoing displacement of Native Americans and the near extinction of the bison upon which they had depended. "In New Zealand, the freezing industry arrived in the nick of time to place farming in that country on a paying basis," Critchell and Raymond concluded, thus permitting European colonization to proceed apace.

Farmers in Europe also got the short end of the stick. Before the arrival of the first successful shipment of frozen meat in 1880, two-thirds of the meat sold at London's Smithfield meat market was produced in the UK. (Much of the remainder was shipped live from Europe.) Within three decades, 45 percent was imported as dressed meat—at which point Yorkshire farmer John Mackenzie told Critchell and Raymond that, "speaking as a sheep owner," he was personally aware that "the importation of frozen mutton has reduced the profits of the rank and file of British flock owners to vanishing point."

Increasingly, Mackenzie and his fellow sheep farmers abandoned their unrewarding efforts to raise animals on the North York Moors. Instead, they began to burn its grassy tussocks to encourage the growth of heather, the preferred food and nesting habitat of red grouse, which aristocrats and the emerging middle classes would pay to shoot. Previously poor, acidic, and overgrazed grassland quickly became a summer idyll, first for fashionable Victorian gentlemen in game-shooting parties and later for poets, artists, and tourists who marveled at the "wild" beauty of the pink-

and purple-clad hillsides. A working landscape was transformed into a national park, in just one of the curious epiphenomena of refrigeration.

A similar transition took place in the Scottish Highlands, where, as Critchell and Raymond reported, "vast areas" had "been cleared of sheep in recent years and put under deer." The river of dead meat flowing into Britain stimulated a countercurrent in live humans, as former farmers emigrated—"in many instances, betaking themselves to Australia, New Zealand, the Argentine, and the United States of America."

In Ireland, which had already lost a quarter of its population due to hunger and emigration during the potato famines of the 1840s and '50s, tenant farmers were all but wiped out by the crash in meat prices, further fueling the rise of the Irish independence movement. "Irish agricultural incomes were dependent on the British price of beef," economist David McWilliams has argued. "The arrival of cheap meat from the River Plate was the biggest change to Irish agriculture ever. It also changed Irish politics forever."

Unable to compete, tens of thousands more Irish tenant farmers emigrated to the Americas, and the ones who remained blamed the British for their losses, contributing to the country's growing anger and appetite for change. In Britain, absentee landlords, alarmed by the fall in value of their Irish holdings, began to support the land reform for which the Irish Republicans had campaigned, seeing it as a chance to sell their estates to the government for redistribution and get out before they lost everything. "Would 1916 have happened without floating freezers?" McWilliams asks, referring to the Easter Rising, which led to the establishment of Ireland as a free state. "Who knows," he concludes. But "the radicalization of Irish tenant farmers and the gradual disengagement from Irish land by all landlords . . . were the direct consequences of the disruptive technologies deployed . . . in Uruguay." In other words, refrigeration can take at least some of the credit for Irish independence.

Meanwhile, the enormous investment required to keep meat cold—and the opportunity to offset some of that by monetizing by-products—meant that meatpacking increasingly made economic sense only at scale. The industry consolidated, leading to the emergence of the "Big Five"—Swift, Armour, Cudahy, Morris, and Wilson—who between them slaughtered more than 80 percent of all cattle in America by 1916. Centralized slaughter led to centralized pollution: famously, Chicago had to reverse the course of its filthy river in order to prevent packing wastes from entering Lake Michigan, the city's drinking water source.

Even animals themselves were transformed. "The New Zealand farmer, the Argentine estanciero, the Queensland grazier, have in mind the fancies of the English meat-buying public in breeding their animals for freezing," explained Critchell and Raymond. The logistics of the dead-meat trade depended on large volumes and year-round slaughtering, so to get more meat to market faster, farmers began to produce "baby beef," pigs rather than hogs, and lamb as opposed to mutton. Animals were increasingly fed concentrated grain and oil seeds rather than forage, so that they grew faster and were ready to harvest sooner. Within a few decades, cattle breeders proudly pointed out that a thirteen-month-old Hereford bull was now the size and shape of a fully grown, four-year-old one, while the selection pressure exercised by intensive feeding measurably increased the length of pigs' intestines.

It's worth remembering that all of these far-reaching and often unexpected consequences of refrigerating meat were spurred in part by a nutritional fallacy: the mistaken conclusion that protein from flesh foods was the only essential nutrient. If chemists had come down in favor of grains and beans instead, the world might have looked very different.

II. Better Living through Chemistry

It was the most talked about meal in the United States. In the weeks leading up to the luncheon, its organizers received so many requests for seats that they switched the venue to one of Chicago's largest banquet rooms. Newspapers from Los Angeles to Pittsburgh covered the rapidly expanding guest list, which included the city's mayor and health commissioner, at least one member of Congress, and dozens of bureaucrats from Washington, DC, New York City, and beyond, as well as many of the nation's most distinguished food and agricultural scientists.

The occasion for all of this excitement was the world's first cold-storage banquet: a meal at which only previously refrigerated foods were to be served. On Monday, October 23, 1911, more than four hundred guests sat down amid the drapery and gilt of the Hotel Sherman's Louis XVI room, unfolded their white linen napkins, and, over the course of two hours of what *The Egg Reporter* later described as "unalloyed pleasure," consumed a five-course meal in which everything except for the olives in their dry martinis had spent between six months and a year in the refrigerated rooms of local cold-storage companies.

Rather than the grower or variety, the menu proudly listed each item's most recent address: the salmon came from a short stay at Booth's Cold Storage, the chicken had resided at Chicago Cold Storage since December 1910, and the turkey and eggs had spent the past eleven and seven months, respectively, at the Monarch refrigeration plant. Addressing a reporter from the *Bulletin of the American Warehouseman's Association*, Meyer Eichengreen, vice president of the National Poultry, Butter and Egg Association, one of the event's sponsors, was happy to provide more detail. "Your capon received its summons to the great unknown along about last St. Valentine's day," he explained. "And the egg in your salad— go right on and eat—well, some happy hen arose from her nest and clucked over that egg when winter was just merging into spring."

"Nothing served at the feast" was "'fresh,'" reported the *San Antonio Express*. Even the pie for dessert was made with cold-stored flour, butter that had been churned in June, and—despite the fact that it was peak apple season—fruit that had spent the past year at Booth's, alongside the fish. "This hotel has never served a better luncheon," said the Sherman's head chef, Lucien Fromente.

Harry Dowie, the Poultry, Butter and Egg Association's national president, proclaimed the banquet a success—edible proof that cold-stored foods were not just perfectly appetizing but, as he put it, "superior to those we style as fresh." "I really believe, as you claim, that there is more flavor to cold storage poultry than the kind that is advertised as freshly killed," Congressman Martin B. Madden agreed, declaring his readiness to spread the word about the wonders of refrigeration back in the nation's capital.

This gala luncheon was the centerpiece of the Poultry, Butter and Egg Association's fifth annual convention. Initially suggested by Henry Brownell, an Ohio egg dealer and cold-storage operator, the idea of serving only foods that had been stored in the city's refrigerated warehouses was seized upon by Dowie as "a form of education"—and a necessary one, given that "there is a disposition on the part of the general public to feel that the cold storage business is against the public's best interests."

At the time, suspicion of refrigerated food was widespread. *The National Provisioner* pointed out that the ladies in attendance at such cold-storage feasts—as well as some of the men—"were in a flutter of expectation at the prospect of eating, for the first time in their lives, a dinner where every item on the bill of fare had been in cold storage." That flutter was not necessarily the thrill of anticipation but rather a tingle of both horror and fear. Indeed, in his opening remarks, Charles E. McNeill, the association's secretary, paid tribute to the diners' bravery: "What better example of courage could we have than their presence today, for it took considerable courage in the face of all that has been written in the newspapers to sit down to such a spread."

The Egg Reporter gathered a sampling of those headlines: COLD STORAGE EVILS; THOUSANDS OF TONS OF FOOD UNFIT TO EAT FOISTED ON PUBLIC BY FREEZER OWNERS; BAD EGGS, POISONED POULTRY, DEADLY FISH, UNWHOLESOME BUTTER AND DECAYING VEGETABLES KEPT TO GET BENEFIT OF HIGH PRICES; and SCIENCE OF KEEPING EATABLES IN GOOD CONDITION NOT KNOWN TO STORAGE MEN were among the highlights. The banquet wasn't a celebration of refrigeration's power and potential; it was a pan-icked attempt to counter this tidal wave of negative publicity. After all, as *The Egg Reporter* concluded, attendees "had demonstrated to them in a forceful way the fact that cold storage foods are really edible and that one could partake of a cold storage meal and still live." Nonetheless, immedi-ately following the luncheon, *The Journal of the American Medical Asso-ciation* published an editorial arguing that one cold-storage banquet proved nothing and pointing to "the condemnation . . . of millions of pounds of cold-storage food" as evidence of "dangerous abuses" in the system.

Fear of refrigerated food was part of a more pervasive anxiety about food safety and food poisoning at the start of the twentieth century. Con-sumers had ample cause for concern. Gastrointestinal infections and di-arrhea added up to the third biggest cause of death among Americans in 1900, just behind tuberculosis and pneumonia and way ahead of today's leading killers—heart disease, cancer, and stroke. Babies and small chil-dren in particular were susceptible to "summer complaint," or fever and diarrhea blamed on a combination of "hot weather, foul air, domestic filth and improper food." During the first few years of the twentieth cen-tury, it was not unusual for a weeklong New York City heat wave to leave an astonishing 1,500 infants dead of the summer complaint.

For many Americans, the chief suspect behind this national epidemic of stomach flu was food—most often cold-storage food. Newspapers, medical journals, and congressional records from the time are filled with stomach-churning stories of otherwise healthy Americans laid low by spoiled food. A series of "cheese poisonings" sickened three hundred people in Michigan in the 1880s. "Ice-cream poisoning" was a common and often lethal affliction: in 1893, the *American Druggist and Pharma-ceutical Record* warned that "deaths from ice-cream poisoning are re-ported with painful regularity, and physicians seem powerless to save the patient once symptoms of poisoning have set in."

Refrigeration had succeeded in delivering perishables aplenty to peo-ple's plates—but was all this cold-storage protein poisoning them?

———

There are things about which it is better not to speculate, and all that I can say is that I hate the smell of ammonia, and grow faint at a draught of unusually cool air," declares the narrator of H. P. Lovecraft's short story "Cool Air." Written in 1926, the story revolves around a mysterious Dr. Muñoz, who uses a state-of-the-art refrigerating machine to cool his apartment down to a frosty forty degrees, even in New York City's summer heat. The narrator confesses to feeling an inexplicable repugnance when he first meets the invalid doctor but soon grows to enjoy his learned conversation, blaming his initial reaction on the "singular cold"—"for such chilliness was abnormal on so hot a day, and the abnormal always excites aversion, distrust, and fear."

When the system fails (as they were wont to do at the time), the true horror is revealed. "A kind of dark, slimy trail led from the open bathroom door to the hall door, and thence to the desk, where a terrible little pool had accumulated," wrote the narrator. Prior to liquefying, Muñoz had scribbled a note of explanation. "The end is here," it read. "Warmer every minute and the tissues can't last." Muñoz, it emerges, died eighteen years ago, relying on the dark arts of refrigeration to halt his decay ever since.

For many Americans in the early years of the twentieth century, the zombie foods that emerged from cold-storage warehouses were similarly horrifying. What kind of unnatural technology could deliver a two-year-old chicken carcass that still looked as though it was slaughtered yesterday? On a fundamental level, cold disrupted what geographer Susanne Freidberg calls "the known physics of freshness," making it difficult for the consumer to tell which foods, if any, were truly fresh. The fridge was an unnerving intervention in life's natural progress toward decay and decomposition, a disquieting suspension of organic matter's inevitable destiny.

Metaphysical implications aside, this new ability to preserve food without changing its external appearance created a perceptual imbalance in the marketplace. A peach might seem fresh from the outside, even as its tissues, like those of Dr. Muñoz, had begun to break down; an egg could look good but conceal a rotten interior. Without the ability to see into an ever-lengthening supply chain, Freidberg told me, "consumers simply had to trust in the merchants and trust in the technology."

Many did not. Frigoriphobie, as the French called it, was only one facet of the average consumer's more general and often justified mistrust of an increasingly industrialized bill of fare. Nevertheless, fear of fridges was widespread. In 1912, a Detroit-based Jewish doctor ruled that meat that had been refrigerated for more than two days was neither fresh nor kosher; after an incident of food poisoning in Agatha Christie's Miss Marple mystery *4.50 from Paddington*, Lady Alice Crackenthorpe is quick to blame the refrigerator, declaring that "people keep things too long in them."

The immature but immortal animals that Gustavus Smith and Charles Tellier had struggled so mightily to ship were greeted with suspicion. Some countries resisted the dead-meat trade outright: the first cold-storage warehouse in Paris was greeted with such outrage that its operator ended up tearing it down, and the French effectively put an end to frozen-meat imports by imposing steep tariffs.

"The idea of eating meat a week or more after it had been killed met with a nasty-nice horror," as Swift's son, Louis, put it. "Eat meat dressed a thousand miles away? No Yankee had ever been served a steak which originated more than a few miles from the stove that cooked it, no sir, not if he knew it!" The British public felt similarly, with an additional aversion to "preserved meat in any form" as a result of "the disastrous outturn of some shipments from South America of 'charqui,'" explained Critchell and Raymond. In 1880, at a legal hearing over plans to lease a London building to a cold-storage operator, members of the public protested. A Mr. Judd told the court "that it was against public morals to do anything which would prevent the getting rid of perishable meat as quickly as possible" and that "the Court should, therefore, be careful not to start or encourage a new industry for preserving that in which decay has taken place."

Although it sounds bizarre today, it was perfectly reasonable to assume that cold-stored food might be bad—for the public's health, if not its morals. In the early days of refrigeration, cooling was subject to frequent lapses: meat might be stored at ambient temperature before or after being loaded into a train car, and the amount of ice in that train car might not be sufficient to keep it cold. Given the long journeys meat now regularly undertook, a bad microbe had plenty of time to reproduce en route, if thermal conditions permitted. Worse, merchants often overestimated refrigeration's magical powers, putting food that was "on the turn" into cold storage in an attempt to halt its decay, so that it could still be sold as "fresh" the next day. As a headline in a Buffalo, New York, newspaper put it, THERE IS NO DEATH, ONLY COLD STORAGE.

For additional peace of mind, packers, dairymen, and meat merchants would often supplement refrigeration's preservative powers with other putrefaction-prevention measures. Formaldehyde, borrowed from the embalming industry, borax, which moonlighted as a laundry whitener, and salicylic acid, better known today as a zit zapper, were among the wonders of modern chemistry that were commonly used to help keep meat, butter, cheese, and dairy good. Such solutions were openly and legally sold under brand names such as Preservaline and Freezine. Less scrupulous individuals found these additives useful to restore the illusion of freshness to food that had aged in transit. Meat that had turned gray could have its rosy blush restored through the application of salicylic acid; promotional materials for sulfur- and formaldehyde-laced Freezine boasted that meat could be "exposed for sale . . . more of the preparation applied, and still look good to the eye."

The industrialization of meat slaughter and processing also led to some sanitary corner cutting, famously exposed by Upton Sinclair in his 1906 novel *The Jungle*. Sinclair intended the book as a labor rights manifesto, but with its graphic descriptions of diseased animals and filthy conditions in Chicago's Union Stock Yards, it was read as a call to action on food safety. Even the Poultry, Butter and Egg Association's Dowie admitted that, on occasion, the iced, dressed chickens he received for sale were not perfectly fresh. "Boxes opened up you could not tell what was in them," he said. "It was as if it was in cotton batting, white with mold, entirely white with mold."

The old certainty that food was good had been based on proximity and appearance—assurances that had been thoroughly disrupted by the introduction of refrigeration. Instead, consumers were being asked to place their faith in an increasingly opaque supply chain and trust that a new technology that they didn't understand would keep their food safe. Ignorance created doubt, which sowed fear, and by the 1911 Cold Storage Banquet, matters had come to a head. Under pressure from an agitated public, the Senate was considering a bill to regulate the nascent cold-storage industry and guarantee public health. Among its provisions were extremely short limits on the duration that meat, fish, eggs, and butter could be stored under refrigeration.

The only problem, as Harry Dowie and other representatives of the nation's fishers and farmers eagerly pointed out in their congressional testimony, was that those limits had no basis in science. Americans were eating plenty of refrigerated beef and chicken, and some were fine while others weren't, but no one knew why. For much of its first half century, refrigeration had been an engineering problem. Now it was time for the chemists to get involved once again.

C hemist Harvey Washington Wiley led the campaign for federal food safety regulation that resulted in the passage of the landmark Federal Meat Inspection and Pure Food and Drug Acts in 1906, and he subsequently found himself in charge of enforcing them. His personal passion lay in the investigation of preservatives: in 1902, he assembled a group of "young clerks, vigorous and voracious," to test the health effects of consuming the freshness-extending chemicals commonly found in America's beef, butter, and milk. The group supplemented their otherwise healthy meals with capsules of borax, followed, over the years, by salicylic acid, sulfites, sodium benzoate, and formaldehyde. They were quickly dubbed the Poison Squad by the press.

The results of Wiley's experiments—his long-suffering volunteers regularly reported dizziness, headaches, stomach pains, and bleeding, as well as nausea and vomiting, becoming so sick that the sulfite study had to be halted early—provided compelling evidence that the unregulated usage of chemical preservatives was bad for Americans' health.

When it came to refrigeration, there was no such research to draw on. If Wiley came across a news story that claimed cold-storage warehousemen were selling turkeys that had been refrigerated for a decade, he had no idea whether to be outraged or impressed. Wiley couldn't tell Americans how long cold-storage turkey was safe or what temperature eggs should be kept at, because no one could. The most basic questions about how best to use cold to preserve food remained unanswered. Meatpackers like Gustavus Swift and Philip Armour had established their own best practices, designed to maximize economic returns, but no one had studied refrigeration from the point of view of how to make sure cold-stored food wasn't deadly.

Wiley decided to establish a new government bureau, the Food Research Laboratory, in order to set evidence-based guidelines for how to refrigerate animal foods. Polly Pennington was his first choice to lead it.

Dr. M. E. Pennington, who was almost always assumed to be a Mark, Michael, or Martin by the farmers, wholesalers, engineers, cold-storage operators, and government officials who sought her professional assistance, first fell into refrigeration by way of chemistry. At the age of twelve, she picked up a copy of Rand's *Elements of Medical Chemistry* in the Philadelphia Mercantile Library and was bowled over by the realization that the world around her existed thanks to the actions and reactions of its chemical building blocks. "Like a flash of light in a dark place, I got the idea of the realness of the invisible world," she recalled.

Chemistry was not considered a suitable pursuit for young ladies at the time: when Pennington petitioned her high school to study the subject, the headmistress flatly refused. A few years later, the University of Pennsylvania allowed her to enroll, but after she completed her studies, the trustees decided that women were not eligible to be awarded a bachelor of science degree and gave her a "certificate of proficiency" instead. Still, Pennington persisted, finagling her way into a PhD by means of a university statute covering "extraordinary cases." Her dissertation, appropriately enough, was on the derivatives of the "refractory metals" niobium and tantalum, so called because of how hard they are to wear down. Following postgraduate research at Yale, Pennington returned to Philadelphia determined to bypass the male-dominated field's closed doors by founding her own lab when she was just twenty-five.

Her start-up, which performed clinical analyses of medical samples for the city's doctors, not only "coasted along prosperously for several years," according to a profile written by *The New Yorker*'s Barbara Heggie, but also put her right at the center of one of the era's hottest topics: the intersection of bacteria, chemistry, and the nation's health.

Pennington came to Harvey Wiley's attention when the city of Philadelphia hired her to clean up its milk supply. She had tackled this daunting task with what was to become her trademark combination of pioneering research, rigorous inspection, and friendly persuasion. In her lab, she conducted detailed experiments to find out what difference the temperature and rate of cooling made to bacteria levels in raw and pasteurized milk. The process she developed to check the health of dairy herds became standard board of health practice nationwide. Despite having no legal powers of enforcement, she convinced the city's pushcart ice cream vendors to boil their equipment by showing them slides of the microbes growing in their "hokeypokey" and penny licks, in an apparently highly effective appeal to their better natures.

Wiley decided that Pennington was exactly the person to take on the challenge of making refrigeration scientific. The only obstacle was that the civil service did not hire women. He enrolled her for the exam without her knowledge. "I was mad as wrath," she recalled, but decided to sit it anyway. He then changed her name on the paper, again without consulting her, so that Mary became Dr. M. E. She scored top marks, was offered the job, and, by the time anyone realized she wasn't a man, had donned a khaki divided skirt and a sensible sweater and begun work on the problems of chickens and eggs.

In 1907, when Pennington started the research that was later published as "A Chemical and Bacteriological Study of Fresh Eggs," America's annual egg output was still concentrated in a few months at springtime—but refrigeration had made seemingly "fresh" eggs available every day. The result was a mess: cold-storage operators could sell April's eggs in November at a premium, claiming they were rare, newly laid specimens; meanwhile, until she got home and tried to fry them for her husband's breakfast, a housewife had no way to know whether the "fresh" eggs she'd bought were as described or rotten. All too often, the latter proved to be the case.

Poultry posed another set of problems. As a 1925 profile of Pennington's work in *The Field*, an illustrated agricultural magazine, put it, "back in 1906, when the food law was passed, a chicken had a rather strenuous time getting to market." Its head would have been chopped off with an axe on a block, and the rest of its body would have spent a couple of days floating in a tank of cold water before being packed in a barrel with ice for shipment. "The soaking of the birds in the melted ice, the dirty heads and feet, and the gradual dissolving out of the soluble parts of the flesh caused a loss in eating quality and induced decay," Pennington wrote, with scientific restraint, next to an illustration depicting the sorry state of these waterlogged and germ-laden chickens.

Pennington hired a small team—including a handful of women—and got to work both in the field and at the laboratory bench. She conducted experiments on such topics as the rate of decomposition in drawn versus uneviscerated poultry and the biochemical reason chickens turned a disconcertingly intense shade of green when held at fifty-five degrees.* She also traveled thousands of miles to put together a detailed survey of the nation's railcar icing stations and to record the temperatures of three sample chickens from each car at regular intervals. Newspaper reports at the time claimed Dr. Pennington rode in the refrigerator car alongside the poultry while making these observations; in fact, she was quick to tell journalists, she rode "in comfort," in a specially outfitted caboose with a built-in laboratory. "That's the way to travel, leisurely jogging along, talking over the crops with the farmers at the crossings, switched into a train yard at night," she reminisced.

Over the next decade, her adventures answered all the basic questions of how long eggs and chickens could be stored, how rapidly they should be cooled, and what humidity and temperature they should be held at in order to guarantee they remained safe to eat. Along the way, she ended up designing a new, standardized refrigerator car with better air circulation, a more sanitary process for slaughtering, plucking, chilling, and packing chicken, the first science-based egg-quality charts (the predecessor of today's USDA egg grades), and a protective carton for eggs.

Unlike her more theatrical boss, Harvey Washington Wiley, who treated the food industry as an adversary to be fought using high-profile court cases and media shock stories, Pennington preferred a low-key, collaborative approach. For starters, the law did not guarantee her entry to packinghouses and cold-storage warehouses. In order for her to gain the access she needed to figure out what was going wrong with America's food, business owners had to trust that she could help them. In return,

---

*"Green struck" chickens were surprisingly common. Pennington quotes the only preexisting scientific paper on the subject, from France, which found that *"la putrefaction verte"* was the reason for spoilage in a third of meat condemned as inedible at the central Paris market of Les Halles between 1904 and 1911. In her own study, she found that carcasses held at warmer temperatures turned "a dull blue-green color," accompanied by bloating and "the characteristic odor of decomposing protein," but that the intense green of a "typical green struck" bird was due to the growth of a particular intestinal bacteria that excreted hydrogen sulfide, which in turn transformed normal red-brown hemoglobin in the blood-filled capillaries of the chicken's skin into bright-green sulfhemoglobin. The dead chickens' blood literally ran green.

rather than warn the public away from cold-stored foods by harping on the horrors of soggy chickens and whiskery eggs, she focused on helping farmers, meat-packers, and warehouse operators reduce their losses and deliver good food. According to *The New Yorker*, she was known in refrigerating circles as "severe but somehow charming." One cold-storage operator, "carried away by admiration, called her 'the voice of conscience in the refrigerating world.'" It should come as no surprise that at the Cold Storage Banquet in 1911 the keynote speaker was none other than Dr. M. E. Pennington, the United States' "supreme authority on matters connected with the refrigeration of perishable foodstuffs."

In an era where refrigeration's detractors blamed the dead-meat trade for everything from cholera to cancer, while cold-storage men advertised their product as "better than fresh," Pennington refused to become an evangelist for either side. "Storage makes nothing better, at least in the line of poultry," she admitted—but she was quick to add that, if handled properly, cooled quickly, and kept at the right temperature and humidity, a "frozen and transported chicken is as near excellence as the vast majority of us Americans can ever attain if we persist in demanding broilers in midwinter and roasters in the summer and early autumn."

Cold's powers of conservation were not magical, she told the American Warehousemen's Association—but they were to thank for the fact that "butter and eggs are now staple foods in the homes of even small wage-earners, and the family that does not buy some poultry at least once a year is poor indeed." What's more, thanks to Wiley's success in banning toxic additives and Pennington's achievements in making refrigeration effective, that chicken was now much less likely to make them sick. By the 1930s, both the protein panic and the subsequent fridge phobia had been assuaged, and adequately chilled, Freezine-free eggs and bacon were on the American breakfast menu year-round, with only high cholesterol levels to fear.

When the Second World War began, Pennington was in her late sixties, still busy solving cold-storage problems for government and industry alike. In a talk she gave in 1939 at the Massachusetts Institute of Technology, she reflected that, just a quarter of a century earlier, "any perishable food subjected to refrigeration for more than a few hours . . . was looked upon askance and its quality discounted." Today, she said, the situation was precisely the opposite: "We are inclined to question the quality of any perishable not put under suitable refrigeration as promptly as possible." In the span of just a few decades, the public's perception of refrigeration had flipped: something that had seemed risky, untrustworthy, and unnatural became instead essential to good health, in that it allowed consumers to consume perishable protein in the quantities necessary to achieve their full potential. This narrative not only appealed to the American enthusiasm for self-improvement but also reflected a widespread optimism about the power of technology to manage and, ultimately, improve upon nature.

The American public's newfound faith in refrigeration was, to a large extent, Pennington's legacy—but it was not sufficient. Refrigeration must do more than assure food safety. "There is a difference discernible to anyone between an egg which is inedible because it is 'rotten' and an egg which has lost its fine flavor and its thick white, but is still an edible egg," Pennington told her audience at MIT. "It is our job, now, to hold the flavor and the thick white." In the future, she told the students, refrigeration must strive to make food *good*, not just prevent it from going bad.

III. *When Muscle Becomes Meat*

Refrigeration's power to preserve is not its only effect on animal flesh. Cold, as Polly Pennington foresaw, can also make it better. Lowering the temperature not only slows the growth of microbes, delaying the otherwise inevitable process of fleshly decomposition; it also orchestrates a cascade of biochemical reactions that transform meat's flavor and texture. To see this second, equally important part of refrigeration's powers in action, I set my alarm for 2:00 a.m. to arrive at Master Purveyors at the same time as a truck full of freshly slaughtered beef.

Outside the Master Purveyors facility in the Bronx is a large painted-concrete cow lurking behind a scrubby hedge. Inside is up to $1.4 million worth of beef, sitting on wire racks in the company's refrigerated dry-aging room. The business, as explained by company vice president Mark Solasz, involves selecting the best-quality beef and butchering quarter-carcasses down into strip loins, short loins, ribs, and more. Mostly, though, it consists of letting the meat sit in a walk-in fridge for three to four weeks, during which time, thanks to a curious alchemy of cold and time, it becomes 15 percent lighter and 20 percent more valuable.

Master Purveyors is located inside New York City's Hunts Point Market, one of the largest centralized food-distribution sites in the world. Between a third and half of the meat consumed in the city comes through Hunts Point, an island of sorts in the southeast Bronx, cut off from the rest of the city by the East River, the Bronx River, the Bruckner Expressway, and a freight rail yard. The city's produce and meat wholesalers moved here in the 1960s, when Washington Market in lower Manhattan was demolished to make way for the Twin Towers. The low-slung buildings, hidden behind a row of semitrailers, are showing their age, and the tile-walled corridors in between the units are chipped and narrow.

Outside the razor-wire fence that encircles the market, the city was dark, the streets empty. Inside, I was nearly run over by a forklift truck and a reversing tractor trailer before I even reached Master Purveyors' unit. Once I was safely in the cold, no one had time to do more than give me a white coat, plastic protective glasses, and a yellow hard hat and tell me to stay out of the way. They were too busy maneuvering enormous quarter-carcasses of beef, hanging from hooks attached to an overhead rail-and-pulley system, out of the back of a truck, onto the loading dock, and through yellowing plastic curtains into a refrigerated room. Each quarter-steer was between 180 and 220 pounds in weight, but the wheels were well oiled. Accompanied by only the occasional screech of metal on metal, the carcasses zip-lined in with a quiet rumble. As they lined up next to each other under the fluorescent lights, suspended by their rear shanks, they formed a grotesque chorus line: frozen in a Rockette-style scissor kick, all fluffy white skirts of fat and blush-pink limbs.

These cattle had been slaughtered in Maryland a couple of days earlier, then chilled. As the room filled up, it began to smell—not unpleasantly but pervasively—of raw meat: metallic, slightly sweet, and a little buttery. The concrete floor was covered in drifts of white powder that crunched gently underfoot. I discovered later that it was salt, put down to provide grip. Next door, in the processing room, the high-pitched hum of a band saw drowned out the radio and the refrigeration system alike, as curls of creamy fat and gobbets of pinkish gristle flew through the air. (When I showered later that day, I discovered little shreds of meat in my hair and one of my ears; the pages of my notebook are still smeared with it.) A dozen butchers had already begun slicing, whittling, and sculpting big chunks of beef into individual rib eyes, fillets, and tomahawk steaks.

The task of breaking each carcass down into those larger chunks—the so-called primal cuts—fell to Juan "Ricky" Bernardez, a Black man from Honduras with an intimate knowledge of beef anatomy. With a big hook wedged between his knuckles like a pirate, and an even bigger knife in his other hand, Ricky showed me how it was done, moving his knifepoint down the ribs—one, two, three, four, five—then holding the hanging carcass steady with the hook as he sliced straight through, running his knife beneath the fifth rib with a swift stroking motion that left the bottom half hanging on by an inch. With what looked like little more than a gentle nudge, he swung the lower half up onto a hook of its own and finished the cut, separating the ribs from both the loin and the flank. "You gotta swing," he told me. "If you lift, you will break yourself. I can cut a hundred a day."

All around the edges of the room, these primal cuts—dozens of briskets, loins, and ribs—were layered four or five high, impaled on the ascending hooks of stainless-steel meat trees that hung from the ceiling. "I wouldn't want to eat that," Solasz said, appearing out of nowhere as I admired the fruit of this upside-down forest. At this point, he explained, the beef is *too* fresh—it has not completed its transformation from muscle into meat. "It would be tough right now," he said. "It needs a little age to be tender."

For most of refrigeration's short history, humans have focused on its ability to stop spoilage by slowing down the reproduction of bad bacteria. But in the much longer annals of meat eating, cool air has been equally, if not more, appreciated for its ability to ripen red meat, turning dry, tough muscle fiber into a juicy, savory steak. This is why, for millennia, farmers have slaughtered their meat in the autumn, when temperatures were falling but not yet freezing. In medieval Europe, the festival of Martinmas, on November 11, was associated with the harvest of pigs and cattle; going back still further, the Anglo-Saxon name for November was *Blotmonad*, or bloodmonth, to mark the slaughter of livestock. The resulting meat glut would undoubtedly keep better in frosty midwinter than summer sun, but, no less important, the milder cool, damp conditions characteristic of late autumn are perfect for tenderizing meat.

"The animal may be dead, but there's still all sorts going on," explained British meat scientist Stephen James, who has spent his entire career studying the peculiarities of postmortem muscle. The steers that Ricky Bernardez was busy dismembering at Master Purveyors had lost consciousness and stopped breathing days ago, but the cells in their muscle tissue were still a hive of activity.

During what meat scientists think of as the first phase, which begins immediately postslaughter, those cells had continued to respire, burning through their stock of complex sugars—but in the absence of breath and blood, this had become an anaerobic reaction, producing lactic acid as a by-product. James likened it to what happens inside an Olympic runner's leg muscles in a sprint: the lungs and arteries can't deliver oxygen fast enough to the legs, so the muscle's energy reserves are no longer being replenished and the lactic acid isn't being removed. The result is a cramp in runners and rigor mortis in slaughtered beef. Paradoxically, James explained, it takes energy to relax muscles. With all its fuel gone, the muscle no longer has enough energy to unlink its fibrous proteins and send them back to their resting positions, so they remain contracted.

For beef, rigor typically sets in within the first twenty-four hours after slaughter. At that point, the muscle is tight and acidic—it would taste terrible. Fortunately, a second phase of cellular activity commences: postrigor. Muscles contain enzymes whose job it is to break down fibers so they can be rebuilt; they're essential to basic muscle maintenance and beloved of bodybuilders. These enzymes don't quit just because the animal whose cells they inhabit has died; instead, they gradually degrade the proteins that became knotted during rigor, weakening their links so that the meat becomes tender.

Cold plays an essential role during both phases, James told me. He and his colleagues have found that the postslaughter cooling conditions have a bigger impact on how meat tastes than most other predeath factors, including an animal's breed, age, and diet. In other words, for all that foodies obsess over grass-fed Wagyu, it is refrigeration that has the power to make meat maximally delicious.

To start with, the speed with which rigor sets in during phase one makes all the difference to the tenderness and juiciness of the resulting meat, and that pace is primarily determined by temperature. Like Goldilocks's porridge, cooling has to be just right: a carcass that remains too warm for too long or gets too cold too soon will likely turn into meat that is putrid, flabby, and watery, or tough.

Today, thanks to ever leaner animals and increasingly efficient refrigeration systems, the latter issue is more likely to be a problem. Starting in the 1950s, scientists in New Zealand noticed that the muscle tissue in lamb and beef that was chilled down to forty degrees in a few hours, as opposed to over the course of a full day, ended up tasting like leather when it came time to cook it. "Cold shortening," James said, shaking his head. "Major headache."

Scientists eventually traced this new "cold shortening" crisis back to calcium. When you turned to this page, your hand couldn't have moved without the encouragement of a little burst of calcium, released from a special pump in the muscle sheath. This calcium pump evolved to function at normal body temperature. At fifty degrees, it fails, which floods the muscle with calcium—the trigger to contract. Earlier, less powerful refrigeration systems had cooled carcasses slowly enough that the muscles had used up their energy reserves long before they reached fifty degrees, which meant that the calcium call to action fell on unresponsive ears. The muscles had nothing left in the tank to contract with. Modern rapid chilling offered meat-packers a lot of benefits—it reduced the chance of spoilage, decreased drip and shrinkage due to water loss, and, most important, allowed them to process meat faster—but it did mean that muscles hadn't yet exhausted their energy reserves by the time they cooled down. Once they felt that calcium rush, they did what they'd evolved to do: they contracted. Cold-shortened muscles lose up to 40 percent of their length, and all that dense, bunched-up muscle fiber makes for very chewy eating.

Time is money, as Benjamin Franklin supposedly said, and rather than give up on faster refrigeration, meat scientists took another leaf out of the founding father's playbook and started electrocuting their carcasses. Among Franklin's lesser-known electrical experiments was a successful, if unwise, attempt to slaughter turkeys by shocking them. "I conceit that the Birds kill'd in this Manner eat uncommonly tender," Franklin wrote to his correspondents in the UK's Royal Society of Arts. (He told his brother about the experiment too, confessing that he had accidentally electrocuted himself in the process, a sensation that he knew "not how well to describe" but that left him "feeling like Dead flesh.")

No one picked up on Franklin's turkey-tenderization insight for two hundred years—but today, if you bought your beef and lamb at a supermarket in the developed world, you can safely assume it was electrically stimulated. The carcasses at Master Purveyors had been zapped; Solasz told me it's a standard step at the slaughterhouses he buys from. Electrocuting meat averts cold shortening because it's a fast track to rigor mortis: it forces the muscles to rapidly contract and relax, burning through the last of their energy reserves. To be effective, the stimulation needs to come within the first hour after slaughter: typically, processors stun the animal, eviscerate it, then shock it in a series of quick pulses over the course of a minute before splitting it into sides and spraying them with chilled water to bring the meat's temperature down as fast as possible.

Different configurations are used—some recommend a probe-based nostril-rectum setup for maximum tenderization, but it's faster to just run the current through the rail from which the carcass is hanging by its leg to an electrified bar that makes contact near the neck. Every aspect of the process of turning an animal into meat at industrial scale is somewhat disturbing, which is why slaughterhouses tend not to welcome visitors, but there's something particularly creepy about seeing a still-warm cow-shaped carcass dangling overhead, flailing intermittently like a piñata at a party.

Adding high voltages to wet and slippery environments is risky for workers, and the equipment is an investment, but for the big companies that slaughter most meat, it's a convenient way to speed up the production process. Electrical-stimulation evangelists point out that shocked beef is brighter red, which consumers prefer. Electrocution also makes the meat spongier, so it holds moisture, and it helps detach the meat from the skeleton, making it easier for "boners," as men like Ricky Bernardez are known, to disassemble the carcass. "The meat is loose! You will not believe it unless you see the faces of your boners when they first get the stimulated meat," as one equipment manufacturer memorably put it.

B y the time the electrically stimulated carcasses arrive at Master Purveyors, phase two—variously known in the industry as ripening, aging, or conditioning the meat—is already underway. Creating the conditions for it to happen in a very particular manner is at the heart of what happens at Master Purveyors.

Swinging open a pair of double doors, Solasz led me into the dry-aging room. It wasn't small—1,600 square feet, Solasz told me—but it was so crammed with meat it was hard to move: metal shelves packed with fifty- and sixty-five-pound primal cuts lined the walls, while, in the center of the room, rows and rows of sharpened hooks hung like candelabras from the railings, bristling with chunky short loins, ribs, and flanks.

To make maneuvering more challenging, an army of standing fans stood at the end and midpoint of each row, sending cool, moist air whipping around the room, over, under, and in between every rib and shank— a couple of thousand pieces of meat in total. The gusts of air smelled malty and mushroomy. "It's not a smell," Solasz corrected me. "It's an *aroma.*" He compared it to an aged Gruyère or Comté cheese—nutty and a little sweet. "It's my life," he said, inhaling deeply. "I love it." Then he disappeared to take another call.

Seeing the meat brought in that morning side by side with the cuts that were already a couple of weeks old was like seeing unflattering then-and-now photos of Hollywood stars: the racks of firm, bright-red-and-white sirloins looked impossibly plump and youthful next to their aged, shrunken, purplish-black future selves. The dried-up surface of the fully ripened meat felt lizardy to the touch, except those parts of it that were coated in a soft, snow-like layer of mold. As I edged my way in, I discovered that Solasz's son, Max, short and serious like his father but less frazzled, was also in there, using a fiberglass hammer to gently tap at the edges of cuts, like a sculptor in the studio. "Sometimes things get a little compressed in transit," he said. Max told me that he'd learned the trick from his grandfather, Sam. "I didn't grow up wanting to be in the meat business," he said. "I'm only here because of him."

Sam Solasz was born in northeastern Poland, in a small town near Bialystok, in 1928. His father was a butcher, and he was the second-youngest of eleven siblings, all of whom later died in concentration camps or were shot by Nazis. In November 1942, when his family was rounded up for deportation, fourteen-year-old Sam escaped by hiding curled up inside a tractor tire—he waved to them but never had the chance to say goodbye. He lived with relatives in the Bialystok ghetto, where he earned the nickname *Malach*, or "Angel," for his ability to smuggle desperately needed food, medicine, and other supplies in and out, using his red hair, his blue eyes, and a Virgin Mary medallion he'd won in a card game to pass as Christian. Finally, in February 1943, the Germans rounded up the city's few remaining Jews and put them on a cattle car headed for Treblinka. As the train slowed to enter a station, Sam jumped and ran; he was one of only three people on that train to survive. He spent much of the rest of the war hiding in the dense forests of eastern Poland, relying on the butchering skills he'd learned from his father as a child in order to prepare wild game to eat—but also to make himself indispensable to partisan fighters and, eventually, Russian soldiers.

When he finally made it to America after the war, Sam was twenty-two years old, had ten dollars to his name, and spoke no English—but his meat-cutting skills got him a job in New Jersey right away. In 1957, he started his own business, Master Purveyors, in Manhattan's Meatpacking District. "This is what kept him alive," said Max. "When he moved here alone, this is what gave him an opportunity to build a family."

Since Sam died in 2019, the dry-aging room has become Max's purview. Sam taught his grandson the art of aging meat, using the movement of cold, damp air to direct a process of controlled decomposition.

"It's about airflow, temperature, and humidity," Max told me. "Not a lot of dry-aging rooms have this kind of air circulation, but this is how my grandpa taught me to do it." It's a delicate balancing act: the room has to be cold enough to prevent bacteria from multiplying but warm enough for enzymatic activity to occur, as well as for pale-gray, whiskery patches of mold to grow on the fatty parts of the meat. The mold is not just harmless but essential: it releases its own enzymes, which also help break down muscle and connective tissues, and is easily trimmed off when the meat is prepared for sale. The room's humidity and air circulation work hand in hand to promote exactly the right rate of evaporation: enough to slowly concentrate the flavor in the tissues and avoid sliminess but not so much that the meat dries out.

After that, Max explained, it's a waiting game as the enzymes transform both texture and flavor, breaking down muscle to create smaller, more flavorful amino acid and sugar molecules and unlacing stringy fibers to create sponge-like pockets that will trap juices during cooking. At the same time the fat marbled throughout the muscle is gently oxidizing, which creates its own complement of aromatic fatty acids.

Each piece of meat goes through this biochemical metamorphosis at a different rate, Max told me, depending on its fat content, surface area, and more. He keeps track of how long the meat has aged using slips of paper, rotating cuts so they age evenly, but he judges when it's ready based on feel. "Somewhere within that eighteen- to twenty-four-day mark, it'll start really hardening up," he said, showing me a piece whose surface felt like wood. "I'm looking for that, I'm looking for the fat covering, and I'm looking at what makes the most sense financially, with the orders my butchers have to cut."

By 5:30 a.m., the rush was on to load boxes of beef onto trucks for delivery to some of the city's best restaurants and hotels. The first light was just creeping over the horizon. One of the men put a plastic bag over his head and poked his arms through the sleeves, ready to start spraying down the rooms. He told me he'd be asleep before 10:00 a.m.

W e never even studied dry-aging," Stephen James told me, laughing, when I asked him whether he or his colleagues had done any research into what was happening in the moist, cold, fast-moving air of Mark Solasz's meat locker. In meat science textbooks, Master Purveyors' "classical" or "traditional" style of ripening is described as "nearly extinct." Muscle still needs time and cold to ripen into meat—but it no longer needs to sit still in a cold room for that to happen.

In the 1970s, with the proliferation of plastic, meat-packers figured out that they could "wet-age" meat instead. Today, beef carcasses are dismembered into steaks, ribs, and loins on a factory production line by poorly paid slaughterhouse workers. The various parts of the steer are vacuum-sealed in retail-ready plastic bags and on their way to supermarkets within a few days of meeting the stun gun. Companies avoid paying to cool a huge room full of beef for three weeks, while the biochemical transformations that make meat tender happen inside the refrigerated trucks, warehouses, and grocery distribution centers that the animal parts move through toward their final date with digestion.

This innovation eliminated most remaining urban wholesale and retail butchery jobs, making the team at Master Purveyors an endangered breed. Wet-aged, precut boxed beef can be unloaded and put on the shelves in the supermarket by minimum-wage employees rather than skilled meatcutters. What's more, because the aging happens in a sealed bag, there's no evaporation and thus none of the shrinkage seen in a Master Purveyors steak of similar vintage—which means even more savings. Cutting the beef at the plant also concentrated additional by-products, which made it easier to find markets for them: before long, the chemically treated, finely textured trimmings nicknamed "pink slime" started making their way into burgers.

By the 1980s, 90 percent of American beef came to market in a vacuum-sealed bath of its own juices. Most of us will never have eaten any other kind. But the question of whether this cheaper, wet-aged meat is any less tasty than traditionally ripened beef is surprisingly difficult to answer. I would swear my Master Purveyors steak tasted better—beefier and nuttier—than the standard supermarket fare, and there are studies to back me up, including a paper showing that dry-aged beef not only has more of the amino acids that are associated with meaty flavor notes and umami taste but also that those amino acids are more concentrated, because some of the beef's water weight has evaporated. But overall the results are inconclusive. After conducting blind-taste-test panels, some researchers have determined that dry-aged beef has a more "beefy and brown-roasted flavor," in contrast with wet-aged beef's "more intense sour and metallic note and strong bloody/serumy flavor." Others have found that consumers mostly can't tell the difference.

What is certain is that refrigeration, combined with vacuum packaging and a quick electric shock, has made deliciously tender beef both affordable and available year-round. Refrigeration, in collaboration with antibiotics, intensive breeding, and corn subsidies, is the reason twenty-five-cent chicken wings and dollar burgers can be found in dive bars and on fast-food menus across the land. Over the past 150 years, cold has proven its nineteenth-century doubters like Wentworth Lascelles Scott thoroughly wrong: it has demonstrated that it can not only stop spoilage but also deliver better-tasting meat at a scale Mr. Scott could barely have imagined.

Nor can most of the rest of us. There are approximately 22.7 billion broiler chickens living out their five-to-seven-week spans on Earth at any moment, compared with just half a billion house sparrows or a quarter of a billion pigeons. Those chickens are also double the size and five times the weight of their preindustrial ancestors, giving them a combined mass that exceeds that of all other birds on Earth. The team of researchers behind these calculations used them to suggest that the layer of chicken bones currently piling up in landfills around the world is, in fact, an ideal marker of the Anthropocene, the proposed unit of geological time in which Earth's systems have been profoundly reconfigured by humans.

Chickens may be a signal to future geologists, but environmental scientist Vaclav Smil suggests that cows might perform that role for aliens. Meat and dairy animals so vastly outweigh all other vertebrates that "if sapient extraterrestrial visitors could get an instant census of mammalian biomass on the Earth in order to judge the importance of organisms simply by their abundance, they would conclude that life on the third solar planet is dominated by cattle." In aggregate, livestock make up 62 percent of all mammals on Earth; humans, at 34 percent, account for most of the rest. Everything else—dogs, cats, deer, rabbits, whales, elephants, bats, and even rats—only adds up to the remaining 4 percent. Livestock takes up nearly 80 percent of global agricultural land; cattle ranching is responsible for the deforestation of an area more than double the size of California in the Amazonian rainforest alone.

Denizens of the oceans have fared little better under refrigeration: in the 1950s, enormous freezer trawlers, capable of remaining at sea for days at a time and equipped with new, plastic-based nets and military-developed sonar, were introduced. Their catch, frozen in large blocks, could be band-sawed up into the conveniently anonymous form of the fish stick. This precooked, crumb-coated, whitefish cuboid was introduced in 1953 and was an immediate hit. Within months of their introduction, fish sticks accounted for 10 percent of all noncanned fish sales—thanks in large part to a government-financed marketing budget secured by none other than Senator John F. Kennedy, Democrat of Massachusetts. Fish are notoriously hard to count, but according to the best estimates, their numbers have decreased by half over the past fifty years.

In 1626, Francis Bacon—statesman, natural philosopher, and snow-stuffed chicken casualty—wrote that "the producing of Cold is a thing very worthy the Inquisition." Heat and cold "are Nature's two hands whereby she chiefly worketh," he explained. For millennia, humanity had controlled only half the equation—a power that alone was enough to change the course of human evolution. Within our first century of domesticating cold, we not only rearranged meat production and consumption, with effects extending from Irish independence to Amazonian deforestation, but also altered the composition of Earth's biomass beyond recognition.

4.

INSIDE THE
TIME MACHINE

I. Sleeping Beauties

"They're alive!" insisted Natalia Falagán, an agricultural engineer at Cranfield University, a postgraduate research institution in the British Midlands. Bananas, apples, blueberries, even the humble potato: despite having been picked, they are still living organisms—and how they look and taste when we eventually eat them depends in large part on what and how fast they breathe.

Falagán led me into her walk in refrigerator, where we were surrounded by racks of decomposing produce. Potatoes, onions, apples, and blueberries were hooked up to sensors and monitors like critically ill patients in an ICU. Fruits and vegetables continue to respire to the bitter end, she explained, frantically burning through their own internal sugars, acids, and vitamins as a last-ditch substitute for their parent plants' life support, until they either are killed by being eaten or simply collapse and die.

On our return from an expedition to the supermarket, most of us would hurry to refrigerate meat, fish, and dairy first; as we've seen, with the honorable exception of beer, the first commercial use to which the new technology of refrigeration was applied was to keep animal products cool. But this implied hierarchy of perishability is wrong, Falagán told me: fruits and vegetables are the most liable to decay. This process can't be stopped; it can only be slowed.

Unfortunately, Falagán's patients are metabolically unique, which means that they all require their own personalized chill-out regime to counteract the stresses of postharvest life. "It's not enough to simply lower the temperature," she explained. "You have to know what the right temperature is for each commodity, and the beauty"—or challenge—"is that that can differ even between different varieties of apples."

Refrigeration's antiaging powers often require supplementation through atmospheric engineering. When it's not photosynthesizing, plant matter breathes in oxygen and exhales carbon dioxide, so tweaking the levels of these gases can slow respiration. Much of Falagán's work consists of experiments in which she applies different "treatments"—varying temperatures and ratios of gases, applied according to varying schedules—to Tupperware tubs full of fruits and vegetables, then monitors their reactions in real time. When I visited, coils of plastic tubing connected dozens of containers of potatoes to a gas-mixing board capable of dispensing atmospheric blends; a respirometer took readings of how fast the spuds were breathing at two-minute intervals. "With this instrument, we can see how well we've put the potato to sleep, basically," Falagán said. "Is it happily dreaming, or is it stressed out, tossing and turning?"

Such sophistication is a far cry from refrigeration's early days, when entrepreneurial warehousemen prescribed cold as a one-size-fits-all treatment that could keep apples, onions, and berries completely fresh for years on end, stored under the same roof and the same thermal and atmospheric conditions. The results were not appetizing: frostbitten strawberries, onion-scented apples, and sprouting potatoes. Produce refrigeration's great leap forward had to wait until the 1920s, when the need to consume vitamin-rich "protective foods" became the new dietary orthodoxy.

As we've seen, refrigeration's mass adoption came about in part because nineteenth-century chemists had mistakenly arrived at the conclusion that consuming large quantities of animal protein was essential for health. Sustaining the productivity of urban workers—essential cogs in newly industrialized societies—required an ample supply of meat, butter, eggs, and milk. As a result, in the late 1800s, refrigeration pioneers focused on shipping beef and lamb, while the fresh fruit that European aristocrats had chilled at such expense in their snow wells and icehouses in the 1600s and 1700s was seen as an optional extra. If they ate them at all, city dwellers consumed fruits and vegetables grown nearby, when they were in season; the most delicate and perishable produce, such as strawberries and asparagus, was a frivolous luxury rather than a dietary necessity.

Although the British navy started to distribute lemon and, later, lime juice to its sailors in 1795, in order to prevent scurvy, the term "vitamin"—and the concept that fruits and vegetables might also contain substances that were essential for life—emerged only in the 1910s. Flush with their new nutritional knowledge, chemists had begun to feed livestock rationally engineered "purified diets" that contained the scientifically recommended amount of protein, fats, and carbohydrates, but nothing else—then realized, as their cattle succumbed to debilitating disease and death, that something was clearly missing. Gradually, researchers narrowed in on these substances and named them, arriving at the alphabet of vitamins we know today.

As these new findings were publicized in the 1920s, produce began to be seen as "protective," and popular media urged city dwellers to shift their diets from meat and potatoes toward leafy greens and daily salads. "Hardly were the results of the laboratory experiments printed than the new heroes . . . were taken up with the enthusiasm of a Lindbergh or a Babe Ruth," wrote journalist Eunice Fuller Barnard in 1930 in *The New York Times Magazine*. "Lettuce, formerly a vegetable Cinderella, within a decade occupied the center of the grocer's stalls. . . . Orange and apple drink stands sprang up on countless street corners, and spinach was irrevocably installed in the menu as childhood's major sorrow."

The only obstacle to satisfying this newfound fervor for fruits and vegetables lay in getting enough of them into the cities, where people lived. In 1916, economist Arthur Barto Adams lamented the enormous economic and nutritional burden created by the fact that some 40 percent of perishable produce never reached the consumer at all, instead succumbing to rot en route. "In the year 1911 the New York Board of Health condemned and destroyed 6,500,000 pounds of fruit and 2,500,000 pounds of vegetables because of decay," he concluded in a landmark report. "In warm weather Florida oranges lose 30 per cent in transportation alone, and if we add the decay after the fruit reaches the consuming center the total loss would be astounding."

Across the Atlantic, in Britain, a densely populated but tiny country that was already reliant on food imports, the submarine blockade of World War I made such losses existential. In August 1917, government officials estimated that if German U-boats continued to successfully sink resupply ships, the country would run out of food within weeks. Under these circumstances, losing a full third of a stored potato harvest to sprouting or half a cargo of Australian apples to "brown heart"—a postharvest affliction that turns the center of an apple into cork—seemed unbearably careless rather than the unavoidable cost of doing business. This wartime anxiety eventually led to the construction of an experimental facility built on the grounds of a Cambridge college. It was named the Low Temperature Research Station, and it was here, over the next decade, that many of the discoveries that still inform the cold storage of fruits and vegetables occurred.*

At the time, the scientific knowledge of what happened inside an

---

*The name was a matter of hot debate. In a scribbled note I found in the archives, the acting director dismissed Low Temperature Research Station as "far too long," writing that "a classical man suggested 'Psychreion,' which seems to hit it off quite well." I consulted my brother, who had the benefit of a classical education; he advised me that *psukhrós* is Ancient Greek for "cold" or "frozen," and the suffix *-eion* means "place of." He also volunteered a suggestion of his own: "Cryonomeion," from the better-known *cryos* ("icy cold," "chill," or "frost" in Ancient Greek), *nomos* ("laws," commonly used as the suffix -nomy, meaning a system of rules or knowledge), and *-eion*. Fortunately, Low Temperature Research Station, or LTRS, it was.

apple or a potato once it was plucked from the branch or shoveled out of the dirt was basically nonexistent; when I flipped through the minute books from the first meetings of the researchers charged with investigating these matters, in the LTRS's seemingly untouched archives, I found an almost dizzying list of unknowns. Why did fruit wrinkle as it aged? What was going on, both chemically and physiologically, when a vegetable ripened, softened, sprouted, rotted, or turned brown? At what temperature would plant tissues freeze? Variously, individual scientists expressed interest in tackling the "woolliness of peaches," "bladderiness of plums," and "breakdown in certain varieties of apples"; as a team, they wished to understand how best to extend "the survival of plant organs" by "prolonging the 'death' processes as much as possible." Appropriately, while their new facility was being designed and constructed, they carried out experiments in food palliative care at the university's medical school.

Apples were chosen as the first object of study, and William B. Hardy, the meat researcher charged with setting up the Low Temperature Research Station, recruited a precocious, pipe-smoking young scientist named Franklin Kidd to lead the effort. As a boy, Kidd had conducted his own amateur experiments into the effect of oxygen and carbon dioxide on seed germination using gas cylinders procured by his father, a civil servant who moonlighted as a philosopher. Kidd continued this research at university, partnering with botanist Cyril West to measure apple growth rates and carbon dioxide production over time. Kidd's Quaker beliefs meant he was a conscientious objector; instead of serving in the trenches, he and West (who had been invalided out of the war after falling from a horse) began attacking the problem of apple storage.

There were already clues that the solution would involve depriving fruit of air as well as warmth. Apples that were not consumed at harvesttime were traditionally packed into barrels and held in the damp coolness of underground cellars, where they seemed to last longer. In Afghanistan, a particularly ingenious device known as a kangina is still used to keep fruit fresh for up to six months by sealing it into an airtight, disc-shaped container made of two clay bowls joined together.

In addition to these traditional practices, a few early researchers had observed that atmospheric modification seemed to help preserve apples. In the early 1800s, Jacques Étienne Bérard, a chemistry professor in Montpellier, was awarded a prestigious grand prize by the French Académie des sciences for his report that ripening could be delayed for weeks and sometimes months by placing healthy apples, as well as cherries and pears, under a glass cloche, sucking out the air, and refilling the vacuum with nitrogen gas. "Would we not be tempted to believe, after these experiments, that fruits . . . when they are separated from the tree, retain a certain vegetative force?" Bérard wrote. "And thus, if we place a fruit in a circumstance such that ripening cannot take place—for example, in a medium devoid of oxygen—then it is possible that this vegetative force may be for some time suspended, and yet be preserved in the fruit, so that when the circumstances again become favorable, the force reclaims its territory and resumes ripening."

By the late 1800s, a couple of Americans had also stumbled upon the same trick and put it into practice with varying degrees of success. In Cleveland, Ohio, a professor named Benjamin Nyce lined his apple-storage shed with sheet iron to make it airtight and cooled it with natural ice before filling it with barrels of freshly harvested apples and sealing the door. In the first forty-eight hours, the apples, breathing heavily from the shock of their removal from the tree, would suck up all the oxygen in the room, at which point, Nyce reported, it would become "so charged with carbonic acid gas [carbon dioxide] that a light would not burn." Thus deprived of oxygen, he found, "the russets keep through the month of May in the most sound condition."

In California, the short-lived Carbonic Acid Gas company of San Jose piped carbon dioxide gathered from an old mercury mine into rooms full of unrefrigerated cherries, apricots, pears, and plums, in an experiment that apparently met with great success. "The carbonic acid gas process holds the fruit in a state of suspense," concluded the company's triumphant report at a fruit-growing convention in Sacramento in 1895. Sadly, when further tests were conducted on ripe peaches in August 1896, the fruit was discovered to have been "baked into a soft mass" after just seven days; without cooling, atmospheric modification could do only so much.

Whether or not Kidd and West knew of these previous observations, Kidd's early experiments on seeds had shown that part of the plant, at least, seemed to become dormant when deprived of oxygen. Perhaps the same might hold true for ripe fruit?

The pair, who worked together for decades, appear to have been chalk and cheese—Kidd gained a reputation as irritable and difficult, while West was renowned for his kindness, as well as for spending hours gazing at plants, enraptured by their beauty. Outside the laboratory, they rarely met. Kidd composed lyrical poetry that expounded on his philosophy of life, while West spent every free moment walking the British countryside, impeccably attired in a dark suit, white silk scarf, and black trilby. One friend, a fellow botanist who authored the definitive five-volume guide *Flora of Great Britain and Ireland*, reckoned that West "probably saw more British species of plants living than any other botanist has ever done" and fondly recalled an excursion during which West became separated from his companions, only to be rediscovered some hours later "sitting on a rock stroking the hairs of *Hieracium holosericeum*," a yellow-flowered weed that resembles a dandelion.

To test the effect of storing apples at different temperatures and in atmospheric blends, Kidd, West, and Kidd's soon-to-be wife, Mary—a bishop's daughter whose previous research focused on the "black spot" mold problem in frozen beef—fashioned a series of boxes, "like small flanged coffins with flat lids." They slathered the interiors with petroleum jelly to make them even more airtight, filled each with a bushel of hardy British apples, and—holding their breath as the apples strained for theirs—sealed them up.

Promising early results led to the construction of a larger experiment: small, glass-fronted huts inside a commercial cold-storage warehouse in London, complete with ventilation ports and tubes to test a range of gas blends. Reducing the oxygen to zero caused the apple to ferment; lowering the temperature too far caused its cell walls to break down and become mushy; letting the carbon dioxide levels creep up too much turned out to be responsible for the infamous "brown heart." It took until 1927, when Kidd and West published their initial results, to settle on a recipe for a synthetic atmosphere capable of doubling the postharvest life of apples: chill to forty-six degrees and open ventilation ports as needed to raise carbon dioxide to 10 percent and reduce oxygen levels to a similar proportion. Removed from this carefully controlled atmosphere in February or early March, the authors noted, the resulting perfectly "green, firm, juicy, and acid" apples sold for more than double what they would have fetched fresh from the tree during harvest-glutted October.

With Vaseline-lined coffins replaced by galvanized-steel panels and rubber gaskets, the first commercial refrigerated-gas storage facility for apples was built outside Canterbury, in Kent, in 1929. When the doors were opened and thirty-three tons of Bramley's Seedlings were removed in late March 1930, the flawless condition of the apples—"perfect in appearance and flavour"—was considered newsworthy enough to merit a mention in *The Times*.

Kidd struggled to communicate with fruit growers, leaving the practical translation to West. Within a decade, more than 80 percent of England's fruit-storage space combined atmospheric modification with refrigeration. Adoption was slower in the US, but today nearly nine out of every ten apples grown in Washington State are put to sleep in sealed storage units every autumn, only to be reawakened when the market is ready for them.

Modern, refrigerated, controlled-atmosphere apple storage facilities are massive. These blank-walled warehouses are capable of being built on short notice to accommodate a particularly abundant harvest, yet they're so high-tech that they can sense the exhalations of the fruit inside and adjust themselves accordingly. Washington State now has the largest volume of controlled-atmosphere storage of any growing region in the world: vast, anonymous cuboids filled with millions of apples hibernating in the dark.

The biggest warehouses contain tens of thousands of fruit bins stacked twelve high; each bin holds some two thousand apples and weighs up to one thousand pounds. Because the buildings have to support such weights while remaining completely airtight, with no cracks caused by shifting or settling, their construction involves pouring concrete floor slabs anchored to deep pilings, under the supervision of geotechnical engineers. Semitrucks loaded with fruit bins can drive straight into the rooms through enormous sliding doors; the walls are precast concrete lined with insulated metal panels that snap together seamlessly. These warehouses represent a very substantial investment for the growers, packers, and cooperatives that own them.

Until the 1960s, the oxygen and carbon dioxide levels in such facilities were largely product-generated: the apples breathed in and breathed out, and sensors opened or closed vents to the outside in order to maintain the desired gas levels accordingly. From early on, lye, also known as caustic soda, was used to help absorb or "scrub" excess CO_2. Then, thanks to the rise of Cold War–era nuclear-powered submarines, new atmospheric-control technology emerged. Whirlpool's Tectrol (short for "total environmental control") and the Atlantic Research Corporation's Arcagen system were both capable of creating and maintaining precision atmospheres using catalytic converters, rather than by manipulating apple exhalations.*

---

*In the second half of the twentieth century, the challenges of maintaining human life in extreme environments—specifically, long-term submarine voyages and space travel—gave rise

With greater control came ever more refined respiratory prescriptions. Today, many controlled-atmosphere storage facilities try to hover just above the point at which an apple will suffocate—somewhere between 0.5 and 2 percent oxygen. This represents something of a gamble for growers. The three-quarters of a million dollars' worth of apples packed into each room might sleep more deeply, emerging firmer and fresher after spending ten months in 0.5 percent oxygen as opposed to 2 percent—or they might stop breathing altogether and ferment into an unsalable mess. After all, as Kidd and West discovered back in the 1920s, two apples of the same variety harvested on the same day from different sides of the same tree can have quite different metabolic rates.

The doors to controlled-atmosphere storage rooms often have a little pressurized glass window built in, to allow anxious growers to monitor changes in the skin color of their apples that could indicate stress, as well as a hatch on the roof through which employees can scoop out individual apples using a fishing net. Humans can't survive more than a minute or two in these ultra-low-oxygen conditions; every so often, industry publications carry a story about an unfortunate worker who died while holding their breath and "scuba diving" for apples.

In the past few years, however, the same technology that Natalia Falagán uses to monitor her produce in real time—plastic tubs bristling with valves, monitors, and sensors—has been commercialized under the moniker *SafePod*. Using this system, a couple of hundred apples can serve as the equivalent of canaries in a coal mine. Growers can probe the metabolic limits of these representative apples from within the safety of a Tupperware bubble and use the feedback from the test fruit to

---

to a suite of curious new technologies, many of which have filtered into commercial food preservation and even creation. For example, in a conversation a few years ago, Morris Benjaminson, the scientist who created the first lab-grown meat (goldfish fillets), told me that he never saw it as a replacement for "the steaks and lamb chops" we eat today but instead as something suitable "to keep people going . . . on submarines that stay underwater for long periods of time." Cultivated chicken cells are now being served in US restaurants and, although this new, no-kill meat still requires refrigeration, it seems to have a substantially longer storage life than conventional chicken, due to the absence of bacteria acquired during slaughter.

automatically tweak the atmospheric controls for the entire room. It is a process that empowers apples to continuously redesign their environment in response to their metabolic needs—an atmospheric architecture reverse-engineered from plant physiology.

What's more, apples that are destined to die young can essentially self-select onto supermarket shelves sooner. "Is the controlled atmosphere making their heart pump fast? That could be an indicator of what rooms to market first," explained the SafePod system's Michigan-based inventor, former apple grower Jim Schaefer. "We're making every season apple season," he promised.

Astonishingly, a century after Kidd and West's groundbreaking studies began, no one knows exactly how the apple time machine works. Cold exercises its standard slowing effect, of course, and scientists know that carbon dioxide and oxygen levels influence the activity rates of various metabolic enzymes, but the precise cascade of decay-inducing chemical reactions that is halted or diverted by atmospheric manipulation remains a mystery.

Still, for economist Dana Dalrymple, it was a rare example of an agricultural technology whose blessings accrued to consumers and farmers alike. In a 1969 essay on the development of controlled-atmosphere technology, he concluded that its widespread adoption meant that "consumers benefited from higher-quality fruit over a longer season; growers received higher net prices."

Today this verdict is debatable. Certainly, no one in 1920 would have imagined being able to eat a fresh apple in June, whereas a century later ten-month-old apples are not only cheap and still relatively tasty but also ubiquitous. Shoppers are admittedly often horrified to discover that their kid's lunchbox staple may well be nearing its first birthday, as opposed to being fresh off the tree, but that's the result of voluntary ignorance rather than deliberate deception: apples are only harvested from July through

November in the Northern Hemisphere, so if you're eating an American apple at any other time, logic would dictate that it must have been stored. But while precision-preserved fruit is undoubtedly of higher quality than fruit left to age in ambient conditions, some flavor sacrifices are nonetheless inevitable.

"The more ripe a piece of fruit is when you harvest it, typically, the less time it will retain good quality in storage," explained Kate Evans, Washington State University's leading apple breeder. But an apple harvested too soon, before it is ready to ripen, will never develop the capacity to produce flavor chemicals in sufficient quantity. Apple growers thus try to harvest fruit in the tiny window where it has reached physiological maturity but hasn't yet begun to ripen—in other words, they err on the early side. Evans, who, like me, grew up in Great Britain, has experienced the disappointing results. "I challenge anyone to have decent imported Golden Delicious in the UK," she said. "It's never delicious, because it's picked too soon, because the aim is to store it for a long time."

Another issue is that although, as Jacques Étienne Bérard noticed two hundred years ago, gas-stored apples retain their "vegetative force" and will happily ripen once returned to circulation, they don't ripen exactly the same way off the tree as on it. As early as the 1970s, scientists knew that after more than a few months of suspended animation, apples don't fully recover their ability to produce flavor compounds. They are still sweet, juicy, and crisp, but they won't taste as complex or unique—in short, as good—as their tree-ripened cousins. The subtle citrus, nutty, or ginger notes of a Cox's Orange Pippin, an Egremont Russet, or a D'Arcy Spice would never survive storage; these older varieties are also rarely stored. Kidd and West struggled to develop a successful atmospheric regime for the Cox—it was and is one of Britain's most popular apples, but after a few months of controlled-atmosphere storage, it routinely falls prey to a host of maladies with such suggestive names as *watercore, bitter pit*, and *diffuse browning disorder*.

"Our supermarkets want life to be easy," said Evans with cheery resignation. "They want varieties that they can consistently have on the shelf for twelve months of the year, and that means varieties that store pretty well." Evolutionary biologists use the term *ecological filter* for processes or conditions that prevent certain species from inhabiting a particular landscape; over the past century, controlled-atmosphere cold storage has proven to be a powerful ecological filter for apple varieties in the marketplace. Evans confessed that when she began her career as an apple breeder, at the Kent-based annex of the Low Temperature Research Station where Kidd and West continued their fruit-storage research in the 1930s, she used to say her favorite apple was the Cox's Orange Pippin. "Now I usually just say: the next one," she told me— meaning the next apple to emerge from her breeding program, which, whatever its other qualities, will be guaranteed to store well.

"Eating quality comes first," she assured me. "But then it's all about storability, because I'm breeding apples for Washington State and we have this massive crop. There's no way that you would be able to get through that crop in terms of sales without having a huge proportion of it in cold storage." Evans doesn't even bother evaluating her new apple varieties for their flavor and crunch until they've survived at least two months in cold storage. Her first launch for the program came in December 2019: WA 38, as she calls it, or Cosmic Crisp, as it has been dubbed in the marketplace, will happily hang out in controlled-atmosphere storage for twelve to fourteen months without any noticeable ill effects.

This means Cosmic Crisp can also be found on supermarket shelves, looking and tasting more or less the same, all year round, which is a prerequisite for any food that aspires to the elusive status of a household brand. What's the point of spending $10.5 million (the marketing budget for Cosmic Crisp's launch) and hiring influencers and even an astronaut to promote your apple if it's nowhere to be found when the consumer goes shopping? "The point is not just producing fruit for people to eat," said Evans with a laugh that turned into a shrug. "It's getting people to buy the fruit—and buy it again."

This is where the other half of Dana Dalrymple's mutually beneficial equation comes in, and his conclusion has not aged well. At first, the expense of controlled-atmosphere storage was worth it for growers, because they could sell their harvest for more in the spring, when apples were rare. Soon more growers invested in controlled-atmosphere storage and planted more trees so they could fill their storage facilities, and that increased supply meant that prices inevitably fell. "Growers are really struggling to make a profit on a lot of our varieties," agreed Evans.

In short, controlled-atmosphere cold storage turned apples into widgets, and the only way for widgets to compete is on price. Growers gave up the distinctive flavors and appearances of traditional varieties to gain an edge in terms of availability, but once June apples became commonplace, there wasn't much left with which to differentiate their produce.

Seen from this angle, Washington State growers' unprecedented investment in branding and marketing the Cosmic Crisp is a somewhat desperate attempt to turn that trend around and make an apple exciting and distinctive enough to spend money on again.* As exciting and distinctive as, say, an Emneth Early—a flavorsome apple that bakes up into a fluffy, snow-like texture and was heralded as one of the first varieties to ripen each summer—or the Roxbury Russet, North America's oldest variety, which was esteemed for its keeping quality—to name just two of the now-almost-forgotten apples that were eliminated by refrigeration's filter.

Novelist John Steinbeck grew up in Northern California's Salinas Valley, describing it in *East of Eden* as "a long narrow swale between two ranges of mountains," where "the topsoil lay deep and fertile," bursting into color after the winter rains before turning "a brown which was not brown but a gold and saffron and red" in the fall. With a large

---

*The first branded apple was the Pink Lady. According to the marketing director behind the Cosmic Crisp's launch, the Pink Lady is "probably the most branded piece of produce in the world."

aquifer for irrigation, abundant sunshine, and a coastal breeze to temper summer's heat, the region's agricultural potential was quickly apparent to settlers. In the newly salad-crazed 1920s, one crop came to dominate the valley: lettuce.

Controlled-atmosphere cold storage transformed the apple; for lettuce, it proved equally revolutionary. Today, the ninety-mile length of the Salinas Valley is known as the Salad Bowl of the World because it produces more than 70 percent of the nation's lettuce. "Nowhere else in the United States does so small an agricultural area produce so large a percentage of a widely consumed commercial crop," concluded geographers Paul F. Griffin and C. Langdon White in 1955, at a time when the valley's growers had cornered only half the nation's commercial market.

Before World War I, America's salad eaters—a rare breed at the time—ate soft-leaf or butter lettuces grown on the outskirts of East Coast cities. Twenty years later, if a woman in one of those cities purchased lettuce, it was most likely iceberg—except from late autumn through early spring, when it could *only* be iceberg—from the Salinas Valley. "Iceberg was kind of like shoe leather," explained Jim Lugg, a lettuce legend. The iceberg lettuce's thick, densely packed leaves were substantially sturdier than the tender, loose head of a Boston Bibb, a quality that has been decried by critics—"scarcely a gourmet worthy of the name would disagree, that iceberg lettuce is the lowliest, most tasteless of salad ingredients," as Craig Claiborne put it in *The New York Times* in 1958—but gave it at least a fighting chance of surviving the long train journey from California fields to East Coast markets.

Salinas Valley's early lettuce pioneers weren't always successful. Part of the first crop was fed to local cows, and a subsequent one took more than a month to reach New York, at which point it had been reduced to green slime. By the 1920s, Salinas Valley growers had figured out that refrigeration could give their lettuce a better chance of success: they shipped it packed in crates with ice, then covered those with more ice, which was topped up by yet more ice en route. To have a hope of remaining intact, each railcar load had to be buried under fifteen thousand pounds of the stuff, giving the crisp-head variety previously known as Los Angeles its now-familiar name: iceberg.* New Yorkers, in the grip of the new vitamin trend and fed up with winter's menu of carrots, potatoes, and cabbages, snapped it up.

Over the next decade, American lettuce consumption doubled. Salinas Valley farmers ripped out their beans, slaughtered their cows, and uprooted their fruit trees to plant lettuce, which had the added benefit of producing two crops a year. By the mid-1920s, more ice was manufactured in the Salinas Valley than in the entire city of New York. Growers whose shipments made it in one piece could earn more than the price of their farmland from a single harvest. Others were not so lucky. "It was really a gambling business," said Robert Kieckhefer, a packaging manufacturer who went on to develop a new carton for Salinas Valley lettuce growers after World War II. "Often, after the rail car arrived at its destination, they would open the doors and have nothing but a bunch of slush in there—terrible stuff."

This fate befalls the fictional Adam Trask in Steinbeck's *East of Eden*. In 1915, inspired by the information that New York City is the biggest market for winter oranges, he begins to wonder whether East Coast urbanites might not also appreciate Salinas Valley produce. "In the cold parts of the country, don't you think people get to wanting perishable things in the winter—like peas and lettuce and cauliflower?" he argues. "And right here in the Salinas Valley we can raise them all year round." His neighbors try to dissuade him—"People in the East aren't used to vegetables in the winter. They wouldn't buy them"—but Trask purchases an ice plant and packs six carloads of lettuce. The result was "a sensation in a year of sensations": snow slowed the train's passage through the Sierras, unseasonable warmth in the Midwest coincided with a delay in

*Company legend at one of the biggest lettuce packers in the Salinas Valley, Bruce Church Inc., holds that the name iceberg originated with kids who greeted the ice-mounded, lettuce-bearing railroad cars with the shout "The icebergs are coming!"

Chicago, and the lettuce arrived in New York City as "horrible slop with a sizable charge just to get rid of it."

Lettuce's path from lottery ticket to big business required another breakthrough in cooling: vacuum pumps capable of generating and maintaining very low pressure. Invented by researchers at Eastman Kodak in 1929 as a way to dry photographic film, this new technology was quickly spun off into a stand-alone business focused on extracting the beneficial but delicate fatty acids in fish oil without the need for heat. Rex Brunsing, a compulsive inventor who worked as a "vegetable traffic manager," wrangling logistics for a Salinas Valley grower, heard about this new technology and immediately saw its potential to free the icebergs from ice.

Brunsing tried to interest his employer in his ideas but was fired for his efforts. "He was always pushing some scheme," explained Bud Antle, a packing-shed boss who went on to acquire Brunsing's Vacuum Cooling Corporation and become the largest lettuce shipper in the valley. Eventually, Brunsing managed to get some funding from the Kieckhefer Box Company, which was hoping to sell its new cartons as a replacement for the wooden crates in which iceberg lettuce was typically shipped. Wood was too thick to allow the vacuum to work well, while cardboard would dissolve if used to carry ice-cooled lettuce, so Kieckhefer's and Brunsing's interests were aligned.

In late 1946, Brunsing put together "a strange maze of boilers and pipe fittings" that passing drivers on the Castroville-Salinas highway described as "a poor man's attempt to build a rocket ship." He put boxes of lettuce in a cylindrical chamber, switched the contraption on, and sucked out all the air. In a high vacuum, water boils at room temperature, and an iceberg lettuce is 96 percent water. Inside Brunsing's tube, water in the tissues of the still-sun-warm lettuce quickly evaporated off as steam, taking heat with it. The inside of a head of iceberg lettuce could be lowered to thirty-four degrees in a matter of minutes, as opposed to days using ice—which in turn added days to its shelf life. "Also, lettuce gets worms in it, and the drawing of the vacuum exploded the worms," explained Kieckhefer. In vacuum-cooled lettuce, he pointed out, "there's no way any housewife could ever find a worm."

"Rex hustled me," said Antle, who traded his packing-shed property for part ownership in the vacuum-cooling plant. At first, the valley's big lettuce growers fought the new technology—after all, they'd invested heavily in ice plants. "Hell, a committee of influential businessmen called on me and told me what was going to happen to the industry, and all the men it would put out of work, and all the ice companies going broke," Antle recalled. "We offered it to everybody in the business," he continued. "We couldn't even give it [a]way. It's pretty near goddamn unbelievable the story!"

Despite this initial resistance, within a few years, the vacuum cooler triumphed—particularly once lettuce companies realized the savings to be had by eliminating not only ice but also unionized packing-shed labor, as the new coolers meant lettuce could be packed in the field by immigrants. Worm-free and perfectly preserved: by the mid-1950s, iceberg lettuce was not only vacuum cooled but also America's favorite vegetable—or at least its most consumed. Iceberg's streamlined scale ensured fresh lettuce was available during the winter months, but the efficiency with which Salinas Valley growers could produce and pack it also ended up squeezing out local varieties in the summer.

This effect, in which a place that can grow vegetables more cheaply forces the price for produce downward, outcompeting growers in other regions, is known among economists as the principle of comparative advantage. Adam Smith formulated it succinctly in *The Wealth of Nations*: "If a foreign country can supply us with a commodity cheaper than we ourselves can make it, better buy it of them with some part of the produce of our own industry employed in a way in which we have some advantage."

Refrigeration allowed the comparative advantage of the Salinas Valley's beneficent climate to exert its force on the market, making it economically inefficient to grow lettuce anywhere else. It also meant that, for most Americans, lettuce *was* iceberg—a perfectly serviceable base for blue-cheese dressing but no substitute for the nutty notes of mâche or the peppery kick of arugula. It took the arrival of Jim Lugg to resurrect the nation's lost leaves, using the same Whirlpool Tectrol technology that had transformed apple storage.

Lugg wanted to be a farmer, but his family had sold their California ranch by the time he got out of school in 1956, so he got a job as a director of research at Bruce Church Inc., a company founded by one of the Salinas Valley's original lettuce barons, reporting to Church's son-in-law and company president Ted Taylor.

"Ted and I spent a lot of time over in Save Mart stores in the San Joaquin Valley," Lugg told me with the laugh that punctuated most of his stories. "We'd see people pick up a head of iceberg and then pick up some carrots or a head of romaine." When Lugg asked how they were planning to use their iceberg lettuce, the customers often said they were going to mix it with the romaine or with some shredded carrot to make a salad. It gradually became clear to Taylor and Lugg that America's housewives weren't shopping for iceberg lettuce, they were shopping for salad, and anyone who could give them that in one convenient, preferably pre washed package would be onto a winner.

The only problem was that, while transporting heads of lettuce intact across the country was a challenge, attempting the same feat for a blend of shredded lettuces was positively Herculean. "For me, a precut salad is the perfect storm," agreed Natalia Falagán. "First you have damaged tissue where all the bacteria can come in. Then you have a mixture of vegetables, so they all have different respiration rates. And then you want it to last two weeks?" She threw her hands in the air. "It's impossible."

"I'm probably telling tales out of school, but Ted was the kind of guy that used to say, 'You know what? You've got the imagination and I've got the money, so go figure it out,'" said Lugg with another guffaw. Whirlpool's research center was just an hour north, in Sunnyvale, and Lugg, who had read about controlled storage of apples, immediately saw the technology's potential for slowing down lettuce respiration rates. The challenge was that Ted Taylor didn't want to store his lettuce for months before shipping it, like Washington State's apple growers—he wanted to slow down how fast his lettuce breathed while it was in motion. Somehow, Lugg needed to shrink a warehouse-sized technology to the dimensions of a bag.

"It was differentially permeable membrane that made the packaged-salad thing a home run," Lugg told me. Differentially permeable membranes are fundamental to the basic unit of biological life: a cell's encircling membrane functions like a bouncer at a club, allowing some molecules in preferentially, while letting others in more slowly or not at all. By the 1960s, polymer scientists were capable of designing films that worked according to the same principle. Different plastic blends were extruded in such a way that they let oxygen or carbon dioxide diffuse through at specific rates. After years of testing, Lugg ended up combining five different layers to get the functionality he needed. "Layers one and five are seal layers," he said. "The center layer had the beautiful printing, then on one side of that was the CO_2 laminate that we needed, and the other side was the oxygen laminate."

Astonishingly, the bag itself—a cheap, disposable plastic lettuce bag—was a miniaturized version of Kidd and West's controlled atmosphere warehouse. The microperforations in its carefully designed films let oxygen in at one rate and carbon dioxide out at another to maintain the ideal atmospheric microclimate around the leaves as they traveled the country and sat on supermarket shelves. Lugg had measured the respiration rates of all the different leaves he wanted to combine. "Iceberg's pretty lazy; romaine's a little more eager to do something," he said. "When you get into the tender leaves, they're like lawn clippings—they're hot." Baby spinach was particularly tough, he told me. "We were packaging spinach when it only had five true leaves on the plant," he said. "That early, it's really breathing hard."

As best he could, Lugg made salad blends that combined leaves in a ratio that was pleasing to consumers but, just as important, balanced out the leaves' individual respiration rates. "We'd match the slower respirers with the high respirers, to minimize the number of different films we needed," he explained. "There was a lot of science behind it." In 1989, Bruce Church introduced the first bagged salad, under the label Fresh Express. "That was the Family Classic, which was iceberg, romaine, and some carrot and so forth," said Lugg. "We did European in 1991, with radicchio, endive, and escarole, and the kits came in 1992 with the Caesar Supreme, which was a hit right out of the gate."

Iceberg lettuce consumption has fallen by half since Jim Lugg's controlled-atmosphere bags changed the face of salad; today, it's a nostalgic taste of the constraints of an earlier iteration of cooling technology. Sales of bagged greens have boomed, however, meaning that, despite drought, food-safety scares, pest problems, and ongoing labor disputes, the Salinas Valley produce business has too—although growers now routinely decamp to Yuma, Arizona, which offers even more sunshine and fewer worker protections, for their winter harvest.

When Fresh Express was sold to a Brazilian orange juice giant, Cutrale, a few years ago, all Lugg's records, including his prototype salad bags, were thrown away. Colleagues rescued the plaque commemorating his achievement, which now sits in his garage. Few people are even aware that their salad bag is a high-tech respiratory apparatus. "I've seen a lot of people who are like, *Oh, it doesn't fit in my drawer—I'm going to open it to remove the air so I can squeeze it in,*" said Falagán. "And it's like, *Oh my God! All this work to find the perfect modified atmosphere for your lettuce, and then it doesn't fit in your drawer, so you open it?*"

II. Oracular Bananas

"A stream of air which has passed over an apple would *appear* to be harmless to other forms of life," said William B. Hardy, director of the Low Temperature Research Station, in a 1932 address to the British

Association of Refrigeration. "The appearance is wrong—the air contains some subtle emanations which profoundly influence other vegetable forms."

As Kidd and West continued their storage research in the 1930s, they noticed something peculiar happening to nearby plant matter as their apples aged. Potatoes exposed to elderly apples became covered in misshapen sprouts that, Hardy continued, were "more like warts than anything else," pea seedlings grew sideways and deformed, bananas dissolved overnight, leaves blanched or fell off, and flower buds refused to open. Younger apples would experience accelerated aging, "as though the elderly apple were jealous of youth, and would destroy it." He passed photographs around the lecture hall as he spoke, in order to illustrate "these queer happenings."

Despite Kidd's and West's best efforts, they had not yet managed to measure or identify the chemical culprit, which, Hardy concluded, must be more potent than snake venom, given that it seemed to be present in the air only in minute quantities. "Of what use is this power to the apple? Why can it so influence its fellow-vegetables? In that and in the actual nature of the emanation lie the biological puzzle," ruminated Hardy, before moving on to meatier concerns.*

Unbeknownst to the intrepid cryonauts at the Low Temperature Research Station, the chemical responsible for this biological puzzle was ethylene, and it had been causing curious phenomena for millennia. A colorless, gaseous, sweet-smelling hydrocarbon, ethylene was first isolated by the German alchemist J. J. Becher in the 1660s, as part of his efforts to demonstrate the existence of phlogiston, a mythical combustible element that he believed was found in everything that burned. (The phlogiston theory was scientific orthodoxy for more than a century, before being disproved in a process that led to the discovery of oxygen.)

---

*The LTRS's work on meat's postharvest physiology was also pioneering—it simply took longer to find commercial application. Much of the early research on the fundamental biochemistry of rigor mortis was conducted there, which formed the basis from which scientists were later able to tackle problems such as cold shortening in meat.

Although many of the great scientists of subsequent centuries—Joseph Priestley, Humphry Davy, Jöns Jacob Berzelius—tinkered with ethylene as part of their chemical inquiries, it initially seemed to have few useful properties outside of being extremely explosive. That perception began to change in 1901, when a seventeen-year-old Russian, Dimitry Neljubow, became curious as to why the sidewalk trees that grew closest to streetlights were often abnormally twisted and swollen. In an ingenious series of experiments, he pinpointed the culprit: ethylene. The glass panes enclosing a streetlamp's gas flame weren't airtight, so tiny amounts of uncombusted hydrocarbons—including ethylene—leaked out and caused nearby plants to grow differently.

Over the next couple of decades, horticultural researchers inspired by Neljubow's insight studied what ethylene did to their preferred fruit or vegetable. In Washington, DC, Rodney True, a government researcher known as the Sherlock Holmes of plant detectives, discovered that ethylene caused green lemons to turn yellow; in Saint Paul, Minnesota, plant scientist R. B. Harvey—another Rodney—figured out how to use ethylene to blanch celery leaves, which had previously been achieved using a fiddly and time-consuming process that involved building little cardboard covers for the plants. "It became evident that the ethylene process was a boon to the fruit trade," Harvey wrote in a University of Minnesota bulletin on the subject.

Not to be left out, two University of Chicago physiologists, J. B. Carter and A. B. Luckhardt, decided to see what ethylene did to animals. In addition to knocking out white mice, guinea pigs, and kittens, they and some friends inhaled it themselves. Their report, published in *The Journal of the American Medical Association*, was euphoric. Carter apparently felt "a sense of well-being and exhilaration"; Luckhardt experienced such bliss that he declared he would be "satisfied to lie there under the influence of the gas for all time"; and their friend Archer C. Sudan "laughed a great deal," failed to notice a safety pin the others stuck into him, and, immediately upon his return to consciousness, "talked excitedly and incoherently of his experience." Ethylene was briefly poised to be the next big thing in anesthesia, until, alas, it was realized that the combination of electrical instrumentation and an explosive gas would likely cause sparks to fly in the surgical theater. Instead, according to toxicologist Henry Spiller, ethylene became the drug of choice at high-end séances in the Roaring Twenties.

Spiller came across ethylene in his work on huffing, or, as it's technically known, inhalant abuse, in his role as the director of the Central Ohio Poison Center. In the late 1990s, he joined forces with geologist Jelle Zeilinga de Boer and archaeologist John Hale to come up with a compelling argument that, in ancient Greece, the oracular pronouncements made by the priestesses at the Temple of Apollo at Delphi might well have been the result of huffing ethylene. De Boer is now dead, but Hale and Spiller told me the story, which began when de Boer was commissioned by the Greek government to study the feasibility of building a nuclear power plant near the temple. Much to their dismay, he identified dozens of crisscrossing geological faults, including one in the limestone directly beneath the subterranean chamber in which the priestesses would sit. The rock's pores contained hydrocarbons, including ethylene, which would have risen toward the surface as the faults shifted, dissolved in groundwater, then bubbled up as vapors from springs such as the one that ran through the priestesses' chamber.

Together de Boer, Hale, and Spiller assembled the geological, historical, and toxicological evidence into a narrative that seemed plausible: a priestess would descend into the adyton, an underground chamber that was the innermost sanctum of the temple, inhale the buildup of what ancient Greek chroniclers described as a "sweet-smelling" gas, then begin to babble in cryptic fragments that typically required external interpretation to make any sense—just like Archer Sudan in Chicago in the 1920s. To put their theory to the test, Spiller, Hale, and de Boer got hold of a tank of ethylene and requisitioned Spiller's garden shed, which, in a stroke of luck, matched the dimensions of the subterranean chamber at

Delphi. Spiller ran a hose from the tank under the gravel floor of the shed so the gas emerged from the ground beneath a lawn chair that was standing in for the stool on which the priestess traditionally sat. Finally, he recruited a female neighbor. "Within about two minutes, we were asking her who was going to win the Kentucky Derby," remembered Spiller. Sadly, the trio did not get anything they could take to the bookmakers out of their garden-shed oracle. "Honestly, she was fairly giddy," said Spiller. "She remained awake and talking to us, but you could have probably done minor surgery on her."

Despite ethylene's booming popularity in séances and horticulture, it wasn't until 1934 that Richard Gane, a colleague of Kidd and West at the Low Temperature Research Station, discovered that the "subtle emanations" coming out of apples were minuscule gaseous emissions of ethylene. Apples were capable of producing this potent oracular anesthetic themselves—as were, it emerged over the following decades, figs, melons, pears, and peaches, among dozens of other fruits.

Subsequent research has shown that for fruits and vegetables ethylene is a multipurpose signaling molecule—a gaseous hormone, or chemical messenger, that helps plants maintain their daily rhythms and communicate with each other about damage, stress, and disease threats. It also seems to function as kind of a life-cycle gatekeeper, either triggering or inhibiting progress toward the next stage of fruit being. Baptize tomato and lettuce seeds with ethylene gas and they will be primed to germinate; the reverse is true of ethylene-exposed onions and potatoes. Spray ethylene-infused water onto a field of pineapple plants and they will all burst into bloom. An elderly apple produces ethylene not out of jealousy, as the Low Temperature Research Station's Hardy speculated, but rather as the warm-up for its next performance: self-sacrificial reproduction. Ethylene switches on the genes that produce enzymes that make an apple softer, less acidic, and more sugary—qualities that encouraged long-extinct megafauna to consume it and distribute its seeds via their digestive systems.

Today, more ethylene is produced than any other organic molecule in the world—more than 200 million tons a year, in vast, energy-guzzling factories known as crackers because they create ethylene by cracking apart the molecular bonds holding a larger hydrocarbon, ethane, together. The same chemical that animated the Delphic Oracle now forms the bedrock of our petrochemical civilization as the primary ingredient in everything from polyethylene plastic bags to polyester fabric. Along the way, it has become an essential element in the ever more refined atmospheric manipulations that have industrialized our fruits and vegetables.

Low-temperature and controlled-atmosphere storage already reduce the ethylene response as part of their overall metabolic dampening. To that a warehouse manager wishing to extend the shelf life of broccoli, for example, or apples, can add an "ethylene scrubber"—a filter lined with a chemical like potassium permanganate that pulls ethylene out of the air by reacting with it to form carbon dioxide and water. The latest trick in ethylene reduction is even more sophisticated: 1-Methylcyclopropene, familiarly known as 1-MCP. It slots into an apple's own ethylene receptors, rendering it deaf and blind to the gas's signals. Under the brand name SmartFresh, 1-MCP was commercialized in 2002 and is already used on more than thirty crops, including seven out of every ten apples harvested in the United States.

For other fruits, temperature-controlled exposure to ethylene is the key to commercial success. A well-timed dose of ethylene ensures uniformity—a prerequisite for any food or drink that aspires to the status of a commodity, let alone that of a brand. It was ethylene, in combination with refrigeration, that turned two unpromising unknowns into global superstars of the fruit world: first the banana, then the avocado.

For years, when Paul Rosenblatt picked up the phone, he said one word: "Bananas!"

When I first visited Banana Distributors of New York, a sprawling facility on Drake Street in the Hunts Point section of the Bronx, Rosenblatt was still in charge, supplying a million boxes of bananas every year to bodegas, food carts, and grocery stores across the five boroughs. He'd married into the banana trade; his father-in-law was a lifelong banana man who got his start working for street peddlers at the age of eight. By the time I returned, postpandemic, Rosenblatt had finally retired, selling the business to his neighbors, D'Arrigo New York, who were looking to expand. This too is a family business: Stefano and Andrea D'Arrigo immigrated from Sicily in the early 1900s. Stefano became Stephen and moved to California, where he began growing vegetables using seeds mailed from Messina by their father; Andrea, or Andrew, ran a produce wholesale business in Boston. In 1926, they were responsible for the first transcontinental shipment of fresh broccoli in an iced railcar—a minor landmark in refrigeration history. Gabriela D'Arrigo, who greeted me cheerfully in the early hours of the morning on my most recent visit, is the third generation of the family in the business.

From the outside, the brick building looked unremarkable, even rundown: the letter *a* had fallen off the end of the word *banana*, and a banner promising "Every Color, Every Day" was so faded as to be almost illegible. Inside, however, Rosenblatt's lead ripener, Juan Luciano, who had remained on under the D'Arrigos, choreographed a complex climate-control regime, monitoring and adjusting the precise levels of ventilation, humidity, atmospheric gases, and, of course, temperature, in order to ripen more than two million bananas a week.

Contrary to popular belief, bananas are the ultimate refrigerated fruit. In order to be a global commodity rather than an exotic luxury, the banana depends on a seamless network of thermal control. This comes as something of a shock to most people. Indeed, it seems to directly contradict the advice issued by one of America's most memorable brand mascots, Miss Chiquita. In 1944, when she first shimmied onstage in a tight red dress and Carmen Miranda–style fruit-salad tiara, this sultry lady banana warned viewers, "Bananas like the climate of the very,

very tropical equator / So you should never put bananas in the refrigerator."

In reality, before the advent of refrigeration, bananas were a rare and expensive treat outside their tropical homelands. Such prominent Americans as Abraham Lincoln and Andrew Jackson likely never tasted one. In the first volume of his Romance of Big Business series, Frederick Upham Adams, an author and political activist who also invented the electric lamppost, described a vivid childhood memory: his first encounter with a banana palm, on display at the Centennial Exposition in Philadelphia in 1876. "It was surrounded by a crowd of spectators," he wrote, complete with its own security detail to prevent theft or illicit touching. "To my young and impressionable mind," Adams continued, "this was the most romantic of all the innumerable things I had seen in any of the vast buildings." His father bought some "nearly black" fruit at dizzying expense to bring home to the rest of the family in Illinois. "Two of the six were in such an advanced state of decay that they were rejected, but we shared the others," he continued, noting that it was many years before he had the chance to see or eat another banana.

As late as 1899, Scientific American lamented that the banana—a fruit with "forty-four times more nutritive value than the potato"—did not travel well, with "the result that a considerable market is closed to them." (The author held out hope that, dried and ground into a flour called bananine, the fruit might yet "furnish the working classes of many countries with wholesome, nourishing food at the lowest possible cost.") Yet just a few years later, fresh bananas had become commonplace—common enough to be sold cheaply to the poor and given away to welcome immigrants to Ellis Island, common enough for their discarded peels to have become a public nuisance. In 1908, the first edition of Scouting for Boys suggested that a daily good deed might consist of removing "a bit of banana skin off the pavement where it is likely to throw people down"; in 1915 Charlie Chaplin inaugurated a comedy trope by slipping on a discarded banana peel.

Apple pie be damned: by the 1920s, bananas had surpassed anything homegrown as Americans' favorite fruit. Today the banana is the most popular fruit in the world, not just the United States, reliably found in even the most unpromising of breakfast buffets, service stations, and convenience stores.

What enabled this transformation? Steamships, trains, exploitative plantation culture, and even CIA machinations—but most of all refrigeration. Bananas are harvested while green and hard; this phase of their existence, preripening, is known to fruit physiologists as the banana's "green life," and extending it through thermal and atmospheric regulation is the secret to the fruit's commercial success. This story—the transformation of a perishable exotic fruit into a high-volume, mass-market, standardized commodity—is emblematic of an entire century's worth of unceasing efforts, creative experimentation, expensive failures, and increasingly sophisticated solutions that lie behind the standard supermarket produce aisle and its seasonless abundance.

In the tropical heat and humidity preferred by the banana plant, a freshly harvested green fruit has between a week or two before ripening begins. From there, the transition from yellow solidity to brown puree is both rapid and unstoppable. But by the 1880s, the shift from sail to steam had helped reduce transit times from the Caribbean to East Coast markets from three weeks in fair winds down to two or less, no matter the weather, and the first banana entrepreneurs emerged.

Without mechanical cooling, a quarter of all those bananas rotted on their way to US markets. Then, in 1901, came the breakthrough. At the request of the British government, two companies—Elder Dempster and Fyffes, which still dominates the British banana trade today—combined forces to set up a new subsidiary. The Imperial Direct West India Mail Service was, as the name implies, charged with transporting letters, parcels, and the occasional passenger between Britain and its West Indian colony of Jamaica; banana shipments provided the business case for an otherwise loss-making service. The company outfitted six steamships with refrigeration machines borrowed from the frozen-meat trade, and the quantity of bananas reaching Europe quintupled in three years.

This caught the attention of Andrew Preston, a Boston janitor turned banana magnate, who promptly decided to refrigerate an entire ship for his new enterprise: the United Fruit Company. He contracted with the Montreal-based American Linde Refrigerating Company to retrofit a standard fruit ship, the *Venus*. According to the reminiscences of Llewellyn Williams, American Linde's chief engineer, it took eight weeks and nearly $2 million in today's money to transform the *Venus* into "the last word in American refrigerating perfection," complete with seventy thousand pounds of cow hair as insulation.

The *Venus*'s first two banana runs were disappointing—a letter from the ship's captain refers to the condition of the cargo as "rather unfortunate"—and Preston summoned Williams to Boston for an emergency meeting. "My boy, I think your piano is all right, but they do not know how to play it," Preston finally told Williams. "You had better go down and show them how." With Williams aboard the third voyage to ensure the refrigeration system was operating correctly, a full cargo of green bananas made it from Costa Rica to New Orleans without breaking color, and Preston promptly commissioned the construction of three more refrigerated banana boats—the foundation of United Fruit Company's Great White Fleet, so called because the boats were painted white to reflect the sun and assist with cooling.

Thanks to refrigeration, green bananas were now able to survive weeks of transit time between their places of harvest and their ultimate country of consumption. The technology's ability to compress both time and space had bestowed upon "the peoples of the temperate zones a fruit and food product denied to their ancestors through all the ages," as the banana-struck Frederick Upham Adams put it. "This is a real achievement. It is a part of the contribution of our age to the sum total of human progress." Just one obstacle remained in the way of the banana's cold road to global domination: ripening. Refrigeration unlocked the ability to get bananas to US consumers, but ethylene was the key to customer satisfaction.

For years, banana importers simply lit a kerosene stove, closed the door on their banana rooms, and hoped the uncombusted hydrocarbons would do the trick. But for Paul Rosenblatt and Gabriela D'Arrigo to deliver on their promise of stocking "Every Color, Every Day," a stove and a prayer were not enough. Banana importers needed to become fruit whisperers, able to cue their fruit onward to the next phase of its existence on the world stage.

When Rosenblatt opened the enormous latch on the front of one of the cream-painted, heavy-duty, morgue-style doors that lined the corridors in the oldest part of the facility, it released with an audible hiss, swinging open on huge hinges to reveal a dingy, single-story concrete room holding about three hundred cardboard boxes of recently gassed bananas stacked together under gray tarpaulins. At the back of the room, two industrial steel fans were inelegantly mounted in chipboard, pulling air through the wall of banana boxes.

I was keen to huff some ethylene, to experience its metabolic magic and oracular influence myself. As I stepped inside the room, the smell was overwhelming—sweet but a little sour, like the mixture of alcohol and vomit in a scuzzy pub carpet. "There are some people that are so sensitive to it, they're like, 'I can't even be near the banana room,'" said Gabriela D'Arrigo. "It doesn't really bother me."

Furnishing New Yorkers with the comforting certainty that a perfectly ready-to-eat banana is always nearby requires a complex blend of temperature control and atmospheric manipulation, orchestrated asynchronously across dozens of different ripening rooms. Lead ripener Luciano kick-starts the process by treating each room to a twenty-four-hour ethylene sauna. The default is to ripen fruit in five days at sixty-two degrees, but to schedule fruit readiness in accordance with supply and demand, he can push a room in four days at sixty-four degrees.

The end of a black plastic cylinder projected slightly from one of the boxes: its tip was a metal probe buried deep inside the flesh of a representative banana. "They're pulping at the right temperature," Rosenblatt said, reading numbers on the digital display. Too cold, and the bananas risk "chilling injury," which imparts a chewy texture and grayish cast to the fruit; too hot, and they will cook into mush.

The design of these rooms, which dates back to the 1970s, is a legacy of the early days of the trade, when bananas arrived loose and had to be piled up in the rooms by hand. In those days, air-handling technology was less sophisticated, which meant that air circulated through the bunches less effectively and the bananas ripened unevenly. "You'd have a guy in there checking about every hour," said Rosenblatt. "You'd throw a bucket of water on the floor to add the humidity, move the bananas around, get the fans going—it was like *The Flintstones*."

Today's banana rooms are more like *Star Trek*. At two and three stories tall, they are loaded by forklift and rely on powerful cooling units with built-in fans mounted along the edges of the room for increased throughput and improved ripening uniformity. Refrigeration is essential because a standard cargo of bananas is, effectively, a furnace. "The energy coming off a box of ripening bananas could heat a small apartment," explained Rosenblatt. In fact, he continued, he'd experimented with keeping his office warm in winter using heat captured from the ripening process—a thermal exchange powered by pure fruit energy.

For a tourist of the artificial cryosphere, a visit to Banana Distributors of New York is particularly exciting because it is the site of the first two-tier, tarpless banana-ripening rooms ever built. Jim Lentz, president of Thermal Technologies, the largest manufacturer of ripening rooms in the world, told me that he built them for Rosenblatt's father-in-law in the 1980s, back when Paul still had red hair and drove a Mustang. "Really, he was ahead of his time," D'Arrigo told me. "Having double the amount of space, having the air circulate from the ceiling, not having to pull the tarp over—that's what you want, because it's just so much more efficient."

Over the years, these too have evolved. Today, according to Lentz, a state-of-the-art, three-tier pressurized room will cost about $170,000 but can ripen two truckloads of bananas at a time, with "total control over the ripening process and uniformity that's guaranteed," for "fruit that looks better, lasts longer and weighs more." (The extra weight comes courtesy of a nifty humidification system and has nothing to do with shelf life, flavor, or even appearance. "Bananas are sold by the pound," Lentz pointed out. "If you're not taking any moisture out of that peel, there's more money going across the scanner.")*

Appropriately, the roll-up doors are bright yellow—a solid 6 on the color chart with which customers can specify their preferred fruit maturity. The chart ranges from 1 (all green) to 7 (all yellow with brown sugar spots). The most popular shades are between 2.5 and 3.5, but much depends on the retailer's size and target market. The grocery chain Fairway, which sources its bananas from Banana Distributors of New York, expects to hold bananas for a couple of days and will therefore buy greener bananas than a bodega that turns its stock over on a daily basis. "Street vendors like full yellow," Rosenblatt said, as do shops serving a mostly Latin American customer base. Personally, he eats only a couple of bananas each week and favors fully ripe 7s.

When the bananas emerge from their ethylene-immersion treatment, they're extremely sensitive. "Once they've been through the ripening, we will not put them back in a cooler," said D'Arrigo. Luciano is responsible for timing the ripening process to coincide with the truck arrival windows so that bananas don't have to wait. "We're in the business of trying

---

*In Japan, where consumers willingly pay a premium for the most exquisite fruit, one company has also experimented with acoustic design. Playing Mozart's String Quartet no. 17 and Piano Concerto no. 5 in D Major in the ripening room allegedly makes bananas taste sweeter.

to sell a dying product," said D'Arrigo with a shrug. "It's always a race to get it to where it needs to be before it dies."

Modern rooms are equipped with injection ports that meter the ethylene into the room in response to fluctuations in temperature; in the older rooms, the ethylene is produced in a low, even flow from portable Easy-Ripe generators. Rosenblatt told me that when he started out, rooms would be injected with a burst of ethylene released from a cylinder, which not only made it much harder to achieve an even distribution among the stacked bananas but also heightened ripening's risk. Ethylene is highly flammable, and in the early days of the technology, fatal banana-ripening-room explosions were not uncommon. The Great Pittsburgh Banana Company Explosion of 1936 blew the roof off the building, shattered windows in neighboring streets, and buried at least one employee in a hail of flying fruit.

Today, D'Arrigo said, Luciano can monitor the status of the most high-tech rooms with an app. "He's no longer having to stay here all hours of the day and night to make sure a room is ripening correctly," she said. Still, she told me, she'd arrived at four o'clock that morning and wasn't going to leave till five or six o'clock that evening.

"Produce is a labor of love," said D'Arrigo. "I tell people that working here is like a face tattoo—you've got to be really sure you want it." Although their existence is almost unknown, the two dozen ripening rooms on which she and her colleagues lavish such care are a vital link in the largely invisible, highly specialized architecture of mechanical refrigeration and atmospheric control that performs the overlooked miracle of supplying New Yorkers with fresh, ready-to-eat bananas every single day of the year.

Bananas were the pioneers, but, D'Arrigo told me, to get a sense of what ripening can do for the marketability of a fruit, I should consider the avocado. For much of the twentieth century, most people outside Central America would have rarely eaten or even encountered an avocado. (Indeed, for most of that time, they were known as alligator pears, as opposed to their current name, which is a bastardized version of the Nahuatl word for both avocados and testicles.) To the uninitiated, the avocado was challenging: it was a fruit but it wasn't sweet, it was hard and sort of slippery, and it didn't cook well. No one knew what to do with it, and, worse, it wasn't ready to eat.

"When my kids were little, I'd take them to the store with me on Saturdays to watch human behavior at the avocado display," Steve Barnard, CEO of Mission Produce, told me. "People would go up to the display and they'd walk off empty-handed if they didn't find a ripe one." The avocado would ripen eventually, but it wasn't ripe when the customer wanted it, and that meant lower sales.

Barnard's big innovation was to introduce ripening rooms to the avocado industry. He wasn't alone—a San Diego County avocado farmer named Gil Henry was also experimenting with ripening methods at around the same time. "The avocado must become as easy to use as the banana, tomato, or any of its other competitors for the consumer dollar," Henry told his fellow growers at the 1984 meeting of the California Avocado Society. Both Barnard and Henry drew on research developed at the University of California, Davis, to adapt banana-ripening regimes to avocados. (This was no simple matter: avocados require three times the refrigeration to pull heat out of the gassed fruit, or they'll explode.)

What Barnard had was the hustle to make ripening avocados work at scale. He started by persuading the produce manager at his local branch of Ralphs grocery store, in Oxnard, to let him run a trial: ten boxes of ripened avocados, displayed right next to the regular hard ones and priced at twenty cents more per fruit. He put them out on Friday and came back in on Monday only to find a pyramid of oranges on sale where he'd left his avocados. "I immediately get pissed off," Barnard remembered. "Then the produce guy says, 'Well, I sold out of the ripe ones by noon on Saturday, and those other lower-priced ones are still right here.'"

Buoyed by this success, Barnard signed up all of the Ralphs stores in Southern California to take his ripe avocados; within a year, the chain's avocado sales went up 300 percent. Next he contracted with Jim Lentz's Thermal Technologies to build avocado-ripening rooms across the United States, in order to supply first Kroger, then Walmart, Costco, Chipotle, and more. Refrigerated, ethylene-dosed ripening rooms—with a little assistance from Instagram influencers and a free-trade deal with Mexico—are the reason for the fourfold increase in the average American's annual avocado consumption since the 1990s. Recently Barnard built his first avocado-ripening room in Shanghai. "Chinese shoppers don't know what an avocado is," he said. "But four chunks of avocado in every noodle soup—if you could pull that off, you would sell seventy-six million pounds a day."

III. Trading Futures

On a blustery March morning, I went to the Port of Wilmington, Delaware, to visit the largest concentrated juice-storage facility in North America, which is operated by one of the world's largest orange juice companies, the Brazilian giant Citrosuco. Although a majority of Americans have drunk juice that has passed through this particular landmark of the artificial cryosphere, very few have ever seen inside. Refrigeration and its refinements—controlled-atmosphere storage and ripening rooms—have, over the past century, enabled a select cohort of fruits and vegetables to escape the constraints of local seasonality to become global staples and intangible brands. But with the assistance of freezers and aseptic storage tanks, orange juice has transcended them all to become a financial instrument.

Wearing a hard hat over a hairnet, a gray company hoodie, and long johns under his jeans, Brian Fogelman took me on a brisk walk through the plant he's been managing for the past thirty years. "The older I get, the colder it feels," he said, rubbing his gloved hands. A gravel-voiced smoker, Fogelman admitted right away that he doesn't drink juice. To accompany his dry sense of humor, he has a tendency to fret; he keeps unsuccessfully badgering his bosses to set up a wireless system so he can monitor the juice remotely when he's off duty on the weekend.

Our first stop was a console that looked like the operations cockpit of a Cold War–era nuclear reactor: a wooden desk with a cutout that housed a fuzzy cathode-ray screen showing a circuit-board diagram of the facility, surrounded by red digital displays and large black knobs. Fogelman's cell phone number was taped to the edge of the circuit board. The dim, yellowish lighting, greige walls, and dingy carpet tiles added to the period feel, but when it was built in 1983, Citrosuco's "tank farm" was a pioneer—the first facility built to receive bulk frozen concentrate and the vanguard of a revolution that made orange juice an American breakfast staple. "We store, blend, and ship juice out on tanker trucks to all over the country," said Fogelman. "The Northeast market is still the biggest, but some trucks go as far as British Columbia."

We put on safety goggles and jackets to walk through the shorter of the two massive, white-walled warehouses arranged in an L shape on the curve of the Christina River, Fogelman yelling technical details at me over the refrigeration equipment's brain-numbing hum. There were twenty-four two-story-tall tanks lined up in the gloom, their silvery stainless steel sides gleaming in the half-light. Inside each tank was 265,000 gallons of viscous brown slush: orange juice, but with its water and volatile flavor molecules burned off. The result is a simple syrup six times more sugary than juice and devoid of any of an orange's characteristic fruity, floral zing. It has been de-oiled and de-aerated, and its flavor at this point—known in the trade as process or pumpout flavor—is largely defined by what's missing. It can be kept frozen in tanks for up to two years, Fogelman explained; rather than cool each tank individually, the entire room is held at twenty degrees.

This juice-preservation technique was originally developed with US military funding during World War II. The Quartermaster Subsistence Research and Development Laboratory wanted to include a portable, shelf-stable citrus juice as part of its ration packs, in order to fulfill soldiers' vitamin C needs. Separately, physicists exploring potential uses for recently developed vacuum technology realized—alongside Rex Brunsing in the Salinas Valley—that it could remove water from food without destroying nutrients. They were working on dehydrating hamburger and, according to contemporary reports, had "arrived at an end product which was at least acceptable" when the military request for powdered orange juice came through.

Orange juice proved even more challenging than hamburger. The final solution required a two-step process in which orange juice was first concentrated under pressure in a steam-heated evaporator before being spray-dried. Frustratingly, the final powdered orange still suffered from "caramelization and off-flavors and aromas." In other words, it was disgusting.

The war ended before the problem was solved, but in 1946 a spin-off, the Vacuum Foods Corporation, commercialized the results of step one—concentrated orange juice—under the name Minute Maid. A 1950s housewife could, and often did, buy a can of frozen, concentrated Minute Maid, plop its contents out into a jug of water, stir, and serve her family "fresh" orange juice for breakfast all year round. "So why squeeze orange juice yourself?" sang Bing Crosby in his advertising jingles for the brand, in which he was also an investor.

Before the dawn of frozen concentrated orange juice—universally referred to in the industry by its acronym, FCOJ, pronounced "eff-coj"—most Florida oranges were consumed as fruit rather than juice, and most Americans who wanted to wash their breakfast down with juice had to settle for canned tomato. (Orange juice couldn't be canned successfully—the resulting bitter, discolored liquid tasted like boiled turpentine.) Afterward, nine out of every ten oranges grown in the state were juiced, and there were more oranges than ever. Less than a decade after the commercial debut of FCOJ, Florida's citrus acreage had nearly doubled. When writer John McPhee visited the state to report on the orange in the early 1960s, a University of Florida citrus expert told him that "the concentrate boom is the boomiest boom since the Brazilian rubber boom."

Frozen orange juice, as opposed to fish sticks, ice cream, or TV dinners, was also the killer app for home freezers. If it meant a convenient, reliably sweet-tasting, vitamin C–packed drink graced their tables every morning with little more effort than opening a can, Americans could finally see the point in acquiring a new appliance. National orange juice consumption expanded from fewer than fourteen glasses per person per year in 1950 to nearly forty in 1960.

In Delaware, Fogelman receives juice from both Florida and Brazil, which entered the US orange juice market in the early 1960s, encouraged by harvest-wrecking hurricanes and frost in the Sunshine State and enabled by the ability to ship frozen concentrate in drums. Today, specialized refrigerated tankers capable of carrying more than three million gallons of juice arrive eight to ten times per year. In the corner, next to the FCOJ tanks, Fogelman pointed out a double-decker trailer, around which were looped huge black rubberized hoses. These connect the tanks to the ship that brings in the concentrate. "It's a beast," said Fogelman, bemoaning the crystals of potassium citrate that form in the concentrate and get caught in the valves and filters. "We call them candy," he said. Although the concentrate is frozen, it's not solid but rather as viscous and sticky as blackstrap molasses, which makes it notoriously hard to handle.

Once the pumping starts, it can't be stopped for the three to six days it takes to unload each ship, so the team work in shifts around the clock, monitoring the lines for pressure drops, air bubbles, and blockages that can, if not quickly defused, trigger a range of blowbacks, burps, and explosions. "We *could* fill a tank in nine minutes," Fogelman said. "But we don't, because it would put too much pressure on the gear." Given his obsessive nature, I was unsurprised to hear that he'd come up with plans for a new, specially engineered rig that would not only cut down on these headaches but also speed up the ship's unloading time, allowing a faster turnaround on dock.

The tanks on either side of us looked identical and all held frozen concentrate, but, Fogelman told me, their contents varied. One tank was filled with early-season orange juice of a particular shade and sweetness, the next came from midseason oranges with a darker color; one tank contained juice from mild Hamlin oranges while another was from sweeter Valencias; one tank was from Brazil and the next from Florida. "We dial in precisely the ratios our customer wants from each tank profile," said Fogelman.

The ratios, which determine the finished juice's color and sugar level, are proprietary, but the process by which it is made to taste like oranges again is even more shrouded in mystery. Fogelman opened a door labeled "Flavor Plus Room" to offer me a brief peek at a set of slightly smaller tanks that contained the oil- and water-based chemicals that are stripped out and recovered during the original juice-evaporation process. "We mix it in as we're loading the trucks," said Fogelman. "It's all done to order," he said, and smiled. "That way, it's nice and fresh." For the first time during my visit to one of the largest juice-storage facilities in the world, I smelled orange.

The problem of restoring lost flavor to frozen juice turned out to be an opportunity: it gave producers the ability to custom blend their beverage, so that Minute Maid would always taste the same whether Valencia oranges were in season or not. "You take Mother Nature and standardize it," as Jim Horrisberger, a citrus-industry lifer who was director of juice procurement for Coca-Cola, told *Bloomberg Businessweek*. Rather than add back the precise blend of three hundred–plus chemicals that were burned off and captured when the juice was first processed, different companies could formulate their own specific recipe, mixing and matching individual chemicals to create a custom flavor profile. Each of the chemicals originally came from an orange, just not necessarily from the same oranges or in the same ratio—and because it is derived "from the named fruit," to use the USDA's terminology, it doesn't have to be listed on the label as an added flavoring. The end product is still 100 percent natural orange juice for labeling purposes, even if it isn't an orange juice you could ever find in nature.

What this meant was that orange juice could be more than a juice—it could be a brand. In exactly the same way that Coke and Pepsi are both colas, yet Coke fans swear by its vanilla flavor notes, while Pepsi aficionados prefer its sweeter, more citrusy taste, Minute Maid and Tropicana are both juices, but Minute Maid (which is owned by Coke) is known for its "unique candy type flavor," according to one Tropicana insider, while Tropicana (which, you guessed it, is owned by Pepsi) has its own signature taste. "We used to call it the pure premium bite," Chip Bettle, the company's former head of technology, told reporters from America's Test Kitchen's *Proof* podcast. "It would wake you up in the morning because it made your cheeks pucker together."

Some of the juice at Citrosuco's tank farm in Wilmington belonged to Tropicana, but it was stored in the newer building next door. Clutching my notebook with numbed fingers, I followed Fogelman into a slightly warmer and much taller white cuboid, where I craned my neck to gaze up at three huge carbon-steel storage tanks, each filled with a million and a half gallons of chilled, not-from-concentrate juice. To prevent settling, each tank was slowly being churned by a pair of paddles, like an oversize ice cream machine; to prevent vitamin C loss through oxidation, the juice was nestled under a blanket of nitrogen gas, like the head on a gargantuan pint of beer.

I felt dwarfed by the sheer volume of liquid looming overhead—it was enough juice to fill seven Olympic-sized swimming pools. The tanks are about six stories tall and equally wide. "The stairs are a workout," admitted Fogelman, pointing at a scaffolding rig running up the side that he and his coworkers have to climb when they need to get into an empty tank for cleaning or repairs. "It takes forty minutes to kit up before you go in," he said. "You have to wear a lifeline and a gas mask, because of the nitrogen." Not long after my visit, one of Fogelman's colleagues died when he fell into the tank he was cleaning.

From where we were standing, safely on the ground, this room, though immense, seemed much quieter, the refrigeration equipment humming along at a more temperate thirty-four degrees. If we'd been inside the tanks, Fogelman told me, he'd have to be shouting directly into my ear: sounds reverberate inside the epoxy resin–lined cylinder in such a way that it's impossible to hear anything but noise. These aseptic tanks were invented in the 1980s by Purdue University food processing engineer Phil Nelson, who grew up working at his family's tomato cannery in Indiana. His goal, as he explained to *Food Engineering* magazine when he won the World Food Prize for his invention, was to "put a big bulge in the supply line" so that tomato processors could manufacture sauce, juice, and ketchup at their leisure, in response to market forces throughout the year. Nelson's solution made tomato processing season-proof; for orange juice, it inspired federal hearings that hinged on the very meaning of freshness.

At the time, in the 1980s, Tropicana was something of a niche player in the OJ business. The company was committed to the wild idea of selling juice that a consumer could drink straight from the container, without having to add water to concentrate. Rather than evaporating freshly squeezed juice like Minute Maid, Tropicana would pasteurize it, then freeze it into slabs to store in tunnels at its Florida plant before defrosting it as needed to package into bottles and cartons. The whole process was expensive and inefficient, although the resulting juice apparently tasted terrific—better than it does today.

Tropicana was on the verge of giving up on ready-to-serve juice and switching to concentrate when Nelson's tank technology came along. Today one of the largest refrigerated enclosures in North America can be found at Tropicana's tank farm in Bradenton, Florida. At the Port of

Wilmington, Citrosuco holds roughly equal quantities of both frozen concentrated juice and the pasteurized, chilled, but unconcentrated kind, but American tastes have shifted decisively in the three decades since the introduction of gigantic aseptic tanks first allowed not-from-concentrate (NFC) juice to be sold at scale.

Although it requires more storage space, Tropicana's flash pasteurized, de-oiled, de-aerated, and chilled juice has overtaken its concentrated competitors: FCOJ now accounts for less than 4 percent of the entire US orange juice market, to NFC's 60 percent.* "I think consumers perceive it to be a fresher product," Robert Behr, vice president of one of Florida's largest citrus grower cooperatives, told members of the US International Trade Commission at a hearing on import duties in 2012. After up to a year of storage, "it's not going to be fresh orange juice, of course," he added. "But it's not bad."

In response, Coca-Cola has invested in additional tank-farm capacity, leasing another forty million gallons of NFC storage over the past decade in order to supply its Simply Orange and Minute Maid bottling plant in Auburndale, Florida. "It is Coke's largest plant in the world, not just juice plant," Coca-Cola's Horrisberger told commissioners at the hearing. With tens of millions of gallons of capacity in Florida supplying Americans with their daily OJ, these refrigerated tank farms form our national industrial breakfast reserve. At any given moment, Horrisberger claimed, he is sitting on nearly nine months' worth of juice in inventory, in order to have enough of each variety and color of juice to make sure he can ship Simply Orange and Minute Maid to supermarkets, vending machines, and fast-food outlets around the country. "I have a brand to support, and that's the key," he told the assembled members of Congress and civil servants. "We cannot be off the shelf." "I understand," replied Commissioner Irving Williamson of the US International Trade Commission. "Because I want my orange juice when I want it."

---

*The remainder is made up of shelf-stable orange juice and reconstituted frozen orange juice.

———

Despite NFC's triumph in the marketplace, it is frozen concentrated orange juice that has transcended the limitations of perishability to become a financial instrument—traded, speculated on, and hedged against on New York City's Intercontinental Exchange. In addition to transforming where, when, and in what quantities OJ is grown and consumed, as well as how it tastes, the market-distorting powers of refrigeration have also created a specialized niche for juice gambling.

Such bets are legal, though many traders skate close to, if not over, the line. As Shawn Hackett, a financial analyst who trades in FCOJ futures and options put it, "it is a high-risk, high-reward game." Hackett was the fourth orange juice speculator I reached out to; the first three turned out to have paid fines, have done time, or currently be in jail.

"This market is structured for someone who likes a little more of an adrenaline rush," he admitted. That's because the orange juice market is much smaller and thus, ironically, less liquid than that for, say, corn. "When you have lower liquidity, you have a greater opportunity for volatility, because less money can push the market around more easily," Hackett said. "Volatility means there's greater risk, but there's also greater opportunity."

The modern futures contract came into being in Chicago in 1865 to trade grains like corn, oats, and rye. Its invention marked a fundamental shift from the specific to the abstract—from exchanging a particular sack of wheat grown by a particular farmer in a particular field for a particular sum of money on a particular day to trading promises to buy or sell a thousand bushels of a certain grade of wheat at a certain price by the last trading day of the month. Growers who were worried that a bumper harvest might mean that the price of grain would go down could sell part of their harvest in advance; buyers who were worried that crop failures might cause the price of grain to go up could lock in some of their future supply at a fixed price. It allowed both sides of the market to hedge their bets and ensure their asses were not exposed in case of disaster. "That's really what this is all about," said Hackett. "It's to try to help you smooth out things you can't control."

Before long, people who neither grew wheat nor had any use for it realized that they could make money by guessing whether its price would go up or down and buying and selling accordingly. No actual grain changed hands; the men with cigars who dealt in these kinds of futures contracts would fulfill them by simply paying the difference in market price on the date the option came due.

Theoretically, the traders were providing a service: taking on and spreading out some of the risks that could ruin growers and grain purchasers, and dampening the wildest of price swings, all in return for shaving off a little profit here and there. In reality, they also quickly found ways to manipulate the market, driving prices up or down to make a killing. Inside the octagonal pits at the Chicago Mercantile Exchange, traders would gesture and yell at each other from raked stadium-style steps, buying and selling and sweating all day. "It was always crazy chaos all the time," said Hackett. "Then we started getting these electronic platforms and everything changed."

Futures markets began with corn, wheat, and soybeans—commodities that could easily be stored for months in silos or sacks. With the advent of refrigeration, these kinds of contracts soon became common in the egg and butter markets too. In 1966, just a few years after Minute Maid had introduced FCOJ to the thirsty American breakfast table, the orange juice futures market joined the fray.

"For a futures market, you have to have something that's storable, that's consistent, and that's transportable in large quantities," said Hackett. "You have to have a product that could at least withstand nine months of stability factor," he added. "Frozen concentrate, you stick it in your freezer, a year later it's just fine." A viable market also needs to be large and global in scope. Thanks to refrigeration's obscure economic alchemy, the ephemeral extract of a seasonal fruit had finally attained sufficient shelf life and scale to fit the bill.*

For a while, at least. Today Hackett fears that the FCOJ futures exchange is "on a slippery slope to an extinction event." On the one hand, the supply side has shrunk: the Florida farmers who planted acres of cloned trees in the 1960s to scale up production have seen them wiped out by citrus greening disease in the past few years. On the other hand, demand for orange juice in the US has been declining for decades, as OJ drinkers defect to Red Bull or other alternatives, abandoning one branded sugar water for another. Refrigeration brought orange juice to the world, then it took it away—just as it put branded apples on supermarket shelves in August but ensured hundreds of older varieties were lost to time and introduced bananas to northern climes while devastating tropical ones.

At the end of the day, Hackett doesn't even drink FCOJ himself. "Well, down here in Florida, we have really, really good fresh-squeezed orange juice, literally right off the tree," he admitted. "I mean, I'm not saying there's anything wrong with frozen concentrate, but it's nothing like the fresh deal."

For the anointed few in the fruit and vegetable world, a century of research has resulted in metabolic regimes that would put an elite athlete to shame. Today we know more about how to lengthen an apple's life span than a human's. Kidd and West had to make do with trial and observation, treating a fruit one way or another and seeing how it responded. A new generation of postharvest scientists, like Natalia Falagán, has the tools to be able to not only see exactly what is happening inside

---

*Other perishable commodities have failed to form a viable market. According to finance journalist Emily Lambert, iced broiler chickens never took off because "the price of chicken, many traders said, was essentially set by Kentucky Fried Chicken." The commodity Hackett told me he would most like to trade is potato futures, a market that used to exist until J. R. Simplot, the billionaire credited with commercializing frozen French fries, defaulted on his contracts—"just because he got on the wrong side of the market." "The exchange allowed him to do it," Hackett explained. "And it ended the futures market right there and then."

an apple cell under different conditions but also understand *why* it's happening, in terms of both the fruit's changing biochemistry and the genes that regulate that response. Falagán uses those insights to design ever more precise storage programs. She excitedly showed me a graph from a recent experiment in which she discovered that lowering the oxygen levels in a blueberry cold-storage unit gradually, over the course of seven days as opposed to twenty-four hours, will extend the fruit's shelf life by a full 25 percent. Another series of charts helped her explain why: it turns out that suddenly changing the atmosphere a fruit is breathing is a nasty metabolic shock. "The blueberry is like, *Oh my God, what's going on? Why don't I have oxygen?*" she said, pointing out a sharp peak of respiratory stress—the equivalent of a fruit panic attack.

In the future, however, she expects that this deeper understanding of the mechanisms behind cellular decay will start influencing how fruits and vegetables are farmed, as well as redesigning them at the genetic level. "We are learning how preharvest factors affect postharvest quality," Falagán told me. One day soon, a blueberry's life-extension protocol will include a schedule for how to taper its parent plant's liquid intake in the hours before harvest, as well as prescriptions for field-applied hormone and mineral sprays. The fruits and vegetables of our imminent future will not have achieved immortality yet—but they will be inching ever closer to it.

"It's amazing any of this stuff works, really," Irwin Goldman, a vegetable breeder at the University of Wisconsin, told me. He thinks *to vegetable* should be a verb, one that describes how humans have manipulated their favorite edible plants to become the horticultural equivalent of Dorian Gray. To Goldman, the verb phrase *to vegetable* encompasses the way in which humans "take a plant and breed it so it can be harvested when it's super immature, so that it's tender and lovely and we want to eat it"—but also how we then reverse engineer its metabolism so that the twilight years of that harvested fruit or leaf will extend indefinitely. "We're kind of asking a lot," he said.

Of course, as in Oscar Wilde's fable of eternal youth, when we receive the thing we ask for, it's often accompanied by a host of unexpected and frequently undesirable consequences. Refrigerated storage allows perishable bananas, apples, and avocados to circulate as commodities; controlled-atmosphere warehouses transform seasonal gluts into accumulated capital; ripening rooms enable demand to drive supply. A refrigerated supply chain is the reason we eat imported bananas rather than North America's own semitropical fruit, the pawpaw; controlled-atmosphere storage created the Cosmic Crisp and its influencer-led launch; ripening rooms fueled avocado toast's ascent to millennial meme; and frozen juice made OJ a mainstay of the breakfast buffet.* Kidd and West began their research with the noble goal of preventing food loss and shortages, while Natalia Falagán told me she is motivated by a desire to make the food system more sustainable, but their findings were and continue to be implemented in a market economy, where they reshape both supply and demand in the pursuit of profit.

Take the banana: Refrigeration allowed it to escape the constraints of geography and become the world's most popular fruit. But the process of fitting the fruit to its new refrigerated ecosystem created both biological and political vulnerability in the Central American countries where it was grown. In the wild, there are hundreds of different varieties of edible bananas—red ones, round ones, and ones that taste like pineapple. Working out how best to preserve and ripen all of them would be uneconomical, so the banana industry chose one: the large and sturdy Gros Michel. Cloned "Big Mike" banana plants replaced native rainforest across Honduras and Guatemala, turning a landscape of incredible biodiversity into a monoculture—an industrial ecosystem that was maximally productive

---

*The pawpaw is not the only fruit left behind by storage science. Just like avocados in the 1970s, mangoes, peaches, kiwis, and melons are often disappointingly hard and unripe at the supermarket. They also respond to ethylene. When I asked Jim Lentz why he didn't build ripening rooms for peaches, his answer had nothing to do with technical challenges or the nuances of fruit physiology. "Bananas account for 10 percent of all the sales in the produce department," Lentz said. "Other fruits don't generate as much revenue, so we don't really spend a lot of time on them."

but also extremely susceptible to pests and diseases. When a deadly soil-based fungus arrived from Southeast Asia, it quickly laid waste to the banana fields. Exports from Honduras fell by more than half. United Fruit contemplated abandoning the region.

In the end, the banana moguls found a replacement variety that was less attractive, less tasty, and more easily bruised but was resistant to the disease: the Cavendish, which we still eat today. The only "solution" that seemed viable was to find a substitute that could slot into the fruit's existing commercial framework—a very specific, corporate kind of banana that could be grown, shipped, stored, and ripened at scale. Today's Central American monocultures and ripening rooms are filled with these Cavendish bananas—and they, in an entirely predictable historical echo, are now threatened with extinction by a new strain of the fungus.

It's not just the fruit that becomes less resilient under these conditions. A decade before the first Gros Michel plantations in Honduras began succumbing to disease, American writer O. Henry coined the term *banana republic* in a short story based on his time in the country. It came to refer to any country with a puppet government whose strings are being pulled by foreign enterprises, but United Fruit's client nations in Central America provided the model. The recent history of Guatemala, where the United Fruit Company was the largest landowner, illustrates both the condition and its consequences. In the 1950s, banana bosses managed to convince the CIA to overthrow a democratically elected government that had expropriated United Fruit's unused land for redistribution to landless peasants and install a military dictator, leading to decades of violent civil war and tens of thousands of "disappearances."

It's neither fair nor accurate to blame the banana industry for all of Central America's political problems, but, as Dan Koeppel writes in his history of the fruit, "throughout Latin America the destabilization resulting from banana-related interventions created a tradition of weak institutions, making it difficult for true democracy and fair economic policies to take hold. The Latin American tradition of governments not supported by the general population, and propped up by overseas commercial interests, was created under the authorship of United Fruit."

"The banana of commerce is one of Man's proud triumphs over Nature," Frederick Upham Adams concluded in his 1914 panegyric to the United Fruit Company's "Conquest of the Tropics." A century later, it is clear that such triumphs are only ever fleeting and their costs are, all too often, unsustainable. The abundance that refrigeration promises is accompanied by a diminishment in both diversity and deliciousness, the stability it brings to the marketplace comes at the cost of greater risk, and its assurance of plenty is undercut by increased vulnerability.

5.

A THIRD POLE

I. Meet the Thermo King

A newspaper photograph of Fred McKinley Jones in his twenties shows a tall, handsome man hunched over to fit into a classic 1920s torpedo-on-wheels race car. He's sporting a leather cap, goggles, and a sweet, shy smile. The caption helpfully points out that Jones was "the only black in Hallock, Minn."

He'd ended up in this tiny town near the border with Canada somewhat by accident: he was born in either Cincinnati, Ohio, or just across the river in Covington, Kentucky (reports vary), to an African American mother and Irish father in 1893. Either orphaned or abandoned as a young boy, he was taken in by a Catholic priest, who put him through six grades of school. It was an unlikely start for another unsung hero of refrigeration: the man who gave mechanical cold wheels.

By the time Jones was a teenager, cars were just starting to be seen on the streets of America—Henry Ford's Model T made its debut in 1908—and these marvels of modern machinery were much more interesting to him than either the church or formal education. He borrowed some long trousers and convinced a local garage owner to give him a job as a mechanic's assistant, starting the next Monday. "But do you think I'd wait three days to get my hands on those cars?" he reminisced later in life. "I should say not. I was down there at six a.m. Saturday, waiting for him to come and open up the place."

In just three years, he had been promoted to foreman of the shop, having revealed himself as possessing an innate genius for all things mechanical. Jones bought his first slide rule from a boy who found it on a streetcar, and taught himself engineering. "I took two or three correspondence courses," he said. "But I learned most from just reading books, and studying, and trying things out." The garage owner, a man named Crothers, would compete in races on the weekend as a way to enhance the allure of his automobiles, and Jones designed and built a couple of racing cars—but when Crothers wouldn't let him travel to the races to watch his creations compete, Jones rebelled. He shut up the garage and went to the races, and his boss was not happy.

As Jones recounted it: "He told me everybody liked me and that I was a good worker, but who the hell did I think was running the place that I could light out that way without permission? Well, Crothers was a swell fellow and I certainly deserved the lesson, but I was a touchy kid then, and I just up and quit and decided to take a trip to Chicago to see the sights."

After his visit to Chicago, Jones decided to carry on to Saint Louis but got on the wrong train and ended up in Effingham, in south-central Illinois. He went to the town's only hotel for a meal and ended up repairing its furnace, an achievement that caught the eye of one of the guests, Walter Hill, youngest son of a railway tycoon.* Hill managed his father's estate in the Red River Valley, near Hallock, Minnesota, and was in need of a mechanic to maintain the farm's machinery—which is how twenty-year-old Fred Jones ended up becoming, as local newspapers never failed to mention, "Hallock's only colored boy."

With a brief break to serve in World War I, Jones spent the next

*James J. Hill, Walter's father, makes a cameo in F. Scott Fitzgerald's *The Great Gatsby*, when Gatsby's father tells Nick Carraway that if Gatsby had lived, "he'd of been a great man. A man like James J. Hill. He'd of helped build up the country."

sixteen years in the tiny northern Minnesotan town, marrying then divorcing a local girl, winning dirt-track races in cars of his own design, hunting and fishing, and playing saxophone in the town band. He and a friend built a radio broadcasting station and put on programs "whenever we had the money"; he also operated the film projector at the Hallock movie theater. By the 1920s, "talkies" were replacing silent movies, and without the cash to purchase a three-thousand-dollar audio-projection machine, Jones designed and built his own out of leftover bits of farm machinery, including the discs from a plow.

News of this particular invention spread, reaching Joseph Numero, a balding, sharp-eyed Minneapolis businessman with a grating voice and rapid-fire delivery. Numero's firm, Cinema Supplies Incorporated, was also struggling with the transition to sound—until he recruited Jones and began manufacturing his equipment.

Later Numero confessed that when he hired Jones, he was nervous that his other engineers would balk at working for a Black man. "He ducked the problem by simply taking Jones into a room with the other engineers and introducing him," according to *Ebony* magazine. "Then Numero fled to be clear of any explosion."

When he popped his head back into the shop after an hour or two, he saw the men clustered around Jones, who was explaining how to solve a particular technical issue that had stymied them for weeks. A few hours later, when Numero checked again, the shop was empty. Anxious that they'd thrown Jones out or quit en masse, he stepped out onto the street and spotted Jones with the rest of the team at a diner on the other side of the road, having a whale of a time drinking coffee and discussing blueprints. Jones, for his part, told *Ebony* he never imagined that there would be a problem: "'Why should they act any other way?' he asks quite innocently."

Jones's adventures in refrigeration began in 1937, thanks to an idle boast. One hot summer day, Numero was playing a round of golf with friends, including an air-conditioning executive named Al Fineberg and

Harry Werner, the president of a local trucking business. At the time, trucks were still refrigerated in the same way Polly Pennington had recommended cooling railway cars: they were insulated and packed with ice, relied on natural ventilation to circulate the cold air, and traveled largely at night. If the ice melted, the meat or produce would spoil—an expensive and not infrequent occurrence, especially in the summer months.

Werner had recently lost yet another load of freshly killed chickens on their way to market in Chicago and, in frustration, asked his air-conditioning buddy, Fineberg, why on earth he could cool movie theaters using mechanical refrigeration but not trucks. Fineberg, on the defensive, replied that it wasn't that simple. Several people had tried, but their mobile cooling machines were not only too delicate to stand up to the constant vibration and jolting of being mounted to a railcar or truck body but also so big that they'd fill half the trailer's storage space.

Numero, apparently just to tease the air-conditioning man, announced, "Harry, if Al here won't make a refrigerator for your trucks, I will!" Everyone laughed, and Numero promptly forgot all about his ill-considered taunt until Werner called in a few weeks' time to say he was sending round a new truck to have a refrigeration unit fitted.

"I expected to give it a quick once over and tell Werner it couldn't be done," Numero told *The Saturday Evening Post* a decade later. "But Jones beat me to it. He climbed into the trailer, made some measurements and calculations, and popped his head out to announce he guessed he could fix up something. And we were stuck with it."

Jones's background in race car design had equipped him with considerable experience in making things as lightweight, streamlined, and shockproof as possible. Still, his first diesel-powered machine weighed 2,200 pounds and had to be mounted under the trailer, where it rapidly became clogged with dust and mud from the road. It also cost $30,000 to build—the rough equivalent of spending $3.5 million on a prototype today. But it did hold up on test runs, keeping the trailer's contents within ten degrees of the desired temperature. Jones had done what cooling experts considered impossible: invented the world's first truly mobile mechanical refrigeration unit.

Numero promptly sold his movie soundtrack business to RCA, borrowed against his life insurance, and formed the US Thermo Control Company. Over the next couple of years, Jones refined his design. He shaved its weight down to just 950 pounds by using lightweight aluminum and moved the unit to the front of the trailer, where it stayed cleaner and was easier to access for repairs.

Werner ordered some of these early machines, as did Armour, one of the big Chicago meatpacking companies, but it wasn't until the United States entered World War II that this new portable, sturdy, diesel-powered refrigerator—christened the Thermo King—caught on. Thermo King units were used to store blood and serums as well as cool field hospitals, but they also kept food fresh and allowed troops to enjoy an ice-cold Coca-Cola on steamy Pacific island bases. "They could land it on a beach-head in half an hour and have it running," Jones explained. "And it was light enough to drop by parachute."

Disappointingly but unsurprisingly given the pervasive racial bias of the era, Jones didn't have an ownership stake in the company built on his brilliance. When the US military invited the nation's leading engineers to Washington, DC, in order to develop a next-generation field refrigeration unit, Jones, as the only Black person invited, "had to stay at a shabby bedroom in an obscure, Jim Crow hotel," while the other attendees were lodged "at Washington's finest hotels at government expense." (The prototype that the group eventually settled on became the standard for army and marine field kitchens; it was also, according to *Ebony* magazine, Jones's design.)

He never became rich from his inventions and didn't live an extravagant life—his second wife, Lucille, bemoaned the fact that he owned only two suits and two pairs of shoes, a minimal wardrobe that he regarded as wasteful extravagance because he couldn't wear more than one at a time. "You get too much money and you lose the touch," he would say.

Jones certainly didn't seek glory—he turned down an honorary degree from Howard University with the excuse that he was too busy to collect it—and he avoided public speaking. He seems to have preferred working to almost anything else: he and Lucille lived in a penthouse above the company building, and in the evening he would put on his bathrobe and sit for hours, turning over engineering problems in his mind. "I knew that brain of his was churning," Lucille reminisced. "But outwardly, he looked like old rocking chair had got him. He was a night owl. He'd be up, doodling on some invention—he took out 50 patents—until 2 or 3 in the morning. Then he'd sleep late."

Jones did, however, share the secrets to his success in 1953, in a short acceptance speech for an award of merit from the Phyllis Wheatley Auxiliary, a social services group set up to assist Minneapolis's small but growing Black community and named after an enslaved woman who was the first African American to publish a volume of poetry. First, he said: "Don't be afraid to get your hands dirty. Don't be afraid to work. You never know when what you have learned will come in handy." His second tip was to read voraciously. "Find out what others know," he said. "You don't have to buy books. Use libraries! You can educate yourself by reading." Finally, he told the African Americans of Minneapolis: "You have to believe in yourself. Don't listen to others tell you you're wrong. Remember, nothing is impossible. Go ahead and prove you're right."

It was not until the postwar period, while Jones was married to Lucille, that the United States truly embraced the Thermo King. During the decades that followed the 1896 introduction of motorized cargo vehicles but preceded Jones's invention in 1937, American farmers adopted ice-filled trucks as a way to get their produce to market on their own schedule. During the Depression and Dust Bowl years, when rural incomes plummeted, struggling farmers often turned to trucking. Buying a used truck and fueling it was relatively affordable, and it offered a way to make a living and stay on the farm, rather than move to the city for a factory job.

By the late 1930s, this burgeoning motorized competition had begun to make railroad companies a little uncomfortable, but before Harry Werner's festering chickens sparked Fred McKinley Jones's ingenuity, trains still held the upper hand. Compared with rail, trucking perishables was limited to relatively short distances: depending on the season and location, trucks could cover only a couple of hundred miles on America's still-rudimentary road system before their ice melted, while railways had invested an enormous amount of capital to build a national infrastructure of ice plants and re-icing stations at regular intervals along the tracks, in order to keep lettuce, oranges, and beef cool on their long journeys east from California and Chicago.

The reliable machine-made cooling provided by a Thermo King unit changed that, leading directly to the emergence of long-haul refrigerated trucking—which, in turn, remade the geography of food production and processing in the United States once again. The Thermo King's adoption was just one of a handful of food-system transformations that took place immediately after World War II, but without it, American farms could never have become as monolithic and productive, nor the contents of American supermarkets so convenient and cheap.

Well before peace treaties had been drafted, let alone signed, the industries that had been mobilized to make tanks and bombs switched to making tractors and fertilizer, and with the government's enthusiastic encouragement, American agriculture became increasingly industrialized. Small, diversified family farms gave way to larger, mechanized, monocropping *agribusinesses*, a term first coined by a Harvard Business School professor in 1955. Meanwhile, throughout the late 1940s and 1950s, around a million people moved away from American farms each year, leaving the countryside for the burgeoning suburbs. Those who remained in rural areas often earned a living by trucking food to shiny new out-of-town supermarkets, which doubled in number between 1948 and 1958. Indeed, the first business loan ever guaranteed by the Veterans Administration under the 1944 G.I. Bill enabled its recipient, Jack Charles Breeden of Virginia, to buy a refrigerated truck in order to establish a wholesale meat business supplying supermarkets in the Washington, DC, suburbs.

Immediately postwar, Jones also adapted the Thermo King unit for use on railway cars: a 1949 article in the *Kittson County Enterprise* reported that the "former Hallock colored boy has added another invention to his list, this time a refrigerating unit for railway cars that does everything but talk." Nonetheless, as late as 1962, Numero complained to *The Minneapolis Star* that only 1 or 2 percent of the refrigerated railcars used to transport perishables were cooled by mechanical units rather than old-fashioned ice. With so much capital sunk into their existing ice-based infrastructure, the *Star* explained, "wooing the railroads from ice to Thermo King has not been . . . easy."

Although ice kept food cold, it couldn't maintain a steady temperature the way a Thermo King unit could. Not all food appreciated the added humidity of traveling with meltwater; meat, especially, arrived in better condition under mechanical refrigeration. Trains were tied to fixed schedules and even more rigid railway tracks, which typically ran in and out of cities; trucks could haul produce from grower to consumer, wherever they were and whenever necessary, on the new network of high-speed freeways.

As we've seen, Americans had been eating a cartographically eclectic diet for decades. As early as 1919, a geographer noted that a Massachusetts man might breakfast on "an orange from California or Florida" and dine on "a lamb chop from a frisky little beast born on the high plains near the Rocky Mountains, and fattened in an Illinois feed lot before going up to Chicago to be inspected, slaughtered, and refrigerated," accompanied by a potato that, in June, came "from Virginia, in July from

New Jersey, in November from New York, Maine, or Michigan." Refrigerated trucking didn't so much lengthen those supply chains as rearrange them, bypassing metropolitan centers and consolidating growers and packers to create today's food-production system, which takes place at vast scale and mostly out of sight.

The shift is particularly clear when it comes to frozen convenience foods and meat. Without refrigerated trucks, it makes no sense to produce 1.5 million pounds of frozen french fries—15 percent of the North American supply—in a town of fewer than nine thousand, or to slaughter 7,200 cattle every day in a town with a population less than a third of that. This, too, is the Thermo King's legacy: grocery store shelves stocked with cheap prepared foods and a lost connection between their producers and consumers.

Although entrepreneurs and scientists had been working on ways to flash freeze food since the 1920s, no one had figured out how to distribute it until the Thermo King made its debut. In 1927, the very same summer that frozen-food pioneer Clarence Birdseye filed a patent for his newly perfected multiplate machine—widely seen as the moment that began the frozen-food industry—he ended up stuck with more than a million and a half pounds of haddock fillets on his hands and no way to get them to fish-deprived customers in the middle of the country.* Without adequate transportation, let alone freezer-equipped supermarkets and home kitchens, frozen foods remained a high-priced novelty. In 1946, the retail tonnage of all frozen foods combined didn't even equal that of cucumbers and cabbage preserved the old-school way, as pickles and sauerkraut.

Once economical and truly temperature-controlled land-based transportation of frozen food became possible, that changed. Frozen food

*The fish chilled in a frozen warehouse, awaiting the invention of retail freezer units in 1928, followed by Birdseye's 1929 sale of his fledgling company to General Foods, before they finally reached customers just a hundred miles inland in Springfield, Massachusetts, where "Frosted Foods" were launched at a handful of pilot stores in 1930.

shipped by truck tasted better and cost less, and Americans responded by buying it in ever increasing quantities. The development of the refrigerated infrastructure necessary to supply fish sticks and TV dinners had what geographer Tara Garnett describes as a "snowballing effect," in which the expansion of the distribution technology prompted the introduction of new frozen goods, and vice versa. Minute Maid's frozen orange juice concentrate went on sale in 1946; the fish stick was introduced in 1953; and in 1954, after an executive at Swanson Frozen Foods came up with the idea to put turkey, mashed potatoes, and peas in a segmented airline dinner tray, individual oven-ready TV dinners made their debut.

All were eagerly embraced. The quantity of potatoes sold to make frozen french fries increased by an astonishing 1,800 percent in the first ten years postwar. The American diet became increasingly seasonless, homogenous, and convenience based, and the shared, home-cooked family meal began its slow decline. Growers who wished to participate in this lucrative new market had to scale up in order to make their investment in expensive quick-freezing equipment worthwhile, which left many smaller farms in less favorable growing areas unable to compete.

Meanwhile, as refrigerated trucks took over from iced railcars, the big Chicago meat-packers and their urban (and unionized) workforce could be bypassed, to be replaced by an entirely new beef economy based in rural Corn Belt towns where cattle could be fattened most cheaply. Before the Thermo King allowed meat-packers to escape the tyranny of rail, cattle were brought to small stockyards near Chicago to be "finished" preslaughter. After the introduction of refrigerated trucking, yearling calves were instead brought to where the corn was, pumped up to slaughter weight as cheaply and quickly as possible, then dispatched in sprawling, low-slung refrigerated buildings nearby. Industrial feedlots, capable of housing tens of thousands of steer, became the norm across Kansas, the plains of Colorado, and Nebraska in the 1960s; by the end of the decade, industrial-scale packing plants, capable of slaughtering thousands of cattle every day, had risen up alongside them—and the notorious Chicago stockyard closed its castellated stone gate for the last time in 1971.*

The ripple effect of this transformation shapes the geography and economics of American meat to this day. Urban stockyard workers had been unionized since the 1930s; employees of the new rural processing plants were not: they were and are paid much less. With the savings on labor and feed costs, the new meat-packers could cut beef prices while still making more money, and Americans responded by eating ever more meat. The real breakthrough, however, came when one of these upstart rural companies, called Iowa Beef Packers (better known today by its initials, IBP), introduced "wet-aged" boxed beef: rather than shipping entire rear and forequarters to a wholesaler like Master Purveyors, the company began breaking down the carcass into consumer cuts in-house, vacuum-packing them in plastic bags, and shipping them out in cardboard boxes.

As we've already seen, this new wet-aging process reduced shrink, eliminated the time and cooling costs associated with traditional ripening, and led to the near extinction of an entire better-paid class of skilled wholesale and retail butchers. It also saved on transportation costs, because refrigerated trucks could carry more precut, bagged, boxed beef in every load. Not only was the proportion of sellable meat much higher with all the muscle carved away from bones and fat caps trimmed, but boxes could be stacked into a trailer much more efficiently than carcasses hung on rails. By 1975, IBP was the largest beef packer in the world. Today the old Big Five has been replaced by a new, even bigger, four: Cargill, JBS, National Beef Packing, and Tyson, which purchased IBP in 2001. Together they control 70 percent of US beef production, operating at a scale that crushes competition.

That scale, allowing for vast concentrations of food production in places where labor, land, water, or sunlight is cheap and readily available,

---

*The gate, now a National Historic Landmark, is one of the last traces of the 450-acre stockyard, which was bulldozed and replaced by an industrial park. The other, often overlooked trace of the city's once world-famous meat-packing district lies in the name of its basketball team: in 1966, the Chicago Bulls played their first season in the stockyards' now-demolished arena.

is a large part of the reason why American farmers are so productive and American food is so cheap. It's also to blame for many of the catastrophic consequences of the American food system: the underpaid laborers enduring inhumane and dangerous working conditions; the drained aquifers under the Great Plains and California's Central Valley; the enormous feedlots in which animals circulate pathogenic and, increasingly, antibiotic-resistant bacteria; and the huge manure lagoons that turn what was once an essential soil nutrient into both poisonous aerosolized particulate and fuel for toxic algal blooms—to name just a few.

The Thermo King unwittingly made that intensification possible; perhaps more important, it also made it possible to ignore its consequences, by allowing production to take place far out of sight—and smell—of urban and suburban consumers.

The implications of refrigerated trucking were enormous and wide-ranging, but they were national, not global, in scope. Before the Second World War, intercontinental refrigerated cargo was almost always transported by ships, which in turn were mostly limited to carrying bananas and frozen meat, along with the occasional cargo of apples and pears or butter.

After the war, air freight was also an option.* Like so many logistical innovations, this was pioneered by the military, which frequently relied on cargo planes to move food to the front lines during World War II. Immediately postwar, they were repurposed to feed the blockaded civilians of West Berlin, where an entirely new airport, Tegel, was built purely to accommodate the steady flow of meat, fish, and dehydrated potatoes. It was christened with a delivery of twenty thousand pounds of cheese.

The eleven-month Berlin Airlift proved that a city of two million

---

*Perishable cargo was rare in aviation's early years, but not completely unknown: the first scheduled flight from London to Paris apparently carried one passenger alongside a shipment of grouse destined for a high-end restaurant.

people could be sustained entirely on food that arrived by plane—at a cost of $4 billion in today's money. The eye-watering expense of that relief effort notwithstanding, commercial air transportation of perishable goods took off postwar, although it was typically reserved for the most delicate and high-value of foods. A 1960 Boeing promotional film showed boxes of lobsters being loaded into a modified military cargo plane as part of the "new pattern of air age distribution." In 1969, a breathless report from the *Chicago Tribune*'s London correspondent, Arthur Veysey, declared that "jets make any day a day for strawberries" at the city's most exclusive dinner parties. "One month they may come from California and another from New Zealand or Kenya or Israel or Chile," Veysey continued. "People who have money to spend are happy to pay the extra cost."

The introduction of the first jumbo jet, Boeing's 747, the very next year, increased the space available for cargo shipment, as did the growing number of passenger flights worldwide. Nonetheless, even today, the high cost of air transportation means that it's reserved for goods whose value is directly correlated with their speedy arrival. Obviously, short-lived luxuries such as live shellfish and fresh-cut flowers fit the bill, but foods as quotidian as lettuce can sometimes make economic sense as airfreight, in order to keep customers happy if a crop failure or supply chain disruption means that supermarket shelves would otherwise be empty. Similarly, a shipment of early cucumbers or carrots might be flown in to take advantage of an interseasonal gap before the bulk of the harvest arrives by boat or truck.

Shifting variables, from the cost of fuel to trade tariffs, create their own curious, often-temporary market distortions. Recently, shipping companies' efforts to reduce fuel consumption by traveling more slowly have nudged perishables back into the air. Bell peppers from the Netherlands used to reach New York City in a week on a cargo ship; now the journey is more likely to take twelve days, so they travel by plane. For the past few years, the largest and most beautiful cherries grown in the Pacific

Northwest have rarely found their way to American supermarkets. Instead, they've been loaded onto specially chartered aircraft and flown to South Korea or China, where they're sold to high-end customers willing to pay upward of ten dollars a pound for an out-of-season, fresh-fruit status symbol that wouldn't be quite as plump and perfect after a long ocean voyage. Recently, however, such exports have been threatened by Trump-era trade wars with China, the rise of cheaper Chilean and Uzbeki cherries, and improvements in the domestic Chinese cold chain that have given homegrown cherries a longer shelf life. The brief window in which it made at least some economic sense to fly planes filled with cherries from Seattle to Shanghai may be closing.

It might seem as though the relatively tiny quantity of refrigerated food transported by air, which adds up to less than 1 percent of the global trade in perishables, could hardly remake diets and landscapes in the same way that refrigerated trains or trucks did. But that would be failing to take into account the global appetite for sushi, and its effects on one of the ocean's apex predators.

As Sasha Issenberg recounts in *The Sushi Economy*, two simultaneous innovations that took place in 1970—"the ability to make tuna available to diners across long distances, and a newly acquired taste for fat among sushi's greatest enthusiasts"—combined to transform bluefin tuna from a worthless trash fish into one the world's most expensive foods. The man responsible for the first breakthrough was a young Japan Airlines executive, Akira Okazaki, charged with solving the company's unique freight problem.

At the time, Japan was at the height of its postwar "economic miracle," with a reputation for manufacturing and exporting high-quality semiconductors, cameras, audio equipment, game consoles, and medical devices. While Japan's trade surplus increased year on year through the 1970s, the US trade deficit was on track to become the world's largest. Practically speaking, this meant that JAL's planes would leave Tokyo for New York with their bellies full of valuable electronics and optical lenses, but the company had a hard time finding suitable cargo for the return journey.

"We had to find something to put in the empty planes," Okazaki explained in a recent documentary. He pored over the United States' handbook of federal trade statistics, concluding that seafood's "value and sensitivity to decay perfectly matched the economics of air freight." Atlantic bluefin were also worth significantly more in newly tuna-mad Tokyo than to New England fishermen, who generally sold their catch for pet food.

The only remaining challenge was finding a way to keep the fish not just edible but fresh enough to serve raw after a journey halfway around the world. Fish is generally preserved using flake ice, but that was too heavy to ship by air. Dry ice made the fish turn black. One of Okazaki's colleagues had heard that carbon monoxide helped preserve the flesh of people who died by inhaling it, an insight he put to the test by transporting cheaper albacore in gas-filled bags. Its flesh stayed an attractive red but rotted nonetheless. The results of an experiment with spray urethane could be smelled from across the room. "We tried many things," Okazaki recalled with a subdued weariness.

Eventually, after three years of trial and error, and with the help of a pair of Canadian undertakers, who helped him construct an entirely new container—known to this day as a tuna "coffin"—Okazaki was ready to ship Atlantic bluefin to Tokyo. On August 14, 1972—"the day of the flying fish"—a handful of enormous, silvery tuna that, just four days earlier, had been swimming in the waters off Canada's Prince Edward Island arrived for auction at Tokyo's Tsukiji market. As the frosty fish steamed gently on the wet concrete floor and the market's tuna traders looked on in disbelief, a small cut above the tail revealed that their cherry-colored flesh was in perfect sashimi condition.

The impact was immediate. The arrival of overseas tuna by air was deemed by at least one Tsukiji wholesaler as the biggest change the world's largest fish market had ever seen. Back in the homeland of the fish stick, the transformation was equally significant. In the 1960s, the town of Gloucester, Massachusetts, was in decline, its once-abundant stocks of haddock, redfish, and hake depleted. The arrival of Japan Airlines executives and Tokyo tuna traders in the 1970s triggered a boom, which only accelerated over the next decade as sushi became trendy among America's newest demographic, the yuppies. In the twenty years following "the day of the flying fish," tuna catches went up by 2,000 percent—and the average price paid for those catches increased by an astonishing 10,000 percent.

"I could not imagine that it would become the starting point for the growth of sushi all over the world," said Okazaki. Forty years after the invention of refrigerated tuna coffins, the Atlantic's previously plentiful bluefin had migrated by air at such a rate that they had become a critically endangered species.

II. Reefer Madness

After graduating from Cornell with a physics degree in 1976, Barbara Pratt spent her twenties living in a fridge. For the next seven years, she traveled and worked inside a refrigerated intermodal container, or "reefer," as they're called in the industry, circling the globe alongside Peruvian asparagus and Mexican mangoes. Her little-known adventures laid the groundwork for today's globalized food system.

We met at her 180-acre fruit farm in Westchester County, an hour's drive north of New York City, on a gorgeous late-September day. The crisp, sunlit air was filled with the happy chatter of families filling bags of apples in the pick-your-own orchard. Pratt is the third generation of her family to farm the land, which boasts more than forty varieties of apples, a pumpkin patch, and a Christmas tree forest. Her great-uncle paid off the mortgage by selling hard cider illegally in New York City during Prohibition, and her father opened the apple orchards to the public in the late 1960s. Today she manages the farm alongside her husband and kids, while still working full time as the director of refrigerated services for Maersk, one of the largest container companies in the world. Reefers currently make up just 6 or 7 percent of global container capacity, but shipping firms can charge significantly more to carry them than their standard, "dry" forty-foot counterparts.

Unrefrigerated shipping containers made their commercial debut on a rainy Thursday in 1956, when Pratt was not yet two years old. These now-ubiquitous metal boxes were the brainchild of Malcom McLean, the forty-two-year-old son of a poor North Carolina farmer who had built his own successful trucking empire and then sold it all in order to buy a struggling steamship line, despite never having set foot on a boat in his life. His flagship vessel, an aging oil tanker named *Ideal X*, was described by a contemporary reporter as an "old bucket of bolts." The innovative twist-lock corner fittings on his "silvery new trailer vans" were supposed to ensure that the boxes could be lifted by crane, secured to the deck, and stacked atop one another, but had only been tested using modeling clay. Similarly, the containers' overall sturdiness was assessed by having McLean and the rest of the team climb onto their roofs and jump up and down. McLean, a man of stoic Scottish ancestry, was observed to betray a "twinge of anxiety" as, over the course of eight hours, fifty-eight containers were lifted onto the deck, and the *Ideal X* set sail from Elizabeth, New Jersey.

A short write-up on page thirty-nine of the next day's *New York Times* missed the point entirely, suggesting that McLean had simply "pioneered a way of making both legs of a voyage pay." In reality, he had set in motion a logistics revolution that would fast-track globalization by knitting together suppliers and consumers around the world in entirely new configurations—but the *Times'* reaction was typical. The response to the debut of the shipping container was muted, to say the least.

Still, when the *Ideal X* arrived in Houston five days later and those fifty-eight containers were swung off the ship, set down on fifty-eight separate trailer chassis, and sent on their various ways the very same afternoon, it represented enough of a success that McLean's fledgling Pan-Atlantic Steamship Company expanded, rebranding itself as Sea-Land Service, Inc., to better describe its trailblazing vision of a seamless, intermodal, truck-to-ship-to-truck transportation network. Over the next few years, other companies took notice and began to follow McLean's example.

In keeping with the general lack of fanfare surrounding the maiden voyage of the first shipping containers, no one seems to have bothered to record what was transported inside them. It's possible that it was beer—that erstwhile early adopter, first of mechanical refrigeration then, perhaps, of container shipping. Certainly, McLean's initial calculations for the new business used beer as the model cargo, demonstrating that it would be 94 percent cheaper to ship it in one of his new containers than using conventional transport. These extraordinary savings came from cutting out the time and human labor required to hoist each individual barrel of beer—or bale of cotton or box of screws—on and off a boat by hand. Before the introduction of the container, a cargo ship would typically spend as much time at berth, being loaded and unloaded, as it did at sea. This was the real significance of McLean's innovation and the *Ideal X*'s maiden voyage: by shrinking both the time and the cost involved in maritime trade, they made today's just-in-time global supply chains possible.

Despite these enormous advantages, container shipping took a while to catch on. Railways and trucking companies were initially reluctant to embrace this new intermodal form of transport: their existing cargo carriages and trailers represented a significant sunk cost. Throughout the sixties, as dockworkers protested at the prospect of losing their jobs, port authorities dithered about whether it was worth investing in the new cranes and facilities needed to handle shipping containers. Meanwhile, the international negotiations to agree on standard dimensions and design, so that containers would be as interchangeable as the commodities inside them, dragged on.

Still, by the 1970s, the container had conclusively triumphed: the number of registered longshoremen on the US East Coast had decreased by two-thirds and nine out of every ten countries in the world had built deeper, crane-equipped ports to handle these enormous new box ships. The rest is the history of today's world, in which 60 percent of everything that is traded globally spends some of its life in a shipping container, and our lives are filled with products that were assembled on multiple continents from parts that have traveled the globe.

At first, fresh food took no part in this container revolution. Early shipping containers were intended only for the transportation of dry cargo—pallets and boxes of goods with no particular temperature requirements. A full quarter century after McLean's box had made its debut, most perishable commodities were transported the same way they had been for decades: as bulk cargo in the hold of specially designed refrigerated ships like United Fruit's Great White Fleet of banana boats. In the 1970s, when bluefin tuna joined red meat and bananas as one of the few foods whose production, consumption, and value had been revolutionized by refrigerated transportation, the amount of perishable food transported between continents remained a relatively tiny fraction of global production.

Although bulk-cargo refrigerated ships had transformed diets, landscapes, and economies by successfully delivering tropical bananas and antipodean rump roasts to northern tables year-round, they weren't necessarily a perfect solution to the challenge of moving perishable foods. Bulk transportation of meat and fruit suffered from the same inefficiencies that pushed the shipping industry toward containerization in the first place: expensive port labor and increased opportunities for handling damage and delay. Bulk bananas offered particular challenges of their own: traditionally, they were shipped loose, still attached to giant stalks, and longshoremen had to wear special gear to protect themselves from the oozing, rubbery sap.

What's more, all bulk refrigerated ships suffered from a larger issue. Because the hold could be kept at only one temperature, most ships were limited to carrying one type of produce at a time. Bananas, as we know, don't care to spend time at the lower temperature preferred by apples, which themselves need to be kept warmer than frozen meat. To fill a refrigerated ship, producers had to be able to slaughter several thousand animals or harvest tens of thousands of bunches of bananas, storing them till they had enough for an entire load; then, when the ship arrived in port, it flooded the market with thousands of pounds of a single commodity, creating a glut that depressed prices and led to waste.

Banana and meat men wanted to take advantage of the price savings offered by putting their product in a metal box to be transported alongside all the other metal boxes. Chilean grape farmers and kiwi growers in New Zealand, for whom access to overseas consumers was still an impossible dream, were even keener to get in on the container revolution. All they needed was a way to cool each box individually.

In the 1960s, shipping companies experimented with so-called porthole containers, into which cold air could be piped from a central refrigeration machine in the ship's hold. This cooling system didn't work particularly well at sea, and it didn't work at all on land. Finally, engineers at Carrier—which with Thermo King still dominates the transport-refrigeration market today—rearranged the components of Jones's truck unit into a thin, flat "picture frame" configuration that could be bolted onto the front wall of an insulated shipping container.* The result was a reefer, short for *refrigerated* container. It was still a standard twenty- or forty-foot interchangeable steel cube, but one equipped with its own built-in cooling machinery, which could be plugged into either an electrical outlet on the ship or the dock or a diesel generator to be hauled by road or rail.

---

*Carrier was founded by Willis Haviland Carrier, the engineer known as the father of modern air-conditioning, based on a system he designed to control humidity at a Brooklyn printing press in 1902.

The first reefers were introduced in 1968, but while their psychoactive namesake was chilling a generation, the logistical version failed to catch on. Each reefer cost four to five times as much as a single dry container, and their ability to protect their perishable contents was frustratingly hit-or-miss. The rigors of the marine environment combined with the extended shipping times—two or three weeks was the intercontinental minimum, compared with four or five days for a cross-country trip—had created an entirely new set of problems.

"Sometimes refrigerated shipments would arrive okay, and sometimes they didn't arrive okay, and nobody really knew why or what was going on," Maersk's Barbara Pratt told me as she monitored farm operations via walkie-talkie, fielding questions about the price of lemonade and the start date for Christmas tree sales. "Fifty percent—even total—losses were not uncommon." Coffee beans out of the Dominican Republic were going moldy en route. Transpacific fruit and vegetable cargoes kept showing up in Asia frozen solid. Flower bulbs shipped from northern Europe showed signs of cold damage when they arrived in the US, while melons from Mexico rotted as they crossed the Atlantic in the other direction.

"We knew so little about refrigerated shipping back then," Pratt explained. "You didn't have reliable portable recorders or satellite communications like we do today, so the inside of the reefer was like a black hole. The only way to know what was going on was to get in there with the cargo and see what was happening."

Pratt had already applied for and been awarded a place to study for a graduate degree in postharvest physiology at the University of Delaware when a former professor got in touch to see whether she'd be interested in taking on the reefer challenge. Her mission, should she choose to accept it: the design and operation of a mobile research laboratory that could study perishable products in transit and figure out where things were going wrong. "I hadn't done a whole lot of travel before, so it seemed like a good opportunity," Pratt said. "I was going to do it for a couple years, and then, of course, I got hooked and never left."

Her first task was to build a laboratory inside a standard forty-foot container. Working with a South Jersey company that specialized in custom motor homes, she decorated it with a black-and-white checkerboard linoleum floor and shiny red cabinets in a nod to Sea-Land's corporate color scheme. There were two bunk beds, a microwave, a shower, and, of course, a refrigerator. There was also a single window—bulletproof, in case things got unpleasant in one of the less salubrious ports. Most important, Sea-Land's mobile research laboratory—Pratt's new home—was crammed full of scientific equipment, including a microscope, a machine capable of identifying different chemicals based on their molecular weight, and a state-of-the-art, closet-sized computer.

As the summer of 1978 approached, Pratt told me, the lab wasn't quite ready for its maiden voyage, but the cacao harvest was about to begin and General Foods was not happy about the number of pods it had lost to rot last season. "We decided to roll the dice and make it work," she said. The laboratory was loaded onto a container ship headed to the port of Santo Domingo; she and a colleague flew down to meet it, and they got to work investigating the moldy cacao bean problem.

At the core of Pratt's research was the computer, which received inputs from 150 different sensors. Out in the field, Pratt would place her probes at different heights and locations inside a couple of empty reefers, then connect them to a terminal on the outside of the box, from which she could later run a long, gray, salt-resistant wire back to her lab. Once the beans were stuffed into their sensor-studded containers, they were transported alongside Pratt's mobile laboratory to the dock for loading. The longshoremen had been instructed to keep them all clustered together, near the center of the ship so the instrumentation was subject to less vibration and motion at sea. "After they were done with the cranes and putting the containers on and all that, then we would lay out the wires for what we needed to reach," Pratt explained. She would carefully thread them along the catwalk between the stacked reefers and bundle them together using zip ties, "so that they weren't just spaghetti sitting all over the place."

Then, perched in her own forty-foot steel box, walled in by hundreds of other steel boxes filled with everything from leaf tobacco to scrap metal, she and a rotating cast of colleagues got to work. The wires relayed all kinds of information about what was happening inside the reefers: tiny variations in temperature, humidity levels, airflow, and concentrations of different atmospheric gases. "For the first couple of tests, we took turns spending a couple nights making sure the readings were all recording," she explained. "Once we knew the reliability of the computer, we didn't need to do that—we slept in the ship's quarters, we ate meals with the crew."

During the day, Pratt conducted experiments on cacao bean samples and analyzed the data coming in from neighboring containers, troubleshooting moisture problems and hot spots in a kind of refrigeration forensics. "We were trying to see what was happening in the different parts of the containers that we were monitoring, and what the relationships were between the different parameters that we were measuring, and how the refrigeration units were functioning," she said. "It was very exciting, because we were pioneering things—we were working on questions that no one had worked on before."

For the next seven years, she traveled back and forth across the globe with problematic perishable commodities. Because the lab was built into a standard container, it could and did go everywhere a reefer would go, from the field to the highway, from the terminal to the ship, and from the dock to the distribution center, from the West Coast of the United States to Asian ports, and from the Mediterranean and Mexico back home to Newark. Pratt traveled in the lab at all times, accompanied by other Sea-Land employees as needed for different research projects. "I probably spent thirty to fifty percent of my time traveling each year, for two to three weeks at a time," she said. "It wasn't totally unreasonable."

During that time, Pratt studied more than a hundred different foods, frozen and fresh, testing and monitoring temperature and humidity levels, mapping airflow in the container, analyzing the respiration rates of fruit and vegetables, tracking fungus growth, and more, in order to arrive at a set of best practices for shipping different perishable commodities. Pratt and her colleagues were the first to map the airflow in a reefer and the first to figure out the maximum density of produce that could be stuffed in before circulation started to suffer, leading to uneven temperatures.

In a dusty set of Sea-Land promotional slides I discovered at the US Merchant Marine Academy archives on Long Island, the shiny white mobile research lab is pictured on the back of a truck in front of the Houses of Parliament in London and a busy streetscape in Tokyo. In reality, Pratt didn't get to do much of the sightseeing she initially imagined— even when the ship was in port, she was working. Her adventures took place in the lab. "There were a lot of 'aha' moments," she said. Based on her findings, the entire ventilation system of a standard reefer unit was redesigned so that the cooler air entered from the bottom rather than the top, as was standard on a truck.

Today, thanks in large part to the seven years Barbara Pratt spent in a container, reefer shipping is so cheap and reliable that even the most fragile of foods can travel by sea. The research that Pratt and her colleagues did led directly to a world where Peruvian asparagus can happily cruise for ten days on its way to American grocery stores, and New Zealand kiwifruit can spend up to seven weeks wending its way around the Indian Ocean and through the Suez Canal and the Strait of Gibraltar to arrive on a British supermarket shelf. This new ability to transport perishable produce cheaply, efficiently, and, above all, intact across the world's oceans left diets, economies, and entire ecosystems remade in its wake.

On the rare days when her reefer units were running smoothly and her experiments could be left unattended, Pratt told me that she enjoyed spending time up on the bridge, learning how to navigate and operate the ship. Like Polly Pennington, she was typically the only woman. "I found out later that part of the reason they hired me was because my manager told the CEO of Sea-Land that I had a black belt in karate and therefore could defend myself on board ship," she told me. This wasn't true, but fortunately Pratt never needed to call on her nonexistent martial arts training. "It was very much a man's world and still is pretty much today, but I didn't have any issues," she said.

S ealed up in their own can of winter, the pineapples, Popsicles, crab cakes, and beef burgers bobbing through the Suez and Panama Canals aboard ships the size of large apartment buildings often end up better traveled than the people who eventually eat them. The wild salmon for sale in your supermarket may have been caught in Alaskan waters, but it will likely have spent nearly two months of its afterlife crisscrossing its former Pacific hunting grounds inside a reefer unit. This final, assisted migration makes sense only when you realize that the cost of refrigerated container shipping is barely pennies on the retail dollar, while fish pin boning is a delicate operation that can't be mechanized—and that the wages earned by Chinese workers are less than a fifth of the amount earned by their American counterparts.

Nearly two-thirds of all fruit and vegetables produced in the world are eaten in a different country from the one in which they were grown. "The situation we're in now is that we as consumers can have the same fruits and vegetables year-round," said Pratt—a situation described by food writer Joanna Blythman as "Permanent Global Summer Time," which, in turn, relies on the existence of an equally unchanging and pervasive artificial winter.

In order to supply the unchanging cornucopia we expect on our supermarket shelves, produce buyers rely on an ever-expanding cold chain. They engage in a kind of counterseasonal resource buffering, rotating through a global roster of climate analogs to supply, say, strawberries cheaply and in sufficient quantity during California's offseason, from July through December.

Steve Barnard, the man responsible for ripe avocados on American supermarket shelves, described how this process, which he calls "filling the calendar," works. "We're in Peru, we're in Chile, we're in Colombia, we're in Guatemala—they all come with different altitudes, different latitudes, and you can determine when your harvest will be," Barnard said. "Our goal is to have two countries of supply at any one time, and we're looking for three in some cases." It is the logic of comparative advantage, expanded to the world stage: a shifting cast of far-flung California stand-ins, each performing their carefully choreographed role in an agricultural mystery play. This performance ought to inspire awe—an unchanging produce section at the grocery store, every day, everywhere across the Western world—yet it's so familiar that we pay attention only when there's a hiccup behind the scenes and shelves are temporarily empty.

Forty years into the reefer revolution, California is increasingly outperformed by its understudies: today more than half of the fresh fruit and nearly a third of the fresh vegetables eaten in the United States are imported. Consumption of mangoes, limes, avocados, grapes, and asparagus shipped from elsewhere in the world has overtaken that of homegrown produce such as peaches, oranges, cabbages, and celery.

Pratt told me she believes that reliable mobile refrigeration units have had more advantages than disadvantages for producers and consumers alike. Certainly, given the fact that only one in ten Americans eats the recommended amount of fruit and vegetables, anything that might help tilt our diets toward produce would be useful. If avocados and mangoes can persuade us to do what cabbages and oranges could not, the health benefits would be enormous. But despite Pratt's efforts to make sure we have the opportunity to purchase fresh, affordable blueberries every day of the year, USDA data reveal that since 1970 there has been only the tiniest improvement in the average American's consumption of fruit and vegetables—and over the past decade those statistics have begun trending downward again.

Among farmers, the benefits seem clearer. "For growers, it's a great opportunity because they have multiple markets they can sell their product into," explained Pratt. Apple farmers in the Pacific Northwest have gratefully sent up to 30 percent of their enormous harvest overseas; the rise in Chilean fruit exports has helped turn the country's economy into the "Latin American tiger," with its per-capita GDP leapfrogging its neighbors'.

On the other hand, these benefits have typically favored larger growers, who have the wherewithal to manage complicated export requirements and meet supermarket expectations for paperwork and quality control. As Pratt admitted, her own family farm doesn't—and couldn't—compete on the global market. As a pick-your-own operation, it survives by selling an experience as opposed to a price-sensitive, interchangeable commodity. Ultimately, it seems probable that, like Washington State's apple growers, Chilean or Peruvian farmers who do scale up to meet international demand might find the rewards are fleeting, as prices drop and produce buyers move on to regions with even cheaper land and labor costs.

The environmental impact of Pratt's reefer innovation is equally unclear. It's logical to assume that shipping food thousands of miles would be worse for the environment than growing it and eating it locally. In reality, though, it depends. If, for some reason, you're determined to eat a fresh tomato in winter, it may well take more energy to grow a hothouse variety locally than is required to ship one from the opposite hemisphere. Marine transportation is typically less polluting than diesel trucking, so more miles can, counterintuitively, add up to less emissions—though as domestic transportation electrifies, that calculus may shift.

These cost-benefit analyses become even more challenging once you take into account local collapses in water availability or biodiversity, as well as socioeconomic outcomes such as land grabbing and population displacement. Peruvian asparagus has the highest per-acre yield globally, but those bright-green spears are grown using water pumped up from one of the fastest-depleting aquifers in the world.

Meanwhile, the reefer—and with it global trade in perishable commodities—is still on the rise. Although her days of traveling the world in a shipping container are long over, Pratt has spent the rest of her career in refrigerated logistics. "Every day, we have new issues, new problems, new questions that come up," she said. "Sometimes it's new commodities, but other times it's trying to move a commodity into a different part of the world or trying to extend shelf life." Today reefer units are all digitized, and Pratt can see what's happening inside them from her desk—but, she said, "we still don't have all the answers."

Over the past decades, reefer boxes have become as high-tech as the bigger, stationary boxes we store food in on land, capable of maintaining ultralow temperatures as well as dynamically managing humidity and atmospheric gases. Ice cream is still a challenge, because the slightest temperature variation will cause ice crystals to grow, ruining its smooth texture, but Pratt told me that it's now possible to reliably ship such delicate loads as sashimi-grade bluefin tuna. According to *Container Management* magazine, bulk-cargo refrigerated ships are expected to go extinct sometime in the next few years as reefer boxes take their place. Even bananas are shipped in containers now: an estimated one in every five reefers is filled with bunches of the green fruit. Chiquita (the descendant of United Fruit) recently sold off the last of its Great White Fleet, and Maersk is beginning to experiment with adding ethylene en route, which may yet send the banana-ripening room the way of the meat locker.

As the afternoon sunlight lengthened and the crowd in the orchard began to diminish, I said goodbye to Pratt and paid for my own bag of apples to take home. Before I left, I asked her what became of the mobile research laboratory. She told me that its final journey took place in 1986 or '87—she couldn't be certain, but she thought it was a test shipment of mangoes from India. After that, it was sold and moved down to Chile, where it may well still exist—an overlooked landmark of the artificial cryosphere, parked somewhere on a South American farm. "I've never seen it since," she said with a small sigh.

III. Building a New Arctic

Just four minutes' drive from the Route 66 exit on I-44, deep in southwestern Missouri, the ground opens up. Every day, more than five hundred semitrailer truck drivers and a train roll down a carefully engineered ramp—the tracks are at maximum curve and maximum grade—into Springfield Underground, a three-million-square-foot cheese cave one hundred feet beneath the Earth's surface.

Most of the best-known cheeses of Europe—cheddar, Emmentaler, and Roquefort—are aged anywhere from two months to a year or more in natural caves. The process is known as affinage, and it is credited in large part with producing a good cheese's distinct, complex flavor.* In Missouri, Kraft also stores its cheese underground—mostly in large yellow drums, stacked five high, alongside Oscar Mayer meats and Jell-O puddings. "Parmesan, cheddar, American, Velveeta—it all starts out the same," said Louis Griesemer, whose father founded Springfield Underground as a mine in 1946. Griesemer greeted me at the top of the ramp and we hopped into his silver Tacoma for a tour of his subterranean refrigerated empire.

"Dad started it as an open-pit mine, in response to the aglime demand," Griesemer explained. After World War II, leftover ammonium nitrate, originally manufactured to make explosives, began to be applied

---

*These underground limestone caves are natural refrigerators, maintaining a constant cool temperature that is typically between forty-five and fifty-eight degrees. Unlike your average refrigerated warehouse, however, a damp, drafty cave also offers cheese the high humidity level (above 80 percent) that it requires. As an added bonus, different caves feature their own unique combination of bacteria, yeast, and mold: the caves of Roquefort-sur-Soulzon in France harbor the particular strain of *Penicillium* that is responsible for the cheese's characteristic blue veins.

to crops as a nitrogen fertilizer. This restored fertility to the region's post–Dust Bowl soils, but it also raised their acidity, so scientists encouraged farmers to put pulverized limestone—known as aglime—on their fields in order to rebalance them. "Then, in the 1950s, with the construction of the interstate, they needed aggregate," Griesemer continued. "I-44 from Rolla on down is built with rock from here."

By the early 1960s, Griesemer Sr. had hollowed out a sizable void: a 250,000-square-foot room, held up by thirty-foot-tall raw rock pillars that look like enormous gray elephant legs. A new Kraft cheese plant had also opened just a few miles south, churning out Cheez Whiz, Kraft Natural Shredded, Sliced and Chunk Cheese, and American Singles. To this day, it remains the nation's principal boxed-mac-and-cheese production center. (The factory has a giant yellow elbow pasta tube mounted on a concrete pedestal outside its entrance, decorated with the words "You know you love it.")

No one recalls whose idea it originally was, but the two companies realized they had the solution to each other's needs: the crates and barrels of processed cheese currently being stored at the surface could be refrigerated less expensively underground, creating savings for Kraft and a new income stream for the mine.

Griesemer drove me over to a recently mined area that he was finishing out for a refrigeration customer. The 200,000-square-foot space will become Springfield Underground's nineteenth subterranean "building," and Griesemer estimated that its metamorphosis from mine to fridge would take eight months in total. We climbed out of the truck to wander around in the dusty, clammy air. Sweeping yellow curtains blocked off unfinished areas from the roadway, and the rough-hewn chamber was mostly in shadow. Water had puddled in the corner, and a condensed cloudlet of fog and particulates hovered above it, spotlighted by temporary construction LEDs.

To turn rock face into warehouse, Griesemer simply pours a concrete floor around the rock pillars, then builds modular walls between them, using blocks of Styrofoam with shotcrete sprayed on either side. Brine piping is installed to bring the cold down from an ammonia plant at the surface, while air shafts are repurposed to pull truck exhaust fumes upward. A laser scan helps fit the racking in between the pillars but also allows their bulk to be deducted from the room's square footage, so that Kraft is not billed for the cost of chilling the shelves on which their cheese sits. "The cost of construction is also cheaper, because you don't have to do a roof—you just put in a sprinkler system," said Griesemer. "And we don't charge for the pillars.

"The more amenities you need, the less advantage I can offer," he continued. "I can build a data center for ninety-five percent of the cost of building it at the surface, but I can build an industrial warehouse for just two-thirds of the cost of aboveground."

These initial savings are augmented by a lifetime of lower utility costs: Kraft estimates that maintaining the temperature in its underground storage space uses just a third of the electricity of a normal facility. While Springfield made the list of "America's Wildest Weather Cities" in *Forbes* magazine, the temperature in the mine stays at a constant fifty-eight degrees year-round. "The energy bill is the same every month—we've just got that much thermal mass," said Griesemer. "That's what we're all about: thermal mass." When the power goes out, mercury levels don't fluctuate.

The flip side is that changing the temperature takes a while: Griesemer says that it will take a month and a half to draw enough heat out of the rocks to bring his new refrigerated space down to thirty-six degrees. Once the rock has been frozen, it can't easily be defrosted without causing cracks that could lead to the collapse of the entire subterranean structure. "This has been frozen for twenty-something years," he said, gesturing toward a room currently filled with gallon upon gallon of Hiland Dairy ice cream. "It's a risk, because you've got to keep it frozen," he said. "Without a client, you go out of business fast."

After a while, every blasted rock pillar and shotcrete wall started to look the same, and I had no phone reception underground. "Truck drivers do get turned around," Griesemer said. One was backing up in front of us, heading straight for a pillar, and I asked whether such collisions were frequent. "We've never had a pillar damaged," Griesemer said, laughing. "We get a lot of truck damage, but the pillars are fine."

Griesemer told me he'd spent more than forty years underground; recently he retired as CEO to make way for his nephew. He confessed to visiting other subterranean spaces for fun. "Kansas City—that's impressive," he said. "There's an awful lot of underground space that can't be thawed in KC." The largest, he told me, is Hunt Midwest SubTropolis, northeast of Kansas City, just across the Missouri River. "But they always have trouble with trucks, because their ceilings are so low—they're only twelve or fourteen feet tall," he said. "So we might actually be the largest, in cubic feet."

The entire Ozark region is riddled with limestone mines and natural caves, which makes it fertile ground for both urban explorers and conspiracy theorists. In 2012, retired pro wrestler and former Minnesota governor Jesse Ventura dedicated an episode of his TV show to investigating the rumor that Springfield Underground is a bunker built by the Illuminati, "a powerful group, bent on overthrowing our government and taking over the world from a command post in the American heartland." As Ventura and his crew enter the same tunnel that dozens of Kraft trucks pass through every hour, the voice-over describes it as "the doorway to a new world, but the end of ours."

"I don't even get that channel," said Griesemer. "The local media called, and I said, 'First of all, I'm Catholic, so I'm Opus Dei. And I could tell you more but I'd have to kill you.'"

An hour down the freeway, in Carthage, Missouri, Americold runs another huge underground cold-storage facility / suspected Illuminati bunker in a former mine. My guide there, Gabe Gary, told me that no one knows how far the tunnels stretch. "We still mine—you can sometimes feel the blast," he said. "And some of it is flooded—I've never seen it but I've heard about it." Urban explorers who have ventured into the tunnels describe crystal clear underground lakes that are home to a thriving ecosystem populated by turtles and fish.

Although the total underground volume is larger at Carthage than at Springfield, and the facilities encompass a pair of tennis courts (players report that the temperature is always pleasant, but the cave-wall echo can be confusing), the refrigerated storage space is smaller. Still, Gary told me that all frozen food found in Walmart stores to the west of us has spent time in this facility, housed next to rooms full of Hostess products— Twinkies, Ding Dongs, and Ho Hos—and yet more cheese, stored in enormous blocks, three feet wide by four feet tall. But although he agreed that underground refrigerated storage offers attractive utility-bill savings, he told me that's not the main reason why so much of America's food is stored underneath the Ozarks. "It's logistics," he said. "You're always looking at the routes." Springfield Underground, Americold Carthage, even SubTropolis in Kansas City: they're all in the center of the country, strung along the cross-continental path blazed by America's mother road, Route 66.

Making cold mobile reshaped the geography of food. The location and design of refrigerated warehouses has, in turn, been transformed by those new supply chains. Like Louis Griesemer, I've made a habit of checking out the local perishable logistics scene on family visits, road trips, and vacations, much to my husband's chagrin. From pink-stained, wooden-walled cranberry tunnels in Massachusetts to the antique, ice-encrusted brine-cooling pipes of a currently defrosting warehouse in downtown Los Angeles, these pockets of cold are honeycombed throughout the American landscape, and their shifting form and organization tells a story.

The earliest refrigerated warehouses were downtown, usually next to both the rail terminus and the wholesale market. They were multistory buildings, as much for the thermal efficiency as to fit in a compact urban footprint. As central, highly visible vaults for food, they were designed by the same architectural firms as libraries and banks. The goal was to produce buildings that looked solid, reliable, and, above all, trustworthy.

The very first one seems to have been built in Fulton Market in New York City in 1865. It would have been cooled using natural ice and salt and ventilated by fan. (No one seems to have recorded its opening, appearance, or exact location, and it is now long gone.) The first mechanically cooled warehouse opened its doors quite a bit later, in Boston, in 1881; this was followed by an even larger neighbor in Quincy Market that sold cold to the businesses and homes around it through a network of subterranean brine pipes.

This business model was called "pipeline refrigeration," and it enjoyed a brief boom: by 1916, some twenty US towns and cities boasted cold networks, with pipes extending more than eighteen miles underground to supply markets, hotels, restaurants, and even "food preservation cabinets" in private homes. This concept—treating refrigeration as a utility, delivered by metered pipeline, just like electricity and gas—implies an entirely different relationship with cooling, but it died out as the machinery to produce cold shrank to more manageable dimensions, in terms of both size and capital outlay.

The next big revolution in warehousing came, like so many others, courtesy of World War II. The wooden pallet, and its partner in logistics efficiency, the forklift truck, were not introduced commercially until the 1930s, and the US military was the first to use the combination at scale.

Triggered by the Japanese attack on Pearl Harbor and the need to supply vast amounts of food and munitions to a new and even more distant Pacific front line, the Quartermaster Corps quickly invented and standardized the now-ubiquitous four-way pallet—so called because its construction allows a forklift truck's prongs to be inserted from any side. It ordered so many that manufacturers struggled to keep up with demand. The son of one lumber mill owner in Virginia recalled that, in an era before automated assembly, the men in charge of nailing the pallets together were treated like elite athletes. "About six hours was all their arms could take," he told *Pallet Enterprise* magazine. "As soon as they quit, they would put their jackets on to keep their right arm warm, just like a pitcher does."

The combination of pallets and forklifts was powerful: in the two years following their adoption across the Quartermaster Corps in mid-1943, the tonnage of freight handled per person in military depots doubled. In a history of Brambles, the Australian company that today manages the world's largest pool of reusable pallets, author David Manuel records the impression the palletized Americans—"marvels of mechanized muscle"—made when they first debuted their newfangled logistical techniques in Australia. "It was a completely amazed and somewhat goggle-eyed collection of Australians who watched the Americans un-batten their hatches, land forklift trucks with ship's gear and then at the most amazing speed bring off their palletized cargo."

Refrigerated-warehouse managers were among the first to adopt this new goods-handling paradigm postwar. Not only did the Lego-like vertical and horizontal regularity of the pallets-and-forklift system increase productivity (and thus reduce labor costs), but it also imposed new constraints on a cold-storage facility's shape and site. No more imposing multistory fortresses downtown: instead, large single-story buildings provided the optimal environment for forklift reach trucks, which could hoist pallets up to forty feet in the air. The racking into which those pallets were slotted held up the roof, with newly developed polyurethane foams serving as insulation in the prefabricated walls: the result was a featureless cuboid from the outside and a steel scaffolding skeleton traversed by narrow aisles on the inside.

Meanwhile, thanks in large part to the Thermo King, these refrigerated warehouses could afford to spread horizontally: they were increasingly built on cheap land next to the interstate, linked by road to the newly rural meat-packers and suburban supermarkets. The new architecture of cold—the one that, until very recently, prevailed throughout the United States—was taking shape.

B ack in the early nineteen sixties, we claimed to be the first to do a rack-supported building," said Adam Feiges, who, until the company's recent sale to Americold, co-owned one of the largest cold-storage providers in the United States, Cloverleaf Cold Storage. "It's gone now—it was in Sioux City, Iowa."

Feiges told me that cold-storage warehouses in the US tend to be situated near either the producers or the people. As I built a map of refrigerated warehouses, tracing their overlooked geography, the patterns that emerged revealed the underlying logic of our food system, as well as the ebbs and flows of the global economy.

The Southern California Americold warehouses in which I worked are located there because nearby Los Angeles is full of people who need to eat. Feiges's family's business was focused on the production side: it was founded in Iowa in 1952 to serve the state's meat-packers. "In a perfect world, the packinghouses would sell everything fresh," he explained. "They can't do it, because they have to operate at full speed all the time." Growing a slaughter-ready steer still takes at least two-and-a-quarter years from the moment of insemination, which makes it difficult for the beef industry to respond nimbly to shifts in supply and demand. "That's where somebody like us comes in," he said. "We'll freeze that hot meat and then store it until they've got it sold."

Springfield Underground offers a similar service to Kraft, banking its cheese in order to help smooth out temporary mismatches between factory production levels and the national appetite for Cheez Whiz. That said, Feiges told me, no cold-storage manager wants product to sit on their racks for long. "We make our money on turnover," he said. "The old joke in our family was you can go broke with a full warehouse."

In the case of meat, dairy, fish, fruit, and vegetables, refrigeration often maps onto production: the largest french-fry factory in the world, in Burley, Idaho, sits next door to an equally gigantic freezer; the fifty thousand residents of California's strawberry capital, Watsonville, live side by side with more than half a dozen cold-storage warehouses, including an Americold facility.

Ports are also major hubs, though perishable foods often bypass the busiest—Newark and Long Beach—in favor of smaller neighbors, where delays are less likely. On the East Coast, the Port of Wilmington, Delaware, handles the largest volume of fruit and juice in the United States. "We used to be number one for bananas in the world," said Vered Nohi as she took me on a quick circuit of the port's facilities. (The current leader is the Port of Antwerp, in Belgium.)

Nohi had worked at Wilmington for fourteen years, but she told me that the port's big breakthrough came before her time, when refrigerated containers of New Zealand kiwis, Chilean grapes, and Moroccan citrus started showing up at the Delaware ports in the 1980s—soon after Barbara Pratt's research made containerized fruit shipment commercially feasible. Today Citrosuco's orange juice tank farm is located in Wilmington; Dole and Chiquita also use the mid-Atlantic port as their hub, while Del Monte is just upriver, at Gloucester Terminal in New Jersey.

"This is our niche," said Nohi. "We can do rapid cooling, cold treatment, fumigation, controlled atmosphere—we pamper this fruit." Wilmington is dwarfed by Newark—it receives just four hundred or so vessels a year, while its northern neighbor's annual total is more than seven times as many—but it boasts that, with just one stoplight between the port and the freeway, fruit can travel as far as Chicago or Montreal overnight.

Recently, Wilmington has begun to face a new generation of competitors to its south: barely a month goes by without the announcement of a multimillion-dollar refrigerated warehouse breaking ground in the other Wilmington—North Carolina—not to mention Charleston, South Carolina; Savannah, Georgia; and Jacksonville, Florida.

"That's all to do with the widening of the Panama Canal," Feiges explained. Until 2016, this man-made waterway connecting the Atlantic and Pacific Oceans was too narrow to accommodate the gigantic container ships that dominate global trade, which meant that all the refrigerated meat and produce that the US sends to East Asia—its largest export market—was funneled through West Coast ports like Long Beach or Oakland. "East Coast US ports can now compete directly," said Feiges. "Doing business in California, in particular, is perceived as being harder because of environmental rules, labor rules, things like that—so the notion is, well, now we don't have to be beholden to these ports where a strike or a new emissions rule could just kill us."

In terms of domestic supply chains, distribution-focused refrigerated warehouses tend to cluster on cheap, interstate-adjacent land, far out on the fringes of urban markets. For example, a logistics park near Allentown, Pennsylvania, where US Foods, Americold, Millard Refrigerated Services, and others all maintain cold-storage facilities, owes its popularity to its location at the intersection of two major freeways, one leading directly to New York City, the other southeast to Philadelphia, as well as several East Coast railway lines.

"Now, there's exceptions to that rule, like Los Angeles," said Marc Wulfraat, a supply chain consultant for large supermarket businesses. "If you look at the traffic and the flow of goods in the LA market, grocery companies like Ralphs and Kroger have their warehouses right near the downtown core." Part of Los Angeles's exceptionalism stems from its curious political geography: what looks on the map to be a single sprawling conurbation is actually made up of dozens of independent cities, each with its own officials capable of making their own land-use, tax, and environmental policies. The city of Vernon, an industrial enclave just a couple of miles south of downtown Los Angeles, is notorious for its corrupt officials (it was allegedly the inspiration for Vinci in season two of the TV show *True Detective*) but also for charging rock-bottom rates for its utilities. It boasts block after block of cold storage—including an Americold warehouse at which I worked a shift—but just thirty homes.

The maximum distance between a distribution center and the supermarkets it supplies is determined by the standard warehouse operating schedule that I experienced at Americold: the daily cycle of receiving truckloads from manufacturers in the early morning, putting those pallets away before noon, and only then turning to the task of picking and loading mixed pallets to send out on trucks to replenish supermarket shelves. Because those outbound loads typically contain products from three or four hundred different vendors, many of which would ideally be stored at different temperatures and humidities, it's important to keep their time in the homogenous environment of the refrigerated truck to a minimum. "That's why we try to put the distribution center within two hundred and fifty miles of the stores, in order to get the driver out there within five or six hours," Wulfraat concluded. "If you're much farther than that, then you run the risk of spoilage."

This system—a centralized distribution warehouse that receives goods in bulk, then stores them, sorts them, and sends them out to supermarkets within its 250-mile radius—emerged in the United States with the advent of refrigerated trucking. Wulfraat credits it with making American supermarket prices the cheapest in the developed world.

By way of illustration, he pointed to the difference in both supply chain and sticker shock at Whole Foods versus Walmart. Although Walmart now supplies at least 15 percent of all the perishables in America, it didn't even sell food until the 1990s. "They hired people who knew the cold chain and how to build these distribution centers and staff them," explained Wulfraat. Walmart started with a single food-distribution center in its home base of Little Rock, Arkansas, and added a grocery section to all the stores within 250 miles, then did the same in Texas. "And they kept leapfrogging out until they covered the nation," said Wulfraat. The result is an extremely efficient logistics infrastructure, which translates into lower prices for consumers. "When you walk into a Walmart store, you wonder how the heck they do it," he said. "That's part of how."

Whole Foods stores, by contrast, are scattered across the nation; there aren't enough clustered within any given 250-mile radius to justify a dedicated distribution center. "So what do you do?" asked Wulfraat, before telling me: "You end up outsourcing distribution to wholesalers." Those wholesalers add their own profit margin to the distribution costs, and Whole Foods adds that to the price it charges customers. "That's why the 'Whole Paycheck' nickname came about," said Wulfraat. Shopping at Whole Foods is more expensive, not necessarily because its food is healthier or better for the environment but because its supply chain is so inefficient.

These unwritten rules shaped the supermarket shelves and refrigerated warehouses of North America throughout the second half of the twentieth century. But the logic of logistics is always contingent. Today it is changing again, at the confluence of demographic shifts and new economic trends.

A ll right, our kid," Dave Priest said to the warehouse operator. "Strap her in—she's writing a book about the cold chain and she wants a ride on the crane." Priest, a genial northerner, is the general manager of NewCold Wakefield, a fully automated frozen warehouse that can hold as many pallets of food as Springfield Underground, all stacked up in a bright-white, twelve-story cube built atop a former coal mine in England's industrial northeast.

"I've told her she's better off switching it to a sitcom," he continued as I climbed into a tiny gray cabin that hung off the side of a floor-to-ceiling steel beam. From where I stood, the scene felt more suited to a dystopian science fiction movie. I was alone, my safety harness hooked onto a railing that held me, suspended, in a narrow canyon no more than three feet wide. Walls of scaffolding and cardboard cubes stretched as far as I could see into the darkness above, below, and ahead. Around me, the building's metallic guts rumbled, groaned, and whirred as it shuffled pallets around, slotting new arrivals into place and shunting older product outward toward the loading docks in an eternal, internal game of three-dimensional

Tetris. The temperature, at five degrees below zero, was cold enough to make my lungs hurt, but the atmosphere also made breathing difficult: because the ceiling is too high to rely on a sprinkler system, oxygen levels are reduced to between 16 and 17 percent as a fire-suppression measure, from the 21 percent found in normal air.*

With an ear-piercing screech, my platform lurched forward. Everything rattled as I was propelled diagonally upward into the void, boxes of microwave-ready french fries and potato waffles blurring into each other on either side of me. Then, thirty seconds after takeoff, I slowed to a halt above a plastic-wrapped pallet of strawberry Popsicles and hovered, nearly 120 feet in the air. Behind me, on the other side of the beam, a square metal platform rolled out into the racking along a set of rails. A few feet in, it stopped underneath a full pallet of two-liter tubs of Kelly's Cornish Vanilla Dairy Ice Cream, lifted it up a couple of inches, and carried it back to the crane, ready to drop off elsewhere in the building. The entire thing happened without human intervention: the building's internal algorithm had issued the instructions and the machinery had followed them, while I watched, dizzy and open-mouthed.

Moments later, the vanilla ice cream and I swooped down through the canyon, coming in to land near a conveyor belt. While the pallet trundled off toward the loading bay, I unhooked myself and stepped triumphantly through the air-lock door back into the world of humans. As I crossed the threshold, wreathed in a halo of white, cold, deoxygenated vapor, Priest applauded, laughing. "You look like you're coming off that stage on *Stars in Their Eyes*," he said, referring to a popular British TV talent show.

My joyride is not part of the usual tour, even for the handful of outsiders—prospective customers and health inspectors—who ever see

---

*Breathing air that contains less than 19.5 percent oxygen is considered unsafe under federal guidelines, but it won't kill you on the spot: the danger lies in the fact that your oxygen-deprived brain quickly starts to struggle, leading to mistakes in judgment and coordination. Extreme cold, as we've seen, exacerbates that. In some jurisdictions, workers who enter NewCold's freezer for maintenance and repairs have to use a breathing apparatus; for a brief exposure like mine, there was no need.

inside this building. Priest added it to the itinerary in an exuberance of enthusiasm for his beloved freezer, which is among the largest and most technologically advanced in the world. He showed me the building's lungs, which vent excess oxygen to the outside with a *whoosh* like a pressure cooker quick release, and its automatic off-loading system—a robotic assemblage of railings, sensors, and conveyor belts that can empty a truck in five minutes, doing the same job that would take a forklift truck driver half an hour. "I used to work in the old-fashioned cold chain," Priest said. "Then, when I came here, it were like *Alice in Wonderland*. The part that I showed you is the part where I looked and thought, *If they offer me the job, I'm taking it.*"

NewCold was founded in the Netherlands in 2012 and has already become the fourth-largest cold-storage company in the world, with four locations in the US and more under construction. Its meteoric rise illustrates several of the shifts currently reshaping the cold chain. For one, multistory buildings are back: Priest's pride and joy may have cost two to three times as much to build as a similarly sized conventional refrigerated warehouse, but it uses 40 percent less electricity, in large part because full automation means that everything can be much more densely packed and vertically stacked.

Emissions reductions are nice, and in terms of return on investment, utility savings are better, but from the industry's point of view, the real charm of NewCold's warehouse is how few people it takes to operate it. Almost everything at NewCold is automated, which means that employees need only to enter the freezer that I took a swan dive through if something breaks down. "Until recently, companies could go out and find a readily available pool of blue-collar labor willing to come in and do hard work in hostile environments like freezers," Marc Wulfraat pointed out. "Those days are over." Not only is the population of the developed world aging out of employment at a rapid rate but, even in the US, the minimum wage has risen sufficiently that the steep initial investment in an automated system now seems much more reasonable.

Energy efficient, high tech, human free, and, increasingly, closer to cities in order to satisfy the instant delivery expectations of a new generation of digital consumers: the American coldscape is currently expanding at a rate that hasn't been seen since the postwar boom. This has, in turn, attracted investment capital: both NewCold and Lineage Logistics, the current world leader, were founded in the same year with private-equity partners. "Until recently, most of the refrigerated-warehousing companies were family businesses," said Feiges, whose own family business was swallowed up in the industry's wave of consolidation. "That is changing relatively dramatically, with a huge influx of money."

The United States already boasts the largest volume of artificial winter in the world at more than 5.5 billion cubic feet of refrigerated warehouse space. Despite the country's tepid population growth, that is no longer enough. The cold-storage industry is on track to grow by nearly half as much again in the next few years, with most of the expansion located in and around coastal urban areas. Without fanfare or notice, a new and improved Arctic is currently under construction all around us.

This development, as astonishing as it is, is dwarfed by another recent phenomenon: the global cold rush. As people in developing countries around the world, from India to Mexico to Nigeria, move to cities and become richer, one of the first things they do is add more meat, fish, dairy, and perishable fruit to their diets. If everyone on Earth were to require as much cold space to eat as an American, millions upon millions of refrigerated warehouses would need to be built, increasing the volume of mechanically cooled space on Earth tens of times over. Seen from this perspective, the world has hardly begun to break ground on its third pole.

I n Szechuan, we're eaters," said Chen Zemin, the world's first frozen-dumpling billionaire. "We have an expression that goes, 'Even if you have a very poor life, you still have your teeth to please.'" He smiled and patted his not-insubstantial belly. "I like to eat."

Chen, who is now eighty, never planned on being a dumpling mogul. Like almost everyone who came of age during China's Cultural Revolution, he didn't get to choose his profession. He was a "gadget guy" during his high school years. "I liked building circuits and crystal radios and that sort of thing," he told me. "I applied to university to study semiconductor electronics." But the state decided that Chen should become a surgeon, so he dutifully completed his studies and amused himself in his free time by learning how to cook, developing a reputation among friends and family for his Szechuan pickles, kung pao chicken, and, of course, dumplings.

Even after he became vice president of the Second People's Hospital in Zhengzhou, a provincial city about halfway between Shanghai and Beijing, Chen remained bored with his day job. "I didn't have enough to keep me busy," he said, blinking earnestly, hands steepled beneath his chin. "I would wander round inspecting the building, and I had meetings, but I felt as if I spent most of my time reading the newspaper and drinking tea." He engaged in lots of Rube Goldberg–like tinkering: jury-rigging the hospital's aging equipment, fixing his neighbors' radios, and even building Zhengzhou's first washing machine. And he cooked. For decades, his Lunar New Year gifts of homemade glutinous rice balls were legendary among friends and neighbors.

In the 1980s, as China began to open up to the West, Deng Xiaoping, Mao's reformist successor, declared that some people "will get rich first." Chen, who not only was bored but had two sons' weddings to pay for, wanted to become one of those people. It wasn't long before he started thinking about, as he put it, giving "my rice balls legs."

Chinese pot stickers and rice balls are traditionally made communally, in enormous batches, in order to justify the effort it takes to knead the dough, roll it out, mix the filling, and wrap by hand a morsel that stays fresh for only one day. Because of his medical background, Chen had an idea for how to extend the life span of his spicy pork wontons and sweet sesame paste–filled balls. "As a surgeon, you have to preserve things like organs or blood in a cold environment," Chen said. "A surgeon's career cannot be separate from refrigeration. I already knew that cold was the best physical way to preserve."

Using parts harvested from the hospital junk pile, Chen built a two-stage freezer that chilled his glutinous rice balls one by one, quickly enough that large ice crystals didn't form inside the filling and ruin the texture. His first patent covered a production process for the balls themselves; a second was for the packaging that would protect them from freezer burn. Soon enough, Chen realized that both innovations could be applied to pot stickers, too. In 1992, against the advice of his entire family, Chen, then fifty, quit his hospital job, rented a small former print shop, and started China's first frozen-food business. He named his fledgling dumpling company Sanquan, which is short for the Third Plenary Session of the Eleventh Central Committee of the Communist Party of China—the 1978 gathering that marked the country's first steps toward the open market.

Today Sanquan has factories nationwide. The largest, in which Chen and I were chatting, employs five thousand workers and produces an astonishing four hundred tons of dumplings a day. He showed me the factory floor from a glass-walled skywalk; below us, dozens of men and women, clad in hooded white jumpsuits, white face masks, and white galoshes, tended to nearly a hundred dumpling machines lined up in rows inside a vast, white-tiled refrigerator. Every few minutes, someone in a pink jumpsuit would wheel a fresh vat of ground pork through the stainless-steel double doors in the corner and use a shovel to top off the giant conical funnel on each dumpling maker. In the far corner, a quality-control inspector in a yellow jumpsuit was dealing with a recalcitrant machine, scooping defective dumplings off the conveyor belt with both hands. At the end of the line, more than 100,000 dumplings an hour rained like beige pebbles into an endless succession of open-mouthed bags.

Scenes like this are being replicated all over Zhengzhou, a smoggy industrial city that, thanks to Chen's ingenuity, has become the capital of frozen food in China. Sanquan's rival, Synear, was founded in Zhengzhou in 1997, and the two companies account for nearly two-thirds of the country's frozen-food market. The city is home to five of the ten biggest Chinese-owned companies in the industry, according to the weekly *Frozen Food Newspaper*, which is also based in Zhengzhou.

Over the past decade, China has offered a unique opportunity: the chance to witness a transition that, in the United States, took place so long ago that no one from refrigeration's prehistory is still alive to notice the difference. In contemporary China, even as skyscrapers, shopping malls, and high-speed trains have transformed life, the refrigerator still represents a significant step forward on an individual level. China's thousands of years' worth of agricultural and culinary history, not to mention its political system, make its context quite different from that of the United States in the 1880s—but watching what happens as an entire country attempts to cool down its food system on fast-forward casts new light on the peculiarities of American-style food distribution, as well as what is lost or gained once perishables are freed from the constraints of time and space.

Almost any Chinese citizen over the age of forty can remember the moment they got their first home refrigerator, with the exception of those who still don't have one. Liu Peijun, a forty-nine-year-old logistics entrepreneur who now owns and operates three refrigerated warehouses on the outskirts of Beijing, told me that one of his earliest memories from childhood was of hanging meat outside the window to keep it cold until the Lunar New Year feast.

Even in the richest cities—Beijing, Shanghai, Shenzhen, and Guangzhou—it wasn't until the late 1980s, as electrical grids became more reliable and families had more disposable income, that refrigerators became a fixture of many homes. When Chen founded Sanquan in

1992, fewer than one in ten of his fellow citizens owned a refrigerator. Fortunately for his fledgling business, that small percentage still represents a large number of potential consumers in a country as populous as China—and that fridge-equipped pool grew quickly, jumping to 95 percent of urban families by 2007.

But while home fridges are now commonplace, at least in China's cities, refrigerated warehouse space and transportation have lagged behind. As a result, imports—such as a perfect air-freighted cherry from Washington State—often arrive in stores in much better condition than domestic fruit. When Tyson began to import chicken paws from its slaughterhouses in the United States into China, where the appetite for these cartilaginous delicacies is much larger, it quickly discovered that frozen feet didn't necessarily have legs, as Chen Zemin would put it, beyond the port cities of Shanghai or Tianjin.* "We'd bring these things in, they're perfect, and three days later, we find out they're in a room-temperature warehouse somewhere with a wet rag thrown over them to keep them 'fresh,'" recalled management consultant Mike Moriarty, who advises US food companies on how to do business in China.

China's recent economic growth has increasingly depended on expanding its domestic market, as opposed to the export-driven strategy that initially fueled the country's development. In practice, that has meant most of the rural peasants previously subsisting on small family plots have been encouraged to move to cities and start buying things, leaving the remaining handful to become agribusiness entrepreneurs themselves. But without a functioning cold chain, there wasn't any way for those farms to scale up their output, or for the former peasants to buy it. Communist authorities thus stepped in to support the country's nascent perishable-logistics sector with both practical and moral support.

Over the past fifteen years, tax breaks, subsidies, and preferential

*Chicken feet include the leg below the feather line; chicken paws are just the feet below the spur. In another example of refrigeration's curious alchemy, demand for chicken paws overseas has turned them into the third-most-valuable part of the chicken for American meat-packers, after the breast and wings.

access to land have been made available to anyone aspiring to build a refrigerated warehouse. Leading up to the 2008 Olympics, the Beijing municipal authorities embarked on an ambitious and partially successful program of "supermarketization," designed to get meat and vegetables out of the open-air "wet" markets—where food is cooled by standing fans and the occasional hose-down from a cold tap—and safely behind sneeze guards in modern, climate-controlled grocery stores. In 2010, the government's powerful National Development and Reform Commission made expanding the country's refrigerated and frozen capacity one of the central priorities in its 12th Five-Year National Plan. "Under the guidance of Deng Xiaoping Theory and the important thought of Three Represents," the document declares, defeating my interpreter's otherwise flawless translation skills, "we must vigorously develop the modern logistics industry." Refrigerated warehouse capacity in China more than doubled over the subsequent decade.

Liu Peijun is one of this new generation of perishable logistics entrepreneurs. In a windowless warehouse next to a deserted restaurant just off Beijing's Fifth Ring Road, Liu showed me around his freezer room. Piled high on four-story racks stood pallet after pallet of shrimp dumplings. In the dimly lit roar, I also spotted boxes of Häagen-Dazs ice cream and Alaskan frozen crab claws. Liu said those were going to be delivered directly to consumers as part of an online promotion for the Chinese New Year.

Liu started out in 1996 as an agent for frozen-food brands like Sanquan when they first reached Beijing. "I did tastings and promotions in supermarkets," he said. "People sort of shunned them in the beginning, but the dumplings caught on really fast. I quickly realized that the real bottleneck was not consumer demand but the lack of refrigerated storage and distribution." Eventually Liu decided to start his own company, Express Channel Food Logistics. He built his first warehouse on the site of a former chicken shed in 2008, storing and delivering chilled and frozen food for grocery stores like Walmart, e-commerce sites like tmall.com

(China's version of Amazon.com), and high-end restaurant-supply companies. That facility is the oldest of the three warehouses Liu now owns in Beijing, in addition to a fleet of city freight vans. He has recently started renting his first warehouse in Shanghai, and he is also building a super-chilled room capable of reaching minus seventy degrees—the temperature needed to store high-end sushi-grade tuna destined for the city's ever-expanding luxury market.

In fact, Liu added, cold-chain logistics is so hot right now that the interpreter he brought with him last year on a fifteen-day research tour of refrigerated warehouses overseas quit as soon as they got back to Beijing, in order to start a cold-storage business of his own.

Meanwhile, just as iced railcars allowed the American meat-packing industry to centralize in Chicago and helped transform California into the nation's fruit and vegetable basket, so too in China. Within a couple of decades, China's pork industry has gone from a handful of pigs on a family farm to twenty-six-story hog hotels housing tens of thousands of animals. Pigs produce three and a half times as much shit as a human every day, so it's perhaps not surprising that the resulting manure lagoons have already become one of the country's largest sources of pollution, ahead of even heavy industry. And while plant scientists at the Beijing Vegetable Research Center are hard at work selecting and optimizing the varieties of Chinese greens that stand up best to cold storage, party leaders have launched a South to North Vegetable Transfer initiative whose goal is to repurpose the country's southernmost province, the tropical island Hainan (otherwise popular with Chinese honeymooners), as the National Winter Vegetable Base, complete with thirty brand-new logistics centers and an express refrigerated rail link to Beijing.

In the United States, some 70 percent of all the food we eat each year passes through a cold chain. By contrast, during my visit to China, I was told that less than a quarter of the country's meat supply is slaughtered, transported, stored, or sold under refrigeration. The equivalent number for fruit and vegetables is just 5 percent. Today, with 4.6 billion cubic feet of refrigerated warehouse space nationwide, China is rapidly catching up to the US in sheer volume of cold. But that translates to less than five cubic feet per person, or less than a third of what Americans currently have—meaning that the Chinese race to refrigerate is far from over.

Wandering Shanghai's traditional alleyway neighborhoods, I found that refrigeration's reach remains uneven. A ten-year-old girl, home alone after school, reluctantly put down her video game controller to show me a bag of her favorite flavor frozen dumplings (pepper beef) in the family's overstuffed freezer. Returning to the couch, she dismissed me with the statement that she preferred McDonald's to any dumplings, fresh or frozen.

A few houses farther along, a middle-aged, henna-haired lady with electric-blue eye shadow took a break from the delicate operation of hanging her laundry on a series of rods balanced atop electrical wires to let me know that she would never buy frozen dumplings. Shaking her damp underwear for emphasis, she told me: "It's the Chinese people's habit to make dumplings at home." Just across the courtyard, a young family shared their tiny single room with a surprisingly large fridge-freezer, which they described as their most important possession. "We have a baby, you see," the wife explained. "I don't have time to shop every day, and freezing keeps food healthy and safe—no additives."

Her attitude is increasingly common among urban populations. Over the past decade, supermarkets have been by far the fastest-growing sector in Chinese food retail, accompanied by a shift among consumers toward larger but more infrequent shopping trips. In rural areas, people still buy ingredients for dinner daily at wet markets, but in the cities sales of chilled meat, dairy, and frozen food have increased rapidly.

Still, not all Chinese people are ready to embrace the refrigeration revolution. Dai Jianjun is the fifty-something chain-smoking chef of Longjing Caotang—a restaurant on the outskirts of Hangzhou, the scenic capital of Zhejiang Province, which serves an entirely locally sourced, anti-industrial cuisine. When I asked him how he liked frozen dumplings, he took off his corduroy cap, rubbed his shaved head with both hands, and finally, in a calm voice that carried a distinct undercurrent of anger, said, "If I may speak without reserve, they're not food."

Over the course of two epic meals, separated only by a short paddle on a local lake to catch fish for dinner, Dai fed me dried vegetables and mushrooms, vinegar-pickled radishes, fermented "stinky" tofu, and peanuts that six months earlier had been packed into earthenware jars. I visited his on-site bamboo-walled drying shed, where salted silvery fish halves and hunks of pork hung in orderly rows. Between courses, Dai pulled out his iPad to show me a series of videos that demonstrated how radish preservation varies by topography, with hill people drying the vegetable in the sun before salting it and flatlanders working in reverse order. After our boat ride, as the rest of the fishermen beheaded and gutted the catch on a wooden block, the fish boss, who went by the name Mr. Wang, presented me with a particularly delicious yellow mud–preserved duck egg, which, he told me, keeps at room temperature for thirty days.

The rest of the ingredients had been harvested or foraged that day. Dai keeps leather-bound purchase diaries documenting the provenance of every chicken, tea leaf, mustard green, and black fungus. Several entries are accompanied by photos of a farmer picking or slaughtering the item in question. Not a single thing I was served that day had been refrigerated: it was the opposite of the Great Cold Storage Banquet of 1911, and I consumed it with equally "unalloyed pleasure." In fact, the food was revelatory: complex but light and offering a more subtle yet diverse range of textures and flavors than I had previously encountered in Chinese cuisine.

Dai himself barely ate, preferring to smoke, drink (first green tea, then baijiu, a clear spirit distilled locally from glutinous rice), and gesticulate expressively while issuing definitive, if óccasionally bizarre, pronouncements on everything from Italian cuisine (too heavy and good only for producing opera singers) to the celebrated former head chef of El Bulli, Ferran Adrià ("antirevolutionary"). Finally, toward the end of the evening, I mentioned that in 2012 Britain's Royal Society had named refrigeration the most important invention in the history of food and drink. With their faces already reddened from the liberal consumption of baijiu, Dai and the other men all convulsed with laughter.

Once he had composed himself, Dai said, "Within our circle, you sound ridiculous!"

6.

THE TIP OF
THE ICEBERG

I. Cold Case

"Your fridge is one of the most date-ready fridges I've seen in a hell of a long time," John Stonehill told me. "Are you married?" Stonehill is the alias used by Jon Steinberg, the world's first and only refrigerator dating expert. Earlier that morning, I had sent him a set of raw, uncensored shots of my own fridge. A few hours later, we met at a fancy coffee shop in his central Los Angeles neighborhood so he could deliver his verdict over a single-origin Americano.

Given that Steinberg's singular talent is helping individuals find their perfect match based on fridge compatibility, the fact that I have a husband and am not currently looking for another one did, I admit, make our meeting somewhat moot. "It changes everything," he said, shaking his head. "Because when I looked at your fridge, I'm like, *Oh my God, this chick is awesome!*"

Steinberg's fridge epiphany didn't arrive until he was in his late twenties, on a date set up by his mother. The young lady in question was a nice Jewish doctor, almost as preppy as Steinberg himself, and they hit it off. Back at her place, she gave him a quick tour, then opened her fridge to get drinks. "It was literally the most disgusting fridge I've ever smelled in my life," said Steinberg. "Look, we all have busy weeks. We all have leftover stale Chinese food in our fridge from time to time. This fridge was so bad, it smelled like a Porta-Potty at Coachella."

The owner of the stinky fridge turned out to be, at least in Steinberg's telling, something of a hot mess herself, and after breaking up with her, he vowed never to ignore the signals emanating from a girl's refrigerator again. "I got to the point where I'd just make a beeline for the fridge," he said. "I'd see Svedka vodka, I'd go, *Okay, she probably makes forty grand a year, she reads* People *magazine, she watches* Keeping Up with the Kardashians."

Steinberg eventually found Ms. Right: invited back to her house on their third date, he noted that she had a GE stainless-steel model with French doors, with just a couple of gourmet condiments, Fiji water, and a nice bottle of champagne inside. "I said to myself, *She's upscale but she's not the nurturing type,*" he said. "And that's exactly how she turned out to be."

A few years later, Steinberg was explaining his process to a friend, who convinced him to launch a blog called *Check Their Fridge*. Readers could submit photos for an analysis of what it would be like to date that particular fridge's owner; if they also sent in pictures of their own, they'd receive a compatibility score. Steinberg had a successful career as a TV producer by this point, so his fridge dating service was just for fun. "Next thing you know, I'm on network TV morning shows," he said. "Then the *Daily Mail* covered me and suddenly I blew up around the world."

Unwittingly, Steinberg had tapped into one of humanity's most broadly shared compulsions: the desire to peek into other people's fridges and judge them based on their contents. "Tell me what you eat and I will tell you what you are," said famed French epicure Jean Anthelme Brillat-Savarin. Show me your fridge, and I'll show you who you *really* are, replied Jon Steinberg and the internet. Bookshelves are abstract and, often, carefully curated; bathroom cabinets are too intimate. It is the humble fridge that offers a window onto the twenty-first-century soul.

A marketing survey by LG Electronics found that 82 percent of

Americans allow the contents of someone's refrigerator to shape their opinion of them—a result that is surprising only inasmuch as there are 18 percent who don't. In between all the doomscrolling, trolling, and ghosting, fridge peeping is one of the more popular online pastimes. Within hours of Twitter launching its short-lived Periscope live streaming app in 2015, the majority of channels were besieged by requests for refrigerator tours. In the early months of COVID-19, TikTok users fell for the "fridge challenge," which consisted of filming yourself taking something unexpected, like a shoe, out of your fridge; today they're more likely to spend hours watching fridge restocking videos. On Instagram, wellness influencers post "fridgies," showing off perfectly labeled Tupperware and color-coded juices. The Fridge Detective subreddit has attracted tens of thousands of commenters who use uploaded fridge photos to speculate freely as to how old each refrigerator's owner is, where they live, what they do, and much more besides, all while passing judgment on the wrongness of uncovered sliced watermelon and sharing tips on the best store-brand hummus.

Undoubtedly there is something to be learned from the fact that Joe Biden requires a fridge stocked with orange Gatorade while the late David Bowie allegedly kept urine in his (to prevent witches from stealing it, obviously). That said, our voyeuristic insights often have as much, if not more, to do with our own assumptions and biases as those of the appliance's owner. In 2020, *The New York Times* tried to bring the culture wars to fridge hermeneutics by creating a quiz that invited guesses as to whether a particular fridge's proprietor was going to vote for Donald Trump. After hundreds of thousands of responses, the newspaper's verdict was that coconut water drinkers are not necessarily Democrats, stocking up on frozen Banquet meatball meals is no guarantee of MAGA tendencies, and that, "as a whole, we can't distinguish people's politics from glances into their fridges much more reliably than if we just flipped a coin."

As he finished his coffee, Steinberg acknowledged the limitations of refrigerdating. "I'm only talking odds," he said. With an apology, Steinberg told me he had to leave soon—he was late for that most Angeleno of activities, a "notes" call to share feedback with the writers of the TV show he produces. But before saying goodbye, he gave me his notes on my fridge.

Despite his initial compliments, I was braced for the worst. "You're obviously not perfect," Steinberg agreed. Still, I earned points for hosting a multiplicity of beers, as well as extra-large-cube ice trays in my freezer—items that, to Steinberg, were evidence of conviviality rather than excessive alcohol consumption. In general, his fridge reading erred on the side of generosity: he interpreted the dozens of tiny containers stacked in my fridge as a sign that I am adventurous and enjoy variety, rather than as a symptom of internalized intergenerational thrift that borders on hoarding. "Ultimately, I think you might be a hipster *and* a foodie," he said with a slight grimace. "But overall you have your shit together."

A century ago, the mere possession of an electric refrigerator would have been more than enough to knock the socks off Steinberg's great-grandfather, regardless of its contents. Today more American kitchens have a refrigerator than a stove. The fridge's ascent from aspirational novelty to household essential has taken place with spectacular rapidity and, while its contents may (or may not) reveal our true selves, its shifting shape and sound reflect much larger social and technological trends, even as its arrival in our kitchens rearranged both homes and cities.

The domestication of cold is often dated back to the early years of the nineteenth century, when Maryland farmer Thomas Moore designed a rabbit fur–lined "refrigerator": an oval cedar tub under whose lid sat ice, salt, and an insulated tin box to hold food. Moore was not the first to come up with this kind of ice chest—among the holdings of the Palace Museum in Beijing are a pair of exquisite enameled coffers that operate on the same principle and date back to the reign of Emperor Qianlong, in the 1700s—but it started a trend. Thomas Jefferson signed off on Moore's patent, purchasing one of the devices himself, and by the 1840s, several companies had started to manufacture iceboxes for sale. By the end of the century, most families in large cities like New York had one.

In the decades following Moore's invention, the icebox *was* the refrigerator. It evolved into a solid wooden cabinet with a sturdy metal latch that looked like a cross between a safe-deposit box and an armoire. A drip tray underneath caught the meltwater, and a separate compartment on top held the ice block, which had to be replenished almost daily. Fortunately, icemen provided door-to-door delivery and, if rumor was to be believed, more: their rippling muscles (from hauling hundred-pound blocks of ice) and their delivery hours, while the gentleman of the house was away at work, made them the subject of innuendo. The title of Eugene O'Neill's play *The Iceman Cometh* riffs on a common contemporary joke in which the husband calls upstairs to his wife to ask, "Has the iceman come yet?" The answer: "No, but he's breathing hard."

Bawdy humor aside, with its high humidity and a fluctuating temperature that was often well above the forty degrees recommended today, the icebox was better suited to storing berries and leafy greens than the meat, dairy, and leftovers for which most families used it. The development of its mechanical replacement took a surprisingly long time. From the 1880s, dozens of individuals attempted to shrink the enormous machines that cooled warehouses and ships, but it wasn't until after World War I that a handful of engineering advances made domestic refrigeration a practical proposition.

In the 1920s, several solutions to the same problem jostled for dominance. The stakes were high: whoever came up with a fridge that could compete with the icebox on price could count every household in the urban United States as a potential customer. By this point, American cities were equipped with both gas lines and electricity, and the country's burgeoning ranks of middle-class women were eager to escape the tedium of emptying drip pans, cleaning slimy icebox interiors, and going grocery shopping on a near-daily basis.

Kelvinator launched the first successful electric refrigerator in 1918.

The company, which was founded in Detroit by automobile-industry veterans, installed its machinery in customers' basements, then cut a hole through the floor to connect the cold air to a box in the kitchen. In his backyard washhouse in Fort Wayne, Indiana, Alfred Mellowes painstakingly hand built the world's first self-contained electric fridges, with the machinery and food storage all housed in a single compact box. After losing $34,000 in two years—hundreds of thousands of dollars in today's money—he sold the company to General Motors, which went on to market it under the name *Frigidaire*. General Electric decided to dive into the market in order to sell more electricity—an energy-hungry appliance that ran 24/7 was their idea of a gold mine. A white enamel closet mounted on dainty cabriole legs, GE's 1927 Monitor Top was so named because the hermetically sealed cylinder in which all its machinery was housed resembled the oversize revolving turret on the Union Navy's first ironclad battleship, the USS *Monitor.*[*]

The Kelvinator, Frigidaire, and GE refrigerators all drew on mains electricity to power the motorized compressor, as did the fridge I built with Kipp Bradford. The Electrolux-Servel employed a different mechanism. Designed by Swedish engineering students, it eliminated the need for a motor by using a gas flame to heat the refrigerant, instead of compression. It was introduced in 1926 and was quickly christened the "common-sense machine," because its lack of moving parts meant that it operated in silence and rarely broke down. Nonetheless, the electric fridge won, even though consumers complained that each time the temperature fluctuated and the motor whirred into action, the noise was sufficient to rattle apartments and disturb sleep.

Although modern machinery is significantly quieter, the Monitor Top's triumph meant that the low-pitched buzz of the refrigerator's electric compressor became the background soundtrack to contemporary life, so ubiquitous that The Velvet Underground used to tune their instruments

---

[*]To modern eyes, the cylinder looks more like the head of a Lego man.

to it. As founding band member John Cale explained, "the drone of Western civilization" is the sixty-cycle hum of the domestic fridge.

Ruth Schwartz Cowan, in her classic parable of how society shapes technology, "How the Refrigerator Got Its Hum," attributes the electric refrigerator's triumph over its gas-powered rival to General Electric's nearly bottomless budget. Flush with cash earned from Thomas Edison's patents, GE spent lavishly on marketing stunts, sending a fridge on a submarine ride to the North Pole and financing an entire commercial Technicolor film, complete with Hollywood stars and a plot revolving around the need for an all-electric kitchen. Even though it cost more than a Ford Model T when it was first introduced, the Monitor Top ended up selling more than a million units.

Cowan's point is that the electric fridge did not beat the Electrolux-Servel fridge because consumers thought it was better—or because it *was* better. In a capitalist economy, the technological choices available to consumers are constrained by profit, and profit depends not just on sales but on a whole host of other variables, including the ease of adapting existing manufacturing and distribution systems, the marketing resources available, and the compounding revenue from elevated utility bills. In the long run, a refrigerator that runs on electricity, which can now be generated renewably, may have been a better choice than one that requires natural gas—but the Monitor Top's triumph cut off a series of other possibilities. Because the Electrolux-Servel lost, a silent, reliable refrigerator powered entirely by biogas from household waste never even reached the drawing board.* GE provides today's consumers with a list of sounds to expect from their refrigerators, including hissing, clicking, gurgling, and a "chirping/woof/howl" created by changing fan speeds. The sixty-hertz hum is not included, having slipped beneath our perceptual threshold to become the white noise of domestic life.†

---

*Helen Peavitt, curator of consumer technology at the Science Museum in London, notes that in the 1930s at least one British household used sheep droppings to power its Electrolux-Servel.
†This low hum actually vibrates around sixty hertz, due to minute fluctuations in the grid as utility companies respond to changes in demand. The result is an ever-shifting symphony of

The fridge's subtle infiltration and rearrangement of the home extends to dimensions far beyond the sonic. At least at first, the fridge was simply an adjunct to existing food-storage methods. Many homes also had a north-facing pantry or larder with a mesh-screened opening to the outdoors and marble or tile shelving to keep food cool while keeping flies out, and even a root cellar to store vegetables. The existence of refrigerators made it possible to transform the built environment in a way that, in turn, made fridge ownership less optional. Well-insulated and centrally heated homes in the suburbs didn't have the kind of drafty, cold corners that were perfect for a pantry. In new urban apartment blocks, it was impossible to put a meat safe outside or keep a root cellar in the basement. In many temperate regions of the world, the indoor climate hasn't just become more homogenous; it has also become hotter: the interior of British homes has become, on average, ten degrees warmer since systematic measurements began in the late 1970s.

Domestic space gradually reorganized itself around the refrigerator and so, too, did shops, cities, and daily lives. More and more American families drove their car to a self-service supermarket (which made its debut in 1916) and filled their shopping cart (introduced in the 1940s) with a week's worth of food at a time, all of which could be stored in their ever-larger fridges. The relationship was symbiotic: refrigeration facilitated mass food production, enabled mass merchandising, and encouraged mass consumption. All these forces combined to make the fridge's ascent from bit part to starring role, to eventually displacing the hearth as the heart of the home, seem inevitable. When the Monitor Top was introduced, fewer than 3 percent of American households owned a mechanical refrigerator; by the time the US entered World War II,

---

frequency variations that London's Metropolitan Police began to record in 2005 for use in audio forensics. Because the hum is so omnipresent, inserting itself into most recordings, law enforcement can match the particular sixty-hertz fingerprint of UK-made recordings to its archive to arrive at an exact time stamp. Every time you open the fridge door (which the average household does 107 times per day, according to research conducted by LG), triggering its compressor to kick on, you're helping create that particular second's unique audio fingerprint.

following a decade marked by the Great Depression, that had risen to more than half the population.*

In the chapter titled "Why Do They Work So Hard?" from their 1925 study exploring changes in the lives of the inhabitants of Muncie, Indiana, husband-and-wife sociologists Robert and Helen Lynd concluded that refrigeration was, at least in part, the reason that twentieth-century Munsonians increasingly devoted "their best energies for long hours day after day" to work—an activity "seemingly so foreign to many of the most powerful impulses of human beings." The need to purchase a domestic fridge and fill it with "green vegetables and fresh fruit all the year round" was, alongside utility bills for such novelties as running hot and cold water, sewage, electricity, and telephone service, one of a "diffusion of new urgent occasions for spending money in every sector of living" that impelled the city's residents to use up their allotted time on Earth in jobs from which they seemed to derive remarkably little satisfaction.

Among those workers were women; here, too, the fridge played a role. The Lynds cite a 1911 US Department of Agriculture report that blamed the limited variety of foods available prerefrigeration for forcing women to slave away in the kitchen for hours, turning sturdy year round ingredients like cabbages, root vegetables, and sugar into preserves, pickles, relishes, and interminable quantities of coleslaw. The local seed-store proprietor told the Lynds that a good 80 percent of Munsonians grew their own produce in 1890—and "they used to *be* gardens in those days"—whereas only half maintained even a diminished sort of vegetable patch by 1925. For many households, working to purchase fresh fruit and vegetables from elsewhere replaced the labor of growing, harvesting, and processing one's own.†

Freed from at least some domestic drudgery, women defected from

---

*The refrigerator was the most frequently financed appliance through Franklin Roosevelt's Electric Home and Farm Authority, a New Deal–era initiative created to increase sales of white goods, including stoves and water heaters.
†Ironically, one of the largest employers in Muncie at the time was the Ball family, whose glass jars are still used for putting up preserves today.

the home toward paid jobs—in small numbers at first, then, following the Second World War, en masse. In 1900, only 5 percent of married women worked outside the home; by 1940 that number had risen to 36 percent, and by 2000 it was 61 percent. In recent years, a number of economists have attempted to tease out the exact contribution of refrigeration to female employment rates. This has proven challenging, to say the least, because the refrigerator was just one of a host of labor-saving innovations—from washing machines and vacuum cleaners to indoor plumbing—introduced during a short span. The best estimate seems to be that household technologies, including but not limited to the refrigerator, were responsible for up to half of the huge increase in women in the workforce in the twentieth century—although some argue that these analyses ignore the impact of medical advances such as the birth control pill and legal abortion, as well as the fact that the unpaid hours spent on "women's work" have not decreased so much as shifted focus. (The latter may bear some responsibility for the more recent finding, by the same man who first argued that fridges were "engines of liberation," that refrigerator adoption also correlates with rising divorce rates.)

These kinds of transitions have occurred elsewhere, but later and with local variations. For example, in 1948 only 2 percent of British households had a refrigerator. Throughout the 1950s, housewives made, on average, three trips to the butcher's and more than seven to the grocer's every single week. In 1970, fridge ownership stood at nearly 60 percent and the first out-of-town superstore—a Tesco—opened its doors. Village and town centers across the US and Britain were slowly eviscerated by the fridge-enabled move to weekly shopping. One government study found that a supermarket on the edge of town shrank business for smaller, more central shops by as much as three quarters.

Just as we saw with the story of our humming homes, the refrigerator made this particular outcome possible—but not inevitable. In 1973 France, which had roughly the same rate of domestic fridge adoption as the UK, as well as an equal percentage of women in the workforce, passed legislation limiting the construction of stores with a floor plan of much more than 100,000 square feet. (For reference, the average size of a Walmart Supercenter is 179,000 square feet.) Italy and West Germany put in place similar regulations. Today many continental European towns and cities retain lively central shopping districts whose walkable, human scale makes them irresistible to Anglo tourists. Meanwhile, the average interior volume of a French refrigerator is less than ten cubic feet; in the US that figure is 17.5, although trendy French-door models typically offer more than twenty-five cubic feet of storage space. As Canadian architect Donald Chong once suggested, small fridges make good cities— and good cities require only small fridges.

The basic mechanism behind the domestic refrigerator hasn't changed much since the triumph of the electric fridge in the 1930s, and as Kipp Bradford assured me, it's simple enough to be extremely sturdy. This presents something of a challenge to manufacturers. In 1932, advertising executive Earnest Elmo Calkins wrote the introduction to a book outlining a concept he called "consumer engineering," which relied on a combination of planned obsolescence and effective design to induce demand. "Goods fall into two classes: those that we use, such as motor cars or safety razors, and those that we use up, such as toothpaste or soda biscuits," he wrote. "Consumer engineering must see to it that we use up the kind of goods we now merely use. Why would you want last year's hand bag when this year's hand bag is so much more attractive?" For those concerned about waste, Calkins had stern words: "Wearing things out does not produce prosperity. Buying things does."

Fridges are hard to wear out. Early Monitor Tops happily carried on keeping food cold for decades; in 2013 the *New York Post* proclaimed an eighty-five-year-old model used by an auctioneer and his wife in Montgomery, New York, to be the oldest working refrigerator in America. But GE, Frigidaire, Kelvinator, and a growing pool of rivals were motivated by the prospect of prosperity, so, following Calkins's advice, they turned to design to engineer consumer enthusiasm. "It's all aesthetics," said Justin Reinke, who, over the course of his career, has been responsible for the refrigeration divisions of Whirlpool, then Samsung, and now Turkish brand Beko, which is one of the largest appliance makers in Europe, although still a relative newcomer to the US market. "How it looks and how it's laid out are the big ones, then the features are the tie-breakers."

Already in the 1930s manufacturers offered fridges with slide-out shelves, a built-in radio, and foot pedal–operated doors. (The latter seemed like a nifty idea, but because a fridge door opens outward, this feature became known as a knee-buster.) The Crosley Shelvador was a huge bestseller thanks to the ingenious inclusion of a hollowed-out space for shelving inside the refrigerator door; the idea was the brainchild of twenty-something mother of two Constance Lane West. Her patent on the invention lasted until 1953, at which point other manufacturers were finally able to add shelving to the inside of their painfully empty refrigerator doors. Door-mounted butter-conditioning shelves with built-in warmers; dedicated thawing drawers that catch drips; special "kid zones" and color-changing panels: they've all been tried, and they'll be tried again. "It just comes back around," Reinke said.

In recent decades, the excitement has mostly been about new configurations. When a new compartment with a separate door was first introduced for frozen food in the late 1940s, it was mounted on top of the refrigerator, where the ice in an icebox used to go, and for the same reason: cold air sinks. The two sections were powered by the same compressor, so putting the coldest part of the appliance at the top made sense. By the end of the decade, the engineers at Amana had figured out how to use fans to circulate the freshly cooled air in a side-by-side configuration, in which both fridge and freezer sections ran the full length of the fridge, but only half its width; in 1957, they managed to flip the standard setup to launch the world's first bottom-mounted freezer.

For the remainder of the twentieth century, these three configurations were the only choices on the market. Mary Kay Bolger, the woman who oversaw the development and debut of the French-door refrigerator—the first new configuration in more than forty years—is now retired, having worked in refrigerator product development for more than twenty years at Amana, which was subsequently acquired by Maytag, then Whirlpool. "It was an incredible ride," she told me, the thrill still vivid.

French-door refrigerator compartments have two doors that open from the center, swinging out on side hinges; ironically, in France, this format is known as a réfrigerateur américain. At the time, Amana dominated the side-by-side market; this new layout was more expensive and didn't even offer ice and water dispensed through the door, which customers increasingly regarded as a prerequisite. "Bringing ice and water through a side-by-side is easy," Bolger explained. "You're making ice in the freezer section and you're dispensing from there, so it's no big deal." Putting an ice dispenser in a French-door refrigerator is an engineering headache: the freezer is at the bottom, so somehow the ice has to get to the upper section, then stay cold enough not to melt and clump. "Once it gets clumped up, it won't dispense and the consumer has to take the ice bin off and get rid of it," said Bolger. "Then they're unhappy because there's no ice and they've got to wait for more to get made."

Bolger told me that she and her team had done a huge amount of consumer research as they developed this new configuration. "Ninety-five percent of consumers said, 'If you think I'm giving up my ice and water dispenser to buy a French door, you're crazy,'" she said. The French door wasn't a response to customer demand; instead, it was something that the company's engineers had suggested. "When we launched French doors in 2001, they gave us a forecast of thirty-five thousand—that's what they thought we would sell," Bolger said. Under these circumstances, she thought the low sales estimate sounded fair. "What happened is we were instantly on back order," she said. "The orders flew in." Amana sold more than 100,000 French-door refrigerators in the first year—triple the forecast—and Bolger estimates that it could easily have sold another 20,000 if it had been able to scale up production quickly enough.

Bolger credited the French-door configuration's astonishing popularity to its visually pleasing symmetry, which tied into another hot trend from the 2000s: the open kitchen. "People entertain in their kitchen now," she said. "That didn't happen back in the olden days." Stainless-steel appliances, inspired by the kind used in professional kitchens and as seen on the brand-new Food Network, took off at the same time and for similar reasons. "People wanted a refrigerator that reflected what they felt was good taste and state-of-the-art design," said Bolger. "And they had credit cards." (By 2000, more than 70 percent of US households had at least one credit card, a concept that barely existed when the side-by-side or bottom-mount configurations were first launched.)

As Earnest Calkins correctly pointed out back in the 1930s, people normally purchase a refrigerator only because their old one has broken. "Our dealers told us that with the French door, people were coming in and saying, 'I'm going to order that, and I can wait for it, because my current refrigerator's working—I just want this one,'" Bolger told me. "That was the first time that kind of conversation happened. Usually it was, 'Oh my gosh, my refrigerator quit. What do you have that I can take home today?'" Finally, the fridge designers had made this year's appliance sexy enough to justify replacing last year's.

II. Freshness Guaranteed

"The freezer is full of stuff. Extra things that are on sale," said the forty-one-year-old mother. "Extra meat. I think there's a whole ham back there because Ralphs gave us a free ham one time, and it's in there. And just different things . . . cheeses and ice cream. Stuff. Extra stuff."

Over the span of four years, from 2001 to 2005, anthropologists at the University of California, Los Angeles, immersed themselves in an unprecedented study of the daily lives and household environments of dual-income, middle-class Americans. Using thousands of hours of systematic observation and videotaped footage, the team analyzed the lives of thirty-two Angeleno families to create what they called an "archaeology of today." The fridge emerged as a central focus—indeed, the researchers found that the families stockpiled food "to such an extent that they overflow into second refrigerators and garage storage areas." Close to half of the families surveyed kept a second refrigerator in the garage; nationally, one in four households has two or more fridges. (Some have dozens: Martha Stewart once told me she probably has fifty or sixty refrigerators spread across her twenty-one kitchens, including two entire walls lined with still-operational fridges from the 1920s.)

The exterior of the typical Angeleno family refrigerator often had "rather dense and layered assemblages of ephemera," the researchers wrote: on average, fridge doors held fifty-two separate items covering 90 percent of the available surface area, though some boasted triple that. "Pictures, reminders, addresses, phone lists that have not been good for years and years," explained the mother of Family 27. A pattern emerged: the more chaotic and overburdened the fridge door, the more likely the family was to be drowning in a sea of possessions—and the more likely the mothers (but not fathers or children) were to have elevated levels of cortisol, a stress hormone, in their saliva. "This iconic place in the American home—the refrigerator panel—may function as a measuring stick," the team concluded: a shortcut benchmark for the intensity with which any given household participates in our shared late-capitalist consumer frenzy.

Abundance, as manifest in refrigerators and homes, is both comforting and stifling: the thirty-two families stockpiled food so as not to worry about running out but then got frustrated when they couldn't find anything in their overstuffed appliances. The mother of Family 16—the one with the extra ham from Ralphs—confessed that the freezer was the bane of her husband's existence. "He hates it, cause it's so full," she told researchers. "He can't make ice because there are chicken nuggets in it."

Big-box stores incentivize this kind of overstock, rewarding shoppers with two-for-one bargains and bulk discounts. "We're a Costco family," said the mother of Family 23, surveying the jam-packed shelves of her garage freezer and storage shelves. "I should just buy stock." Much of the food hoarded in this manner is premade and microwave ready—frozen pizza, fish sticks, and burritos, valued for their ease of preparation and convenience. The UCLA researchers found that the families saved only a few minutes of "hands-on" time by preparing a convenience meal for dinner, as opposed to cooking from scratch. "The real difference is the effort needed at the planning stages," they conclude. "The family chef can invest less time thinking about the week's meals."

Frozen ingredients and entrées are, after all, reassuringly predictable: they take the same amount of time to cook and taste the same every time. They are also unlikely to rot before they're consumed—although, hidden from view behind a spare ham, they may have shriveled up, taken on a grayish tone, or suffered freezer burn. Other perishables are more subject to decay. "We eat a lot of fresh vegetables and fruits, too," said Family 16's mother. "Well, I should say we have them on hand," she corrected herself with a laugh.

This combination of bulk buying, aspirational purchases, and pride in maintaining an amply provisioned home has helped fuel the expansion of the average American fridge—a trend that shows no signs of slowing. "Customers will typically describe capacity as the number one thing that they want," agreed Beko's Justin Reinke. "But here's the thing: if you go out and ask somebody if they want more money, their answer is going to be, 'Yes, of course I want more money.'" It's the theory of induced demand: just as adding another lane to a freeway will only increase congestion, extra fridge space will inevitably be filled. More is never enough.

The real problem with huge, overstuffed fridges is that there's almost no way the average family can consume that much food before it goes bad. What's more, as geographer Tara Garnett explained, there is a "safety net" syndrome associated with refrigerated storage. In the refrigerator, "the food can always keep longer, goes the thinking, except that suddenly one finds it has gone off." Schedules change, ambitious plans to cook from scratch are jettisoned when everything becomes too hectic, and the result, for many families, is that their refrigerators simply "serve as cleaner, colder trash bins," as food waste expert Jonathan Bloom puts it. The fridge has become a schizophrenic status symbol: on the one hand, it's a public display of sufficient good taste to warm refrigerator dating expert Jon Steinberg's heart, and, on the other, it's all too often a shameful food graveyard, filled with moldy berries and long-expired yogurt.

Garbologist William Rathje became familiar with this uncomfortable tension between the values and ideals associated with the fridge and the very different reality it typically embodies. "The study of garbage reminds us that it is a rare person in whom mental and material realities completely coincide," he wrote in the introduction to *Rubbish!*, his book describing the University of Arizona's thirty-two-year-long Garbage Project, which he ran for most of that time. (Rathje, whose nickname in the field was "Captain Planet," died in 2012.) "The aha! moment for Bill was that when he was studying Maya culture, nearly all of what was being excavated was what those people had thrown away," explained his friend sociologist Albert Bergesen. "He thought, *Why can't we use these techniques to learn about our own culture?*"

Among the many illuminating studies Rathje and his students carried out was an intensive survey of food waste—specifically, formerly edible food that was thrown away, as opposed to inedible discards such as peels, rinds, bones, and skin—in which each participant's self-reports were compared with the sorted and weighed contents of their trash bags. His first finding was that many of his subjects had what he described as a "tenuous grasp" on their own behavior. No one thought they wasted much, and all of them threw away a considerable amount—up to a quarter of a pound of food per person a day. Middle-class households typically wasted more food than either poorer or, perhaps more surprisingly, richer ones. During a recession, Rathje came to the counterintuitive conclusion that economic stress led people to buy *more* perishable food—particularly meat—and then throw more out when it spoiled before they could consume it.

Even for someone willing to put on rubber gloves, get the appropriate booster shots, and sort through trash cans in the hundred-degree heat of an Arizona summer, arriving at an exact measurement for household food waste is tricky. Some disappears into the kitchen sink garbage disposal and some is thrown away at school or work. Rathje's estimate—that 15 percent of the food Americans purchase ends up in our trash cans rather than our guts—is relatively conservative; the US government's assessment of the scale of the problem is substantially bleaker. According to its data, Americans send more than half a pound of food straight to the landfill every single day of the year and, once retail waste is included, squander more than 30 percent of our total food supply.

The issue is not just that we can't possibly eat everything stored in our gargantuan refrigerators—we can't even see it. Who hasn't purchased a fresh block of cheese or another lemon because they didn't spot the one already lurking in their deli drawer or crisper? "When we talk to consumers, they will always say that they want help organizing," Mary Kay Bolger told me. One idea Bolger tested in consumer research was building in a lazy Susan on one of the shelves; consumers said they liked the concept—but not as much as they disliked the idea that they'd lose usable space at the corners. As it happens, in the 1940s General Electric did introduce a lazy Susan, which was quickly retired when it emerged that children had a tendency to spin it so quickly that jars and bottles would fly off in all directions. Bolger laughed when I told her this story. "We played around with a pull-out pantry-style unit, so you could have easy access on both sides and you wouldn't have to move anything around inside your fridge," she said. "People were receptive, but right away they worried about what happens if the kids start hanging off it."

Recently the internet-connected "smart fridge" has been touted as a solution to the food-waste issue, capable of keeping track of our fridge contents and alerting us as items approach their expiration. "Nobody's really figured this one out yet," sighed Justin Reinke. Most people simply aren't willing to take the time to scan or log everything they put in their refrigerator, and, Reinke told me, supermarkets have thus far proven unwilling to share purchase data with fridge manufacturers—after all, if their customers end up buying too much and throwing some away, that isn't hurting Walmart's bottom line. "Even if they did hook that up, how does the checkout know that half that product isn't going to my dad's house because I bought his groceries?" said Bolger.

"People just felt it was the next logical step for innovation," Bolger added. She has strong feelings about the stupidity of smart fridges. "The consumer never really said that that's what they wanted. They want help keeping food fresher longer."

I n the United States, the initial feeling that refrigerated food wasn't truly fresh gradually faded. As Polly Pennington pointed out in 1939, it had taken them more than a quarter century, but American consumers had finally embraced refrigeration as a guarantor of quality—a shift that came about thanks in no small part to her efforts. Today the association between fridges and freshness is so strong that sturdier shelf-stable food is often perceived as less healthy, simply because it doesn't need to be kept cold.

The existence of this perceptual bias did not escape Steve Demos, a Boulder-based hippie who became the millionaire founder of White-Wave. When he launched Silk soy milk in 1977, he made the calculated decision to pay supermarkets a substantial premium to display it in the refrigerated section of the grocery store. At the time, soy milk, which has an unrefrigerated shelf life of at least a year, was sold in squat, rectangular packages on the center-aisle shelves, and soy was rated by Americans as the second-most-hated food in the nation, just below liver. Demos's gamble was that, by packaging soy milk as if it shared the fragile freshness of milk, he could convince Americans to make the switch—and it paid off.

Freshness—a quality that used to be bounded by constraints of time, seasonality, and geography—was increasingly determined by association: fresh foods are foods that require refrigeration; refrigerated foods are thus, by definition, fresh. And yet, with the traditional parameters of perishability disrupted, an uncertainty persists. The sell-by or best-before date printed on some food packaging is perhaps the most obvious response to consumer concerns over how to know whether refrigerated foods are fresh. Whereas a simple sniff or squeeze might have sufficed in an era when most people knew where their food came from and, likely, the person who produced it, for consumers at the end of cold-extended food chains, freshness is a belief system. They crave the guarantee of a printed date.

Despite the evident charms of the urban myth crediting Al Capone with inventing the sell-by date, the very first label with a printed freshness date seems to have been introduced by a San Francisco beer company in the 1930s.* Shelf-life indicators didn't become widespread, however, until the 1970s. Each supermarket and food producer had its own system to keep track of how old its product was and to help manage inventory, creating a cacophony of printed dates on packaging. Dozens of alphanumeric codes representing "pack date," "pull date," and "sell-by date" were in use, and they all meant something subtly different for different

---

*There are multiple versions of this yarn, the most common of which has either Capone or his brother, Ralph, moved to lobby for date labeling on milk in order to protect Chicago's children following the illness and/or death of an unnamed friend or family member after drinking spoiled milk. The more cynical version points out that the Capone family had the equipment to print such labels at their own dairy-bottling operation and thus could have benefited from both the expanded business and the reduced liability for sickness among their own customers. Most of these stories seem to have either originated with or been disseminated by park rangers at Alcatraz, where Mr. Capone, also known as Public Enemy Number 1, spent time following his conviction for tax evasion. Sadly, there's zero evidence that anything the Capone family did or didn't do led to the introduction of milk-expiration labels. Still, Ralph's granddaughter, Deirdre Capone, swears the story is true and that her grandfather's efforts earned him the nickname "Bottles."

foods in different locations. None of them offered reliable guidance to consumers, who were, apparently, desperate to know how fresh the items in their refrigerator really were. When New York State's Consumer Protection Board spent nearly a year gathering and deciphering major manufacturer and retailer inventory-management systems, publishing the results in a free booklet titled "Blind Dates: How to Break the Codes on the Foods You Buy" in 1977, the board's director told newspapers she received more than fourteen thousand requests for a copy in the first week.

While government committees and commissions pondered the matter, supermarket chains instituted their own voluntary labeling schemes—in the process, sidestepping the minefield of clearly stating how old the "fresh" food on their shelves was in favor of their own estimate of its future longevity. In 1972, the British supermarket Marks & Spencer hired supermodel Twiggy to launch a marketing campaign that repositioned its own internal warehouse sell-by dating system as a promise that its foods were fresh. Giant Food, a big chain in the mid-Atlantic, and Ralphs supermarkets in California also introduced sell-by dates in the 1970s. Today the sell-by date is still not mandated by law in the UK or at the federal level in the United States (except on baby formula), although twenty states plus the District of Columbia have rules on dating food—many of which seem capricious, if not downright ridiculous, such as New Hampshire's requirement that cream carry a sell-by date, but not milk; or Montana's insistence that milk cannot be sold more than twelve days after pasteurization, even though it would be fine for at least another week in the rest of the country.

The result has been widespread confusion. In a 2020 study evaluating the reasoning (or lack thereof) behind these labels, University of Maryland researcher Debasmita Patra found at least fifty different types of date labels in regular use. As one FDA spokesman put it, in a delightful semiotic koan, "the dates mean so many different things, they end up not meaning a thing." The final best-by or sell-by date is little more than an educated guess at how long a food can be kept under standard conditions before it starts to taste noticeably worse, with a little padding built in.

Evidence that food date labels help keep consumers safe—or even whether shoppers in the twenty states with rules suffer from food poisoning less often than those in the other thirty—is in short supply. Meanwhile, an ever-increasing body of research has found that these labels, and the resulting doubt as to an expired food's freshness, cause people to throw away a great deal of perfectly safe and edible food. In the United States alone, according to the FDA, 26 billion pounds, or more than \$32 billion worth, of food is discarded uneaten each year simply because the date on its label has passed. In an unfortunate irony, an appliance designed to preserve food has led to an enormous amount of waste.

Designer Jihyun Ryou describes her mission as to "save food from the fridge." It began with an observation: upon the advent of the mechanical refrigerator, not only did once-common food-storage and preservation spaces like the root cellar or pantry vanish, but so, too, did our appreciation of food as fellow organic matter—an awareness that each carrot or egg is also living tissue, with its own metabolic processes and peculiarities. Seen from this perspective, although it might be a convenient one-stop solution for us, for much of the food we eat, the fridge does not necessarily represent an upgrade.

The diversity of household food-storage options in the past allowed our predecessors to keep different foods in the environment that best fit each item's unique needs. Soft cheeses like Brie or Camembert prefer conditions that are more humid and warmer than the standard fridge; a marble slab in a cupboard beneath the stairs was ideal for them. The cool, dark damp of a root cellar suited the storage requirements of potatoes perfectly; in the refrigerator, they tend to darken and become unpleasantly sweet, while exposure to daylight at room temperature causes them to turn green and sprout. Perishable foods don't necessarily play well stored together: the ethylene emitted by apples causes bell peppers and cucumbers to soften and rot, while milk and eggs can absorb the aromatic chemicals emitted by nearby cabbages or mangoes.

Food producers know this perfectly well. Think of the vast controlled-atmosphere apple warehouses of Washington State, or Master Purveyors' meat locker in the Bronx, with its carefully regulated temperature and humidity and its battery of fans to manage airflow. Fruit and vegetable wholesalers like Gabriela D'Arrigo wouldn't dream of storing avocados under the same conditions as berries. Postharvest physiologist Natalia Falagán told me she shudders whenever she sees a peach in a home refrigerator, where the temperature is smack in the middle of what she called "the stone fruit–killing zone."

Pondering how to make this understanding of fresh food—how we should treat it and how long it should last—common knowledge, rather than the preserve of experts, led Jihyun Ryou to her solution: a set of ingenious wall-mounted and countertop units that draw on traditional, prerefrigeration food-storage techniques. Fruits, vegetables, and eggs—all would be freed from our monolithic fridges and instead distributed in a series of carefully tailored environmental niches around the kitchen. After all, beer and ice cream need to be cold, but produce doesn't. It just has to be preserved.

Ryou, who is currently at work on the perfect housing for tomatoes, explained that she bases her designs on scientific research but also on "conversations with farmers and grannies." Her root vegetable unit is a U-shaped shelf made of beeswax-treated maple. A glass panel holds the damp sand in which carrots and leeks are buried, alongside a little funnel to top up moisture levels as needed. It almost looks as though the vegetables are still in the ground, ready to be harvested, with just the orange tops and feathery green fronds of the carrot visible above the sand's surface—and that is precisely the point: Ryou has found that root vegetables last longer and taste better stored upright in slightly damp, loose sand, because it mimics their growing conditions.

Other units include a beautiful marble dish carved into a circular step well, in which cabbages and Romanesco broccoli can sit with just their stems in a thin layer of cool water, almost like a birdbath for brassicas. An enclosed potato drawer cleverly vents to an apple-storage shelf above, taking advantage of the fact that the ethylene emitted by apples inhibits sprouting in potatoes. Rather than throwing everything in the crisper drawer of their refrigerator, a shopper returning home to their Ryou-designed kitchen would simply slot carrots into their sandy shelf unit and place bell peppers on another, specially humidified, shelf, put their apples away on top of the potato drawer, and sit their cauliflower on its marble throne.

Ryou's wall-mounted egg unit, complete with freshness-testing device— a glass of water, to see whether the egg sinks, which means it's still good, or floats, which means it's not—would require a shift in food law, not just household practices, at least in the United States. The issue begins about an hour before an egg is laid, when a hen's shell gland squirts on a protective coating made of protein, lipids, phosphorus, and more. Up until that point, the thousands of pores in the eggshell remain open to allow oxygen and carbon dioxide to pass back and forth to the developing embryo (assuming, as the hen does, that the egg might be fertilized). This final layer, which chicken people call either "the bloom" or "the cuticle," blocks those pores to protect the egg from bacteria it might encounter once it exits its mother's body. In the US, as well as Japan, Australia, and a few other countries, egg producers wash freshly laid eggs in soap and hot water, which gets rid of bacteria but also removes the protective cuticle, so that the egg has to be refrigerated. In the rest of the world, where the cuticle is left intact and flocks are vaccinated against salmonella, eggs are sold unrefrigerated, in cartons stacked alongside such shelf-stable foods as sugar and flour.

Proponents of the American system point out that refrigeration gives the eggs a longer shelf life: the cuticle will keep unrefrigerated eggs good for about three weeks, but washed, refrigerated eggs last up to 105 days.

Critics point out that the energy used to wash the eggs and then keep them refrigerated all the way along the supply chain is an unnecessary waste, given that most eggs get eaten within that twenty-one-day window and that salmonella vaccination, which the FDA has chosen not to make mandatory, is much more effective at reducing outbreaks than washing and refrigeration. Both supply chains produce "fresh" eggs, but through alternate—even opposed—means. Ryou's egg-storage unit not only puts the ability to determine freshness back in the hands of the consumer but also highlights the existence of other ways to define and maintain this elusive attribute.

Leaving aside the potential food-preservation benefits and possible energy savings, perhaps the most important difference between Ryou's food shelves and the fridge is that her exquisitely designed wall-mounted and countertop units would force us to look at our food. The result of this daily confrontation, she hopes, is that we would eat more healthily, waste less, and—intangible but important—rebuild our relationship with these equally biological and perishable, if slightly less animate, fellow organisms.

When we toss food into the fridge, Ryou said, "we hand over the responsibility of taking care of food to the technology." Adopting Ryou's designs wouldn't eliminate the need for a fridge—but they could reduce the need for such a large one, or for multiple fridges. And the new connection to and respect for the fragility of food that they foster would likely also extend to the contents of our refrigerators.

Fridge designers aspire to make their appliances more hospitable to food too. Mary Kay Bolger said that Maytag developed additional drawers that tweaked the atmosphere to suit leafy produce and stone fruit. "We did a lot of tests, and it worked beautifully," she said. "But that technology added three hundred dollars to the cost of the fridge, and people were like, *I'll just buy a new head of lettuce.*"

Justin Reinke told me that some manufacturers had introduced special filter cartridges designed to suck up the ethylene produced by apples, avocados, and mangoes before it turned green beans yellow or wilted lettuce. "As businesspeople, we like recurring revenue," he pointed out. "We'd be more than happy to keep selling you cartridges, but the business never really took off." The problem is something he's seen again and again in market research: "The food-preservation stories are sometimes hard ones for consumers to understand." He and his colleagues have found that Americans don't understand the science behind freshness, and they don't appreciate the potential for shelf-life extension.

"All you really want is for things to be in the condition that you want them in, when you want them," said Reinke. When he asked consumers how they felt about fridge technology that might, say, allow strawberries to stay good for sixty days, their response wasn't necessarily enthusiastic. "A lot of times people would say, 'Sixty days? That's gross! I don't want to keep something that long.'"

Undaunted, Reinke and his colleagues at Beko continue to try to solve freshness from within the fridge, installing hospital-grade filters that remove any fungal spores from the air and engineering ever more precise temperature control. Their latest technology uses LED lighting on a timer to simulate the sun's daily cycle in the crisper drawer. In a Beko HarvestFresh fridge, vegetables will sleep in the dark for twelve hours every night, before waking up to a blast of blue morning light, followed by a midday break for a couple of hours of green light and then a bath of red light through the afternoon and evening.

Whether or not this system works—postharvest specialist Natalia Falagán has her doubts—it is not being marketed as a way to keep food fresh or prevent waste but rather, in a return to 1920s-era rhetoric around protective food, as a way to maintain nutrient levels. "That's what we have to tell them with HarvestFresh," said Reinke. "It saves your vitamin A and vitamin C." Even this health-focused message appeals only to a limited audience, Reinke admitted. "It would be a different story if I was at Whirlpool and I had a gigantic market share and had to sell to a lot of people," he said. "For us, because we're still relatively new to the US, if we can find the people who care about this stuff, that's enough."

In the end, the heavily hyped and equally maligned smart fridge might provide the zero-effort solution that most consumers are looking for: built-in spoilage detection, so that your refrigerator can finally answer the eternal question of whether any given food is still good to eat. Amazon recently filed two related patent applications for image- and scent-based systems that use fridge-mounted cameras and a variety of sensors to detect surface blemishes, changes in fruit and vegetable weight, and volatile chemicals produced by overripe produce or spoiled meat. If the data suggest that a particular food item is on the turn, it will notify the fridge owner accordingly. Target is training scanner systems in its warehouses to build a data set that could ultimately power consumer-level freshness detectors, and Grundig, a German brand owned by Beko's parent company, has already incorporated an odor-based traffic-light system in some of its fridge drawers, which flash red when meat and fish have gone bad or green if they can still be eaten, regardless of what date their labels might say.

Little more than a century ago, refrigeration disrupted traditional assurances of freshness, casting our understanding of perishable food into doubt. Today, in lieu of our reestablishing that knowledge ourselves, the smart fridge seems poised to step in with an answer: a technological solution to a problem caused by technology.

III. The Taste of Cold

In 2010, open-data activist Waldo Jaquith decided to make a cheeseburger from scratch. He and his wife had just built their own off-grid home in rural Virginia, and as soon as they got settled in, they began raising chickens and planting an extensive vegetable garden. Flush with an all-conquering sense of self-sufficiency, he outlined the steps required: bake buns, mince beef, harvest lettuce, tomatoes, and onion, and make cheese. Then he realized that he wasn't being nearly ambitious enough.

To really make a cheeseburger from scratch, he would also need to plant, harvest, and grind his own wheat, grow his own mustard plant, and raise at least two cows—one pregnant, for the milk and the rennet to turn that milk into cheese, and another that could be slaughtered for its meat.

At this point, Jaquith gave up. It wasn't the prospect of milling wheat or slaughtering cattle that put an end to his homemade-cheeseburger plan—it was a scheduling conundrum. His tomatoes were in season in late summer, his lettuce ready to harvest in spring and fall. According to the prerefrigeration agricultural calendar he was trying to follow, Jaquith would have needed to make the cheese in springtime, after his dairy cow had given birth: her calf could be slaughtered for the rennet, and the milk intended to feed it repurposed. Meanwhile, the steer would have traditionally been slaughtered in the autumn, as soon as it started to get cold. If he turned the tomatoes into longer-lasting ketchup and aged his cheese in a cellar for six months until the meat, lettuce, and wheat bun were ready, he could maybe, possibly, make a cheeseburger from scratch. But practically speaking, he concluded, "the cheeseburger couldn't have existed until nearly a century ago." (As, indeed, it did not. Though there are a handful of competing origin stories, they all date to the 1920s and '30s.)

Intended as a celebration of local ingredients and food independence, Jaquith's thought experiment had inadvertently revealed that a cheeseburger is so logistically difficult as to be nearly impossible outside of a highly industrialized and, above all, refrigerated food system. But cheeseburgers are not the only gastronomic delight that manufactured cold has added to our culinary repertoire. No exploration of how refrigeration has reshaped the way we eat would be complete without a look at how we first fell in love with it, seduced by the sensory pleasures of cold: the delightful anticipation of pouring a crisp, ice-cold beer at the end of the day, the refreshing clink of ice cubes in a soft drink or cocktail, and, of course, the sheer joy of licking an ice cream cone in summer.

As we've seen, brewers such as Fred Pabst and Adolphus Busch were among the first to invest in natural-ice storage, followed by mechanical refrigeration. Without it, American-style lager beer was impossible to make year-round or at scale. In California, without access to these expensive new refrigerating machines, resourceful nineteenth-century brewers found a strain of lager yeast capable of fermenting at slightly higher temperatures; they also cooled the hot wort by pumping it into shallow pans outdoors, creating clouds of condensing vapor that may have led to the resulting beverage's name: steam beer. (A similar style of beer has apparently emerged in North Korea, also due to a lack of refrigeration. According to aficionados, it tastes something like the San Francisco–based Anchor Steam.)

Elsewhere in beverage-based restoratives, the introduction of affordable, year-round supplies of ice led directly to the invention of refreshing juleps, cobblers, swizzles, and rickeys. Historian of alcohol David Wondrich has traced both the name *cocktail* and the custom of drinking a blend of spirits, bitters, and sugar back to Britain—but it wasn't until the mixed drink met abundant American ice in the late nineteenth century that the art of mixology was born. Similarly, ancient Chinese, Romans, and Persians all mixed snow or ice with fruit juice or dairy products to make early chilled desserts and sharbat, or sherbet, but it wasn't until the mid-1800s that ice cream became popular outside elite circles.

The opportunity to consume frosty drinks and desserts opened up an entirely new vocabulary of oral sensation. Some found the cold shocking at first. "Lord! How I have seen the people splutter when they've tasted them for the first time," recalled a London ice cream vendor in 1851. One customer—"a young Irish fellow"—took a spoonful, stood stock-still like a statue, and then "roared out, 'Jasus! I am kilt. The coald shivers is on to me.'"*

---

*Cold beverages can kill: take the unfortunate Thomas Thetcher, who died aged just twenty-six in 1764. His headstone, in the corner of Winchester Cathedral's graveyard, reads: "Here sleeps in peace a Hampshire Grenadier, Who caught his death by drinking cold small Beer, Soldiers be wise from his untimely fall, And when ye're hot drink Strong or none at all." The medical literature on the topic is slim, but there are a handful of recorded cases of both sudden loss of consciousness and cardiac arrest following the ingestion of a cold beverage.

The earliest recorded description of brain freeze seems to have been published by Patrick Brydone, a Scotsman traveling in Sicily in the 1770s. The poor victim was a British naval officer who got a nasty shock after taking a big bite of ice cream at a formal dinner. "At first he only looked grave, and blew up his cheeks to give it more room," wrote Brydone. "The violence of the cold soon getting the better of his patience, he began to tumble it about from side to side in his mouth, his eyes rushing out of water." Shortly thereafter, he spat it out "with a horrid oath" and, in his outrage, had to be restrained from beating the nearest servant.

Scientists are still somewhat at a loss to explain the cause of the ice cream headache, to which not everyone is susceptible, but the leading theory is that the sudden, blinding pain is caused by a rush of blood to the head and the resulting pressure of brain on skull. A few years ago, in an attempt to understand some of the mechanisms behind blast-related headaches in soldiers, researchers in the Veterans Affairs New Jersey Health Care System induced brain freeze in thirteen subjects while monitoring their blood flow. The onset of pain coincided with the rapid expansion of a major artery, pumping extra blood upward in a panicked attempt to keep the brain warm; as soon as the artery contracted, the headache passed.

Shock and cranial constriction aside, the taste of cold is not without its pleasures. Although consuming chilled food and drinks makes little to no difference to one's body temperature, it is remarkably refreshing. Again, scientists aren't sure exactly why, but it seems as though when the temperature receptors in our mouths feel cold, they tell the brain that our thirst has been quenched. The body has other ways to monitor hydration levels—primarily by checking how concentrated or dilute our blood is—but, the theory goes, the cooling sensation caused by water evaporating from the warm tongue is an early alert that liquid has been ingested. (By way of proof, one study found that water-deprived rats,

---

Curiously, the founder of Alcoholics Anonymous saw the inscription on Thetcher's headstone as an "ominous warning," quoting it in his 1939 book, and the grave has become a minor site of pilgrimage for those in the recovery program.

mice, guinea pigs, and hamsters would all lick a cold metal tube repeatedly, instead of a hot or room-temperature one—presumably because the cooling sensation triggered an illusory sense of quenched thirst.)

Cold may also have made food and drinks sweeter—particularly in the ice-obsessed United States. At least three of our basic taste receptors—sweet, bitter, and umami, or savory—are extremely temperature sensitive. When food or drinks cool the tongue to below fifty-nine degrees, the channels through which these three taste receptors message the brain seem to close up, and the resulting signal is extremely weak. This is why a warm Coca-Cola or a melted ice cream is so sickly sweet: because they're intended to be consumed cold, they *have* to contain too much sugar to boost the signal and register in our brains as tasting sweet at all.* In 1929, the president of Coca-Cola set up the Fountain Training School to ensure the drink was being prepared and served properly: salesmen were told, "It's gotta be cold if it's gonna be sold." Washing down your food with ice-cold water or a soft drink, as Americans often do, will have the same effect—which may explain why extra sugar finds its way into so many savory packaged foods, from hamburger buns to pizza sauce to salad dressing, in the United States. Everything simply has to be a little sweeter to taste right if your tongue is cold.

For fridge manufacturers, *Cooking with Cold* (the title of a recipe pamphlet distributed by Kelvinator in 1932) meant three things: make-ahead foods for the servantless hostess, eliminating any last-minute preparation so that the lady of the house could spend time with her guests; Jell-O, aspic, and other gelatinous concoctions, including dozens of mayo-laden molded "salads"; and leftovers. Many of the most peculiar recipes of the 1920s and '30s—"Peanut Butter Salad" stands out for its combination of green peppers, celery, whipped cream, and the titular ingredient—are perhaps best understood as status signifiers. In other

---

*This is also why warm beer tastes more bitter. (Which some people enjoy: in the UK, traditional ales—"a pint of bitter"—are not served chilled.) No one is yet sure how salt and sour tastes are affected because the receptors for those tastes use a different signaling channel, but there's some evidence showing that cold makes food taste saltier.

words, jellied foods were popular less because they were especially delicious and more because they demonstrated refrigerator ownership.

Several of these kinds of recipes appeared in *Cooking with Cold's* "Refrigerator Remnants" section, alongside the promise that with "a little bit of this and a bit of that . . . the left-over foods disappear, and are replaced by delightful combinations, well blended by the Kelvinator cold." According to food historian Helen Veit, the term *left-over* was first coined in the early years of the twentieth century; before then, perishable foods had to be eaten too quickly to become their own culinary category. Dinner scraps would be fed to animals, warmed up for breakfast, or added to a simmering stockpot. Refrigeration changed that, but it created a new challenge for cooks. Now that last night's meal could be stored in a recognizable format till the next evening, the pressure was on to liven things up, repurposing dishes as an ingredient in something else to avoid repetition. As home fridges became more common from the 1930s through the 1950s, they were accompanied by an endless stream of advice on how to transform leftovers into new dishes—often by gluing together cheese, fish, and a random vegetable or two with some mayonnaise and gelatin.

Leaving aside its suggestion of "Molded Lamb with Fruit," Kelvinator was not lying when it said refrigeration could make leftovers taste better—at least in some cases. After all, chemical reactions continue in the cold, albeit slowly, and some of them improve flavor. *Cook's Illustrated* investigated this question a few years ago by serving fresh bowls of beef chili, as well as French onion, creamy tomato, and black bean soups, alongside portions that had been made two days earlier. Testers preferred the fridge-aged versions, describing them as "sweeter," "more robust-tasting," and "well-rounded."

By way of explanation, the magazine's science editor pointed to the fact that while the soups and stew were sitting in the fridge, the lactose in dairy would have time to break down into glucose, as would some of the carbohydrates in onion, while the protein in meat would separate out into amino acids such as glutamate, which boosts savory flavor notes. Other researchers note that the collagen in meat, which is released during the first cooking, sets to a jelly in the fridge, then melts to create a silky texture when a stew, lasagna, or chili is reheated. Heavily spiced dishes, like curries, fare well in the fridge, because the flavor molecules in many spices are soluble in fat, and the more time they have to disperse, the more evenly they'll be distributed through the dish, creating a well-balanced whole. Water in a dish also tends to soak into starch over time, bringing flavor with it, which means that the black beans in Cook's Illustrated's refrigerated soup would have had more of a chance to absorb the umami flavor notes in the broth. That said, for all reheated lasagna's sensory delights, not all of the taste transformations wrought by refrigeration are as welcome.

Refrigeration was the first food-preservation method that didn't utterly transform food in order to extend its life. A raisin is never going to be mistaken for a fresh grape, but to many, a refrigerated grape *is* a fresh grape. In reality, however, the taste of a refrigerated grape sits firmly in the uncanny valley of freshness: not different enough to be something else, but definitely not the same as eating a grape straight off the vine.

Today so few people in the industrialized world have experienced the latter that the gap has largely ceased to matter. When mechanical refrigeration was new, that was not the case. The fabulous Chicago cold-storage banquet of 1911 was quickly followed by disapproving editorials, including one that bitterly predicted the eventual triumph of the taste of cold. "There seems to be only one consolation," concluded the Chicago Inter Ocean in its report on the meal: those Americans who remember what food tasted like before refrigeration will eventually die off, while "the generation growing up doesn't know the difference and may be happy in its ignorance."

This prediction has come true. Few of us today can taste the difference between wet-aged and dry-aged beef; even fewer would know to miss the taste of different pastures and seasons in milk fresh from the cow. In communities that have not acquired the taste of refrigeration, however, preferences for traditionally stored or freshly harvested food often remain strong. Among many Alaskan Natives, whale and walrus meat tastes as it should only if kept in naturally cooled underground ice cellars. As the permafrost warms, meat stored in the walk-in freezers that are being imported as substitutes for failing cellars just isn't the same. "There's nothing that tastes better than ice cellar food," one resident of Kaktovik, a tiny village in Alaska's far north, told the Canadian Broadcasting Corporation. After eating beef stored in a traditional yukimuro, or snow cellar, in Niigata, Japan, I would agree: something about spending a few weeks in an insulated room alongside a gigantic pile of snow made the resulting steak as deeply savory and tender as dry-aged beef but even more mellow and a little sweeter.

Meanwhile, in China, frozen dumplings may have reached a certain level of acceptance, at least among the younger generation, but a fish is not fresh unless it was purchased while still alive.* On the third floor of a university building in the otherwise unremarkable city of Jinan, between the Time Temperature Tolerance Lab and the Small-Size Instruments Room, I visited the Room of the Sleeping Fish: four rows of five blue plastic barrels, each filled with flat, gray turbot and hooked up to a system of brine pipes that gradually cool the water in each barrel to just above freezing. A small, bespectacled man in a white coat told me, very quietly, how he uses refrigeration to send fish to sleep. Gently lifting a corner of the netting on the nearest barrel, I looked down on a lone sleeping turbot. It lay unblinking and motionless but for a slow and hypnotic rotation around the bottom of the barrel with the flow of the chilled water. In this sluggish state, a fish can be rolled up, popped in a

---

*Even Walmart sells its fish live in China, from scoop-your-own tanks, relying on a ramshackle system of chilled, anesthetic-laced Styrofoam cubes that act as a sort of mobile ocean to transport seafood inland from the coast.

clear plastic poster tube, and mailed to anywhere in China. As long as it arrives at its destination within three days, the researcher explained, it will simply wake up and start swimming again as soon as you slide it out of the canister—"fresher than fresh," he claimed, as this extended nap apparently helps wind down the stress chemicals released when the fish is first caught. "We invented this method here," he said, fluttering his hands like a revived turbot's fins.

In the United States, where convenience and value are often prioritized over taste, and a refrigerated supply chain has prevailed for generations, hard, acidic supermarket peaches and starchy, cardboard-flavored peas are accepted, if not relished. But even the least flavor focused among us usually admit that one refrigerated fruit in particular is a pale shadow of its freshly harvested self: the tomato. Chef James Beard lamented the "tragic decline" that had turned "this most glorious of fruits" into "an almost total gastronomic loss"; critic Craig Claiborne of *The New York Times* did not mince words in his description of the shop-bought variety as "tasteless, hideous, and repulsive." "Hard," "plastic," "watery," "blah"—Harry Klee has heard all the descriptions of the supermarket tomato, and he agrees. "They have no flavor at all," he said. "They're awful." When I first spoke to Klee, he and I both, coincidentally, had just come in from tending our own backyard tomato plants. "I only grow our own varieties," he told me, which makes sense: Klee has spent much of his career trying to design a tomato that can survive our refrigerated supply chain and still come out tasting of something.

Klee started out as a psychology major, before falling in love with chemistry. "Thanks to a psychology class about drugs and human behavior," he explained. "In the seventies, of course, you took any class on drugs." He switched academic tracks and started studying biochemistry, which led him to molecular engineering, which led him to become the first person on his campus to clone a gene, back when the technology was still new.

At the time, most plant scientists thought that the reason American tomatoes tasted of nothing was that growers in Florida—where up to half of all the country's tomatoes are produced, including almost everything you'd find in an East Coast supermarket between October and June—pick them when they're hard and green and gas them with ethylene to make them ripen and turn red. As with bananas, this has made it possible to ship an otherwise soft and squishy fruit all over the country without enormous losses. And as with Golden Delicious apples, it has meant that growers harvest the fruit too early, before it has even developed the potential for flavor.

Klee wondered whether it might be possible to use genetic engineering to slow down tomato ripening so that the fruit would stay green on the vine for longer, accumulating the nutrients that become the building blocks of flavor molecules. "You'd have a bigger window that you could let them ripen on the plant and still ship them without losing them," Klee explained. "It would be a win-win for everybody."

Klee wasn't the only one trying to create a GMO tomato that could hold up under the rigors of refrigerated shipping. In the 1980s, scientists at Calgene were tweaking a gene that regulates pectin metabolism in an attempt to slow the softening process in ripe tomatoes. Their efforts resulted in the Flavr Savr tomato—the first genetically modified crop sold in the US. It wasn't a success: the variety had low yields, and the resulting fruit was still slightly too squishy to survive industrial handling. But the bigger issue, Klee realized, was that while he and the Calgene scientists had been focused on ripening, they'd missed the fact that commercial tomato varieties had almost no flavor even when they were allowed to redden on the vine.

"The commercial tomatoes have been selected for being very hard, for having great shippability, great shelf life," said Klee. "They don't have the genetic capacity to have great flavor." In 1977, when *The New Yorker* sent a man to Florida to investigate the problem of tasteless tomatoes, the state's growers were excited about an experimental new variety, the MH-1—but not for gastronomic reasons. "You could take an MH-1, stand twenty paces away from another man, and play catch with that tomato without hurting it," the head of the Florida Tomato Committee boasted.

At the start of the new millennium, Klee moved from Monsanto to Florida State University, and from his efforts to slow ripening to trying instead to untangle the chemistry of tomato flavor. To understand which molecules mattered, he and his colleagues grew hundreds of different tomato varieties and measured exactly how much of some seven hundred different chemicals—acids, sugars, and an array of volatile organic compounds in varying combinations—they contained. Then he took more than 150 tomatoes—the ones with the largest variation in flavor chemistry—and gave them to consumer panels to find out which ones tasted the best.

The process took years, but by matching the tomatoes that scored highest in taste tests to their molecular makeup, Klee was able to narrow in on the twenty-five volatiles, plus some sugars and acids, that make up the recipe for the Platonic tomato—or tomatoes. "There are multiple answers to what's the perfect tomato," Klee told me. He's found that, in general, younger people respond to the sugar content, while older consumers and women like their tomatoes more complex, with a greater array of volatiles. Curiously, the most-loved flavor chemicals are all derived from vital nutrients, including essential fatty acids, amino acids, and antioxidants—which implies that the best-tasting tomatoes may also be those that are best for human health.

He also found that breeders needn't have sacrificed flavor in their quest for a more robust tomato—the two qualities are not mutually exclusive. Flavor loss was an accidental casualty of a market that valued a cheap, round, red tomato that was available all year round over one that tasted of something. "I still remember when I was at a tomato conference in California, and one of the biggest growers stood up in front of this audience and said, 'I've never lost a sale to a great-tasting tomato,'" Klee told me. "Growers are paid by the pound and by the size of the fruit," he continued. "They don't get one penny more or less if they produce a tomato that tastes great or tastes lousy."

For the past four years, Klee has been working on breeding a tomato that combines the seven most important flavor genes back into a tomato that has the same yield, disease resistance, and ability to be shipped under refrigeration around the country as a commercial variety. "It's like putting a jigsaw puzzle together," he said. Using only traditional breeding techniques, he crosses a modern commercial tomato with varieties possessing good-flavor genes, then checks to see whether their progeny end up with the best genes from both parents. "Then we repeat the process again and again and again," said Klee.

At the end of August 2022, more than twenty years after he began working on the case of the tasteless tomato, Klee told me he and his team had finally cracked it. "Just last week, we got to the point where we had something that had all seven of the genes we wanted in," he said. The consumer panels loved the flavor, the yield was great, but Klee worried the tomato was still a little too small for growers to embrace. "I think we're going to have to back-cross it a little more," he said. "The problem is that sugar is a direct trade-off with fruit size—the bigger the fruit, the less sugar it has, and vice versa."*

Klee, who is now semiretired, told me that his quest is far from over. Spending more than four days below fifty-five degrees alters a tomato's DNA in such a way that the fruit's ability to make volatile flavor chemicals completely shuts down. "It doesn't affect the sugars and the acids, only the volatiles," Klee said. "But the tomatoes taste worse, no question." Tomato packers and trucking companies can keep the tomatoes at fifty-five degrees or above, but home fridges are typically set at forty degrees, and supermarkets that only have one cold room will set it at thirty-four degrees to protect meat and dairy. "Ten years from now, we could probably understand the genetics of that response and figure out how to prevent it," Klee said. "It's probably beyond the scope of my career."

---

*This is true for fruit in general. Many shoppers deliberately choose the largest apples or strawberries, under the assumption that they provide better value, but they are, almost without fail, significantly less delicious than smaller ones.

Back in the field, the issue of when to harvest the tomatoes remains. Klee explained that tomatoes that are harvested at the "mature green" stage—full-size and ready to start ripening—will still develop the full complement of sugars, acids, and volatiles when they're exposed to ethylene. (In taste tests, consumers can often tell vine-ripened tomatoes from those that are picked at the mature green stage and gassed, but Klee told me it's a close call.) The problem is that it's tricky for pickers to judge maturity visually in the field, which means that up to 40 percent of tomatoes are harvested at the "immature green" stage—they aren't ready to ripen and thus they never will, even after turning red. In other words, scientists also need to develop a field test for maturity in order for commercial tomatoes to taste good.

Klee's approach—reengineering the tomato to taste good in our refrigerated food system, as opposed to redesigning the supply chain to accommodate the tomato—is expensive and time-consuming, but the results are promising. America's future might hold many more delicious salads and sandwiches thanks to his single-minded efforts over the past two decades. His process offers a road map to breeders working on fruits like mangoes and strawberries, whose flavor has been similarly diminished in favor of cold hardiness.

Klee's fruit are, of course, the tomato equivalent of Minute Maid and Tropicana orange juice: a carefully designed, scientifically standardized taste profile that salvages flavor from the wreckage of refrigeration. The logic of industrial orange juice, with its vast and complex infrastructure—chilled aseptic tanks full of different varieties and captured flavor molecules that are blended to match consumer preferences—has simply been internalized in the genetic machinery of these tomatoes. And like supermarket OJ, they offer a solution to a problem that need not exist. We could just limit our fresh-tomato consumption to the summer months, when local specimens, in all their juicy, savory, tangy deliciousness, are available.

"The reality is, most people do not want to eat like Alice Waters," Klee said, referring to the Bay Area chef known for her commitment to serving only local, seasonal produce. "No matter how bad the quality, people are still going to want to buy a tomato in January, and without refrigeration and postharvest handling, that's impossible." Rather than save food from the effects of the fridge, Klee wants us to have our refrigerated cake and enjoy it too. It's a more pragmatic approach than Ryou's refrigeration-liberation movement, and given human behavior, it's probably more likely to succeed.

Although I remain firmly in the Alice Waters camp when it comes to tomatoes, I wanted to taste the sturdy-but-still-delicious tomato of the future. Klee invited me to Florida that fall, but a fungal infection wiped out his experimental plots, so he sent me a packet of seeds instead, and I carefully sowed Klee-lab cherry, plum, and globe-shaped tomatoes alongside some of my favorite noncommercial varieties. Within days, they burst out of the potting mix, sending out shoots in all directions and perfuming the air with that delicious spicy, green scent unique to tomato leaves. In a few months, my diligent watering and applications of compost tea were amply rewarded with a profusion of fruit. They looked like supermarket tomatoes—bright-red Ping-Pong balls compared with the flame-orange, purple-and-yellow stripes, and bulbous ridges of my heirloom varieties—but they were so juicy, tart, savory, and flavorful that they never even made it into my fridge.

Tassos Stassopoulos, a London-based portfolio manager who focuses on investment opportunities in emerging markets, has been using fridges to predict the future for more than a decade. His insights depend on the fact that as a society refrigerates, it doesn't simply consume the same foods, swapping a Cox's Orange Pippin for a Cosmic Crunch, or Marmande tomatoes for MH-1. Instead, its entire diet changes—in ways Stassopoulos can both foresee and exploit.

By the time of his refrigeration revelation in 2009, Stassopoulos had already gained a reputation in the industry for his immersive, ethnographic research process; where other investors typically relied on Bloomberg data or forecasts from big consumer-products companies in order to deduce what people in India might start purchasing in the future, Stassopoulos spent days traveling around the country, asking people himself. He found the process fascinating and threw himself into it, visiting informal settlements and working-class neighborhoods to chat with people for hours—but he still wasn't getting the insights he wanted.

"The problem is that I was asking people, 'Okay, assume you get a salary increase. How will your diet change?' They'd all say, 'I wouldn't change anything,'" Stassopoulos explained. "But we know that as people get richer, their diets change."

One afternoon he was in the city of Aurangabad, a couple of hundred miles inland from Mumbai, interviewing a woman who had just told him the exact same thing. Her family was quite poor, and what little food she had in the house was very traditional—pulses, rice, and pickles. On a whim, Stassopoulos asked her if she'd mind taking him shopping. He gave her some rupees and followed her to the corner shop, where she bought Cadbury chocolate bars, Coca-Cola, and some packaged savory snacks—items that Stassopoulos had cataloged in the fridges and cupboards of people one socioeconomic class above her.

"I realized that the answer is the fridge!" he said. "The fridge could tell me how people would behave once they had some extra money—before they even know it themselves." While Jon Steinberg scopes out fridges to assess romantic compatibility and UCLA researchers analyze them to understand family life in the twenty-first century, Tassos Stassopoulos has figured out how to use refrigerator data to make money.

Stassopoulos started grouping his photographs of fridges by income, in order to see how their contents evolved. What emerged was a journey, starting with a poor family's acquisition of their first fridge. "For them, it's an efficiency device," said Stassopoulos. They use it to store either the ingredients to make traditional dishes or the leftovers from those dishes. Upon their ascent into the middle class, the fridge starts to include treats and international brands—soft drinks, beer, and ice cream. "You have some disposable income for the first time," said Stassopoulos. "You want to provide all these things that your family was previously deprived of, and you want to show off while doing it."

Once a family becomes truly affluent, their fridge will shift again. Where one brand of ice cream in the freezer was an indulgent treat for all the family, multiple brands of ice cream reveal that frozen desserts are now normal enough that individual family members can dislike each other's preferred flavors. "Before, it was just, *Yes, we can get ice cream*," he said. "Now it all becomes about *me*: I *like chocolate and* I *don't like strawberry*." Ingredients from different cultures as well as items marketed as healthy—fat-free, diet, or probiotic foods—also show up on refrigerator shelves at this income level, reflecting, in Stassopoulos's rubric, a desire for self-improvement and, beneath it, a transition toward individualistic, Western values.

The pinnacle of his pyramid is reached once a fridge contains foods that express collective virtue: fair-trade, organic, cruelty-free products in reusable packaging. "This is where the Nordics are," he said. "India is mostly in this efficiency stage, China is at the indulgence stage, and Brazil is already on the healthy stage."

Stassopoulos told me that a middle-class fridge in China is not much help in predicting the future contents of Indian fridges; instead, he looks to middle-class fridges in India and affluent fridges in China to see where the future of each country's consumption lies. Based on Indian fridgenomics, he decided to invest in dairy processors: companies that turn milk into butter, cheese, yogurt, and ice cream. He predicted that these were the items Indian families would add to their diets as their incomes increased—and recent data showing double-digit growth in sales of value-added dairy products, not to mention his above-benchmark returns, have proven him correct.

Fridge signals can also tell him where he needs to divest. "In the past, I used to invest in Yum China," he said, naming the parent company of KFC, Pizza Hut, and Taco Bell. "And when I was looking at fridges there in 2014, I started getting really worried." Buckets of KFC had been replaced by green curry and sushi: for upper middle-class households in China, fast food had already been replaced by international cuisine as a mark of sophistication, and the less affluent were bound to follow in their wake.

Beyond food, he's found that fridge acquisition is a reliable herald of growth in a country's insurance and private-tutoring markets. "With a fridge, women can work outside the home, and that's when they get a say in the household finances," he told me. "Women tend to think more long-term than men—they think about education for their kids, they want to be prepared for a rainy day."

This kind of research may pay off in the end, but it is certainly not fast or cheap. Stassopoulos estimates each study takes six months of planning, days of interviews and documentation on the ground, and weeks of analysis upon his return. The immersive nature of his process has also led to deeper investments in his sources: he has maintained WeChat conversations with many of his interviewees for years and funds philanthropic solutions for needs that investment-worthy companies have failed to address.

But when it comes to the dietary and lifestyle changes his analysis predicts and his investments then enable, his motivations are purely financial. His fridge-based research reveals the dietary shifts that accompany economic development—changes that surely have both environmental and health implications. Stassopoulos is a self-described health-conscious foodie and former vegan, but, he told me, he makes no judgment on the changes in consumption that a fridge both enables and reflects. "If they want to eat it, I don't have a problem with it," he said. "The important thing is, where is the trend going?"

IV. *The Fridge Diet*

Jelena Bekvalac's skeletons are stored beneath a roundabout in central London, just steps from the former site of Smithfield Market, whose blood and gore so revolted Pip in *Great Expectations*. Long dead, their bones nonetheless hold at least the beginning of the answer to one of today's most important questions: Has refrigeration made us healthier?

In both historical and contemporary commentary, received wisdom holds that whatever economic, environmental, and even culinary costs have accompanied the spread of mechanical cooling, it has at least been a blessing in terms of nutrition and human health. That assumption, although widespread, turns out to be based on surprisingly thin evidence.

First of all, it rests on another underlying but largely unproven argument: that, thanks to the fruits of human progress, we live healthier lives than our forebears. This *seems* logical, but it hasn't been easy to verify, in part because arriving at an accurate measure of public health before the dawn of modern medicine is challenging. Birth and death rates, where they exist, can reveal changes in a population's average life span, but they don't provide much of a guide as to whether those individuals' time on Earth was enjoyed in good health or blighted by disease. Meanwhile, not only was recordkeeping spotty, but today's maladies—heart disease, cancer, obesity, type 2 diabetes, and autoimmune diseases—rarely match the descriptions used on historical death certificates, making it hard to understand what exactly ailed our ancestors.

In 2015, Bekvalac realized that the scope of her skeleton collection might allow her to verify whether the optimistic story of development was actually true: whether the rapid technological progress of the past 250 years was a net win in terms of human health, or not. In the absence of a contemporary diagnosis, bones can reveal valuable clues to an individual's underlying health conditions. "For example, there's one gentleman, John Paul Rowe, who clearly had metastatic carcinoma, and the parish record just says 'decline' for the cause of death," she said. Rowe died in 1834, at the age of fifty-nine, and was buried at St. Bride's, in central London; his skeleton was disinterred when the church was destroyed during the Blitz.

For the past two decades, Bekvalac has been curator of human osteology at the Museum of London's Centre for Human Bioarchaeology. As we chatted over tea and biscuits in the terrace café, Bekvalac gestured out the window toward the nondescript garden that sits above her subterranean crypt. Deep in the bowels of the museum, in a dimly lit concrete basement, her bones are kept in numbered cardboard boxes on steel shelving, grouped according to the site where they were found. "Overall, we look after about twenty thousand individuals," said Bekvalac. "They're not all complete, but it's one of the largest collections in the world."

Bones typically make their way into the collection during the archaeological excavations that take place before development projects in the city, from an intact Roman sarcophagus unearthed beneath St. Pancras during the construction of the Channel Tunnel to medieval plague pits discovered during post–World War II reconstruction. They range in antiquity from a Neolithic skull dating back to 3600 BCE, found on the Thames foreshore, all the way to a cache of 274 individuals buried between the Great Fire of London in 1666 and the Burial Act of 1852, which closed the city's cemeteries and graveyards for good and thus marks the latest date for Bekvalac's skeletal acquisitions.

During that span, London grew from a marshy valley containing a few huts to the largest city the world had ever seen, complete with 2.3 million inhabitants, railways, factories, coal-burning fireplaces, and gas-powered lamps. Bekvalac and a colleague, Gaynor Western, decided to use digital X-rays and CT scans to analyze the skeletons of almost 2,300 individuals, just under 1,000 of whom died before 1760, when Britain's Industrial Revolution began, in an attempt to reveal how that transformation affected their health.

Although many people have studied Bekvalac's skeletons, nothing had been done on this scale before—in part, she said, because it wasn't technologically possible until recently. "Without digital radiography, we would have all been skeletons by the time we'd finished it," said Bekvalac. As it was, she and Western spent the best part of three years underground, imaging the crania, lumbar vertebrae, pelvises, left femurs, and the second metacarpals in the hands of up to thirty skeletons a day, then analyzing them one by one for the obscure traces that disease can leave on the bones.

Among the clues Bekvalac and Western were looking for were the dark brown streaks of new bone that build up on the ribs of individuals with chronic lung inflammation, as well as the raindrop-like holes and neoplastic lesions that are the skeletal tell for myeloma and metastatic cancer. "Prostate cancer tends to get this quite florid, fluffy bone buildup, for example," said Bekvalac. "Then we can see wear and tear on the joints, HFI in women, and DISH in men, all of which might be linked to obesity." *HFI* stands for hyperostosis frontalis interna, a condition linked to metabolic syndrome that causes a reef-like, undulating overgrowth of bone in the front of the skull; *DISH* is diffuse idiopathic skeletal hyperostosis, a form of arthritis linked to type 2 diabetes and an elevated body mass index, which hardens the tendons and ligaments around the spine into what Bekvalac and Warner describe as a "thick, dripping, wax-like" substance.

These kinds of clues left on the bone are evidence, not verdicts. If an X-ray of a woman's skull shows the characteristic bony lobules of HFI, Bekvalac and Western can't conclude with any certainty that she was overweight and suffered from a metabolic disease that led to her eventual demise—just that it's more likely than not. "She might have choked on a raisin," said Bekvalac. "We just don't know."

Nonetheless, Bekvalac and Western had a large enough sample size to pull out patterns with some confidence. Although economic status and location made a difference, the traces of chronic lung inflammation, cancer, and obesity were all more common in the recent skeletons than their preindustrial peers. Their conclusion was that "for the most part, the industrialization of the city has been a grotesque assault on the health of Londoners."

"There are all sorts of lovely variables and caveats," said Bekvalac. "But in the end, that's what the skeletons are telling us."

Bekvalac and Western's research confirms that, contrary to conventional belief, the benefits of modern civilization came with a substantial price tag. "It's called the antebellum puzzle," said Lee Craig, a professor of economics at North Carolina State University, whose research belongs to an obscure subgenre of the field called cliometrics: the application of econometric analysis to history.

Antebellum refers to the decades prior to the American Civil War, which began in 1861. The "puzzle," according to Craig, is that during these decades in the United States and Western Europe, "biological measures of the standard of living erode, even though the standard economic measures seem to be going up." In other words, just as Bekvalac's skeleton scans suggest, "progress"—in the form of urbanization and industrialization—seems to have made matters significantly worse.

The next question is, how is refrigeration in particular implicated in that decline, as opposed to the many other ways in which lifestyles changed as Britain and the United States mechanized and modernized? Unfortunately, teasing out that answer is even more ticklish than establishing that public health went downhill following the onset of industrialization. Refrigeration's adoption, and thus also the consequences of that implementation, took place over time, unevenly, and in tandem with dozens of other public health breakthroughs and calamities, from a rise in pollution to the construction of municipal water treatment and sewage plants, and from a decrease in physical activity to the introduction of vaccines and antibiotics. Nonetheless, Craig told me, he'd arrived at an answer: 0.02 inches.

C raig and his coauthors, Barry Goodwin and Thomas Grennes, are the first and only researchers to quantify the impact of mechanical refrigeration on human health, disentangled from all the other lifestyle changes that happened at the same time. Craig's interest in refrigeration emerged from the struggle to solve the antebellum puzzle: the American version of the unexpected industrial-era decline in health that Bekvalac and Western observed in Londoners' bones.

The antebellum puzzle itself first surfaced in 1979, when Robert Fogel, who went on to win a Nobel Prize for his research linking changes in human physiology to economic growth, noticed that the average American male seemed to have shrunk in height by as much as an inch in the fifteen years preceding the Civil War. This was odd, because economists have found that wealthier parents always have taller children, everything else being equal—yet military recruits born during the 1830s were clearly shorter than their peers born before 1820, even though average incomes had risen steadily during this time. Later work showed that the nadir, at least in terms of stature, seems to have been the cohort born in the 1880s—which happens to be when Gustavus Swift's refrigerated railcars began to transform urban meat supplies. After that, adult height slowly bounced back and was eventually reunited with GDP.

Cliometricians argued among themselves about the cause of this sudden dip and subsequent recovery for a few decades. Fogel concluded that the shrinkage was due to "a considerable decline in diet after 1840," which he speculated was caused by a delay in agricultural productivity catching up to population growth. Others have tied it to a rise in infectious disease as people crowded together in cities in an era before germ theory was fully understood.

When Craig dug into the data, he arrived at the conclusion that refrigeration deserves part of the credit for helping Americans bounce back from the antebellum slump. In fact, Craig calculated that by 1900 mechanical cooling allowed Americans to scrape back at least 0.02 inches in height, and likely more. Although this figure sounds underwhelming, Craig pointed out that it actually represents a not-so-insignificant 5 percent of the total average increase in height that decade. (The other 95 percent—0.38 inches—can likely be credited to improvements in public health and sanitation.) What's more, 0.02 inches is an average across the entire population, but Craig referred me to a stack of additional research showing that the poor would have seen substantially greater gains in height.

The ability to arrive at such a precise figure for a phenomenon that encompasses so many variables seems extraordinary. "That's the magic of being an economic historian," Craig acknowledged with a laugh. "You can learn how to tease a lot out of some probably fragile data." As he walked me through his calculations, Craig told me that the mathematical techniques he used were developed only a couple of decades ago. Wrangling the data from which to arrive at the final figures took a similar span of time. In one paper, he and a colleague used military-recruitment records and local agricultural surpluses to arrive at a statistical measure connecting the amount of protein consumed as a child with an individual's eventual adult height: twenty-two pounds per additional half inch. In another, he used government records to conclude that refrigeration increased butter consumption by 5.5 ounces per person per year. Finally, he drew on yet another colleague's estimate of the relationship between height and per-capita income.

Extrapolating from all of this—and excluding everything but the additional calories and protein from butter, cheese, milk, pork, and beef—Craig and his colleagues found that, during the 1890s, the increased consumption of perishable foods enabled by refrigeration delivered an extra 5,500 calories and 400 grams of protein to the average American every year. It was this nutritional boost that augmented the average adult's height by a whisker and boosted GDP by fifteen bucks per household—an increase whose benefits then compounded annually. Better nutrition means greater resistance to other diseases: underweight individuals are significantly more likely to die from tuberculosis, for example, which had become epidemic in Europe in the 1700s and 1800s.

"You know, refrigeration has been called one of the great inventions with good reason," said Craig. "That extra nutrition was an important part of turning around the decline."

Bekvalac's bones tell us that industrialization damaged human health; Craig's calculations reveal that refrigeration helped ameliorate that: both reveal that the timespan under examination matters. The individuals whose skeletons are stored at the Museum of London all died before 1852. By then, Edward Jenner had invented the smallpox vaccine (1796), and John Snow had traced a cholera outbreak to a contaminated water pump (1849). But the City's cemeteries closed before the Great Stink of 1858, which prompted the construction of London's pioneering sewage system, and well before the first shipment of frozen beef and mutton arrived in the UK from Australia in 1879. Alexander Fleming wouldn't discover penicillin until later yet, in 1928.

In other words, it's entirely possible that Bekvalac and Western's post-industrial bodies were unfortunate enough to live in the precise window when rapid urbanization and industrialization had begun to poison their air, water, and food, but before new infrastructure, regulations, medicine, and technology—including refrigeration—emerged to address those problems.*

Similarly, Craig's analysis only examines a decade at the very dawn of mechanical refrigeration, in the context of the preceding century's worth of height data. What the trend looks like depends on where you start and end, which means that Craig's extremely specific findings can't serve as a conclusive answer to whether refrigeration has been a blessing in

---

*This is not entirely a matter of luck, of course: as Simon Szreter, historian of public policy, has pointed out, politics and class relations shaped the impact of urbanization and industrialization, exacerbating its disruptive effects in some countries and mitigating them in others (notably Sweden).

terms of human health over the long term. That question has proven too thorny a challenge for researchers thus far. "Our profession is geared toward the small questions," Craig admitted apologetically. Recently, another cross-disciplinary group of academics, including historian of technology Jonathan Rees, tried to zero in on the effect of refrigeration on mortality over time. He told me that they gave up when the various correlations proved too tenuous.

Given the seeming impossibility of disentangling all the possible influences on human height to arrive at a sense of cold's specific contribution—let alone the limitations of using bone scans and average height as a proxy for a population's health—it might seem simpler just to focus on how refrigeration changed what we eat. Unfortunately, that's not an easy question to answer either.

Rees pointed back to the Lynds' study of Muncie, Indiana, which found that, "in 1890, the city had two distinct diets"—winter and summer. The winter diet, a local housewife told them, consisted mostly of meat, pastry, and potatoes, with only pickles, preserves, and root-cellar staples such as turnips, cabbage, and apples available to add variety and "make the familiar starchy food relishing." "We never thought of having fresh fruit or green vegetables and could not have got them if we had," she said. Without what the Lynds refer to as "the ubiquitous cole slaw of 1890," a winter diet would have been all but bereft of fresh produce. On the other hand—and in contrast to the plant-forward focus of the summer diet—a typical Munsonian's winter menu featured "meat three times a day," according to a local grocery store owner. "Breakfast, pork chops or steak with fried potatoes, buckwheat cakes, and hot bread; lunch, a hot roast and potatoes; supper, same roast cold."

Following this winter diet came "spring sickness." According to an Indianapolis pharmacist, by April, "nearly everybody used to be sick because of lack of green stuff to eat." Similarly, food historian Lizzie Collingham has concluded that, by spring, most pre-refrigeration northern Europeans "were pre-scorbutic, even if they were not suffering from full-blown scurvy." Once the first beans and tomatoes began to be shipped from the South each May, Indiana housewives were urged by physicians, recipe books, and newspaper advice columns to treat spring sickness with the urgent and generous application of "salads of all sorts."

By 1925, however, with the rise of iced railcars and refrigerated ships, all but the poorest of Muncie's citizens were increasingly able to supplement their winter diet with oranges and lettuce from California, as well as bananas from Central America. Over the thirty-five-year period encompassed by the Lynds' study, the average Munsonian consumed less homegrown fresh and preserved produce but also enjoyed more refrigerated fruit and vegetables year-round. Their protein consumption would have likely increased too, as Lee Craig's data suggest.

Again, changes in the diet of the residents of a small town in Indiana between 1890 and 1925 can tell us only so much. Were the precolonization Native Americans who inhabited that land also subject to "spring sickness"? In the UK, some historians have argued that average diets before 1750 were much more diverse and nutritious than we might otherwise imagine. Pulses, berries, foraged greens and herbs, and wild game added variety and vitamins to the starchy staples and dairy on which peasants depended, and it wasn't until industrialization and urbanization really accelerated, in the nineteenth century, that the British diet was reduced to meat, wheat, sugar, and dairy.

It's difficult to know for sure, however, because the information we have on prerefrigeration diets is based largely on bits and pieces gathered together and extrapolated from a series of equally patchy sources: cookbooks, newspaper articles, diaries, household surveys, trade associations, records of the food allocated to widows, and so on. Even where data sets exist, they rarely go back earlier than 1850, by which time industrialization and urbanization were well underway in the US and UK. What's more, most of them track market availability and sales rather than actual intake—which excludes food that was wasted, as well as homegrown produce. Trying to quantify shifts in consumption levels over time thus feels like trying to survey a cloud: nothing can be pinned down with any certainty.

Based on the data that do exist, all we can say is that fresh fruit, cheese, butter, and egg sales increased decade on decade in the United States during the late 1800s, at least in part due to the introduction of cold storage and refrigerated transportation. Red meat availability per capita increased after the introduction of the refrigerated railcar in the 1870s, through the 1960s, but has since declined. Poultry has taken its place, meaning that overall meat consumption has gone up—again, since records began. Data show that fruit sales also increased slightly in the early years of the twentieth century. Vegetables are the neglected stepchild in terms of record keeping, with only the potato considered worth accounting for in the past.

Matters become murkier still when you factor in changes in the nutrient levels of fruits and vegetables. According to an analysis conducted by the USDA, today's commercial tomatoes are not only less flavorful than the varietals they replaced but also contain 30 percent less vitamin C, 30 percent less thiamin, and more than 60 percent less calcium. This pattern of nutrient depletion has been observed in everything from asparagus to oranges—indeed, one study concluded that you'd have to eat eight oranges today to get the same amount of vitamin A that your grandparents would have ingested from one. For phytochemicals such as lycopene or flavonoids, which are increasingly understood to be beneficial to human health, no such data exist: the USDA doesn't measure them because the government hasn't set recommended daily allowances for their consumption.

Scientists suspect that, in developing varieties with higher yields and the sturdiness to be shipped and stored under refrigeration, breeders may have accidentally lost not only flavor but also essential vitamins and minerals. What's more, although refrigeration allows us to keep food edible for longer, it doesn't usually get more nutritious. Spinach left in a fridge for a week will lose three quarters of its vitamin C and 13 percent of its thiamin, while broccoli will retain only a third of its vitamin C and beta-carotene and less than half of its phytonutrients. It's entirely possible that refrigeration's meat-and-dairy-based boost to the American diet has been accompanied by and perhaps outweighed by losses in other essential nutrients over time.

Not all is doom and gloom, however. One of the key ways refrigeration has been shown to affect consumption patterns, and thus our health, is by reducing our dependence on salt as a preservative—which, in turn, has been credited with a dramatic reduction in rates of stomach cancer. (High dietary salt intake combines with *Helicobacter pylori*, a microbe that experts estimate one in two of us carries in our guts, to increase the risk of gastric cancers.) Until the 1930s, this was the deadliest cancer in the US; it's now not even in the top ten, in what epidemiologist Ernst L. Wynder called "an unplanned triumph" for public health. Similar declines following the introduction of cold storage have also been documented in Portugal, Japan, the UK, and beyond. And salt was hardly the worst of the preservatives people turned to in an attempt to keep food fresh before refrigeration, as Harvey Washington Wiley's unlucky Poison Squad volunteers could testify. Swapping formaldehyde and salicylic acid for cold storage was undoubtedly a step forward.

A t the microbial scale, the blessings of refrigeration seem as though they should be a sure bet. Diarrhea, dysentery, and other gastrointestinal upsets were a leading cause of death among Americans at the end of the 1800s, and food poisoning was at least partly responsible.

Food poisoning statistics are, alas, another data set that is widely acknowledged to be incomplete and unreliable at best. The question of whether the incidence of food poisoning has decreased in the United States or the UK over the twentieth century does not have a conclusive answer. What is clear from the data is that *deaths* from food poisoning have sharply decreased—by more than 90 percent between 1900 and 1980, according to one study. To what extent that reduction in mortality is due to cold's ability to slow the reproduction of pathogenic bacteria in our food and drink or to the ability to treat bacterial stomach upsets with antibiotics is, again, yet to be determined.

At the level of straightforward common sense, it seems likely that refrigeration has helped. In 2007, after China had urbanized but before the country added building a cold chain to its Five-Year Plan, the average Chinese person experienced some kind of digestive upset twice a week. When I visited Shanghai in 2014, the pork processor that supplied a fifth of the city's demand still managed without mechanical cooling. It simply did all its slaughtering at night, when the temperature was slightly cooler, in a shed with open sides to allow for a cross breeze. The freshly disemboweled pigs would then hang, steaming, for hours in the smoggy air as their bacterial load multiplied exponentially.

On the other hand, there's evidence that the complexity of a US-style refrigerated food system—one in which the concentration of food production has vastly increased while the distribution has expanded to global scale—has allowed different problems to emerge. The recent spread of new, more deadly, and, increasingly, antibiotic-resistant forms of pathogens such as *E. coli* and *Salmonella* has been encouraged by keeping hundreds of thousands of animals confined to crowded barns and feedlots; those bugs are then given the opportunity to spread more widely as food is moved through long supply chains with multiple handling opportunities. According to Madeline Drexler, visiting scientist at the Harvard T.H. Chan School of Public Health, "this leads to a new, insidious kind of epidemic: one with low attack rates . . . but huge numbers of dispersed victims."

Still, diarrhea no longer even ranks in the top ten causes of death in the United States. With the exception of deaths from COVID-19 and accidents, most Americans now die of heart disease, cancer, cirrhosis and chronic liver disease, diabetes, and Alzheimer's. These maladies are frequently related to what we eat: the sugar-, fat-, and red meat–centric "Western pattern diet." It's easy to see how refrigeration would bear some responsibility for the red meat part of the equation. There's also a reasonable argument to be made that at least one of the reasons so many Americans seem to prefer junk food over fruit and vegetables is because industrialized supermarket produce tastes so bland. As we've seen, it's also possible that our predilection for ice-cold drinks has inadvertently led us to prefer sweeter food.

More recently, researchers have begun to link these kinds of diseases to chronic inflammation. That, in turn, has been connected to the depleted state of the Western gut microbiome—which may be due, at least in part, to refrigeration. "This might be the microbial bargain that we've unknowingly struck," said Stanford University's Justin Sonnenburg, whose research has focused on unraveling how diet affects our intestinal bioflora. "We've focused on reducing acute pressures like diarrheal disease, while at the same time compromising this community of microbes that lives within us and that people haven't been aware of until very recently."

Hardly a day goes by without a new study connecting our gut microbes with yet another aspect of our physical or mental health. Skeptics might wonder whether such tiny organisms can really have such outsize effects, especially because we still have a limited understanding of the mechanisms through which microbes exert their influence, as well as what determines which microbes take up residence in our guts in the first place. But over the past decade, more and more evidence has accumulated that directly ties changes in gut bacteria to a whole host of maladies, including heart disease, diabetes, some cancers, and depression. What's more, it seems increasingly clear that the modern Western lifestyle has completely remodeled our guts, changing which microbial species predominate, as well as wiping many out altogether. These shifts wouldn't necessarily have immediate, acute effects, but researchers increasingly believe that they may well have, as Sonnenburg put it, "longer-term health effects over a person's lifetime, driving towards chronic inflammatory diseases as we get older."

Along with advances in hygiene and sanitation, refrigeration has lowered our everyday exposure to microbes. It has also reduced our dependence on fermentation as a means of food preservation. Fuchsia Dunlop, a British cook and author who writes about Chinese cuisine, told me that she has seen traditional food-preservation skills of the sort practiced by chef Dai Jianjun die out over the past two decades as refrigeration gained ground. "When I first lived in China, in 1994," she said, "everything was dried, pickled, or salted. In Chengdu, they would hang sausages and pork under the eaves of the old houses to dry, and there were these great clay pickle jars in people's homes." Today, she said, most of those old houses have been demolished, and in the new apartment buildings, people put their food in the fridge instead.

Sonnenburg told me that his research has shown that it's surprisingly difficult to create lasting change in a person's gut microbes by changing their diet. "Fermented foods do that," he said. "When you finally see an intervention that actually changes the diversity of people's microbiome, it really is a startling moment." He and his colleagues found that the increase in microbial diversity that followed the consumption of fermented foods was also correlated with a significant reduction in blood-based markers of inflammation.

"We still don't know *how* fermented foods did that, and we also don't know for sure that the increase in diversity is linked to the decrease in inflammation—they just happened at the same time," Sonnenburg cautioned. He and his colleagues are now testing various hypotheses, in a series of elegant experiments involving mice in bubbles and sterilized sauerkraut juice. Thus far, Sonnenburg said, it seems clear that the new microbes in people's guts didn't come from the food itself. Instead, he thinks that eating fermented food did something to people's guts that either encouraged microbes that were previously present at undetectable levels to flourish or allowed beneficial microbes encountered in the course of daily life to establish residency rather than pass through. One hypothesis currently under investigation holds that it might be the molecules that the microbes in fermented food produce as they go about their daily metabolic routines, rather than the microbes themselves, that make the difference. Sonnenburg's colleagues have shown that, when given to mice, these molecules boost a particular type of cell in the gut that in turn dampens inflammation.

In general, Sonnenburg concluded, our refrigerated, hygienic lifestyles seem to be lacking in the kind of low-grade microbial exposure that we evolved alongside—and without that background stimulation, our immune system can end up spiraling into an inflammatory state. Refrigeration rose to glory as a solution to a crisis in nutrition: the difficulty of feeding the expanding populations of newly industrialized cities both adequately and safely. But it's possible that, by refrigerating and sanitizing our way out of the bacterial infections that killed so many city dwellers in the 1880s, we may have inadvertently created the conditions for another, less acute but equally deadly, set of diseases to emerge.

Ultimately, it seems as though the short-term health benefits of refrigeration—a decrease in illness and deaths from bacterial infections, a reduction in cases of stomach cancer, and some additional animal protein to grow a little stronger and taller—might yet prove to have been offset by its downsides. The consequences of any technology are rarely limited to what we hope and imagine they might be when it is first introduced, and refrigeration is likely no exception. A solution to one problem brought with it—and even brought into being—a host of others. The promise of health—protein, vitamins, and an end to ptomaine poisoning—was leveraged to sell mechanically cooled food to wary Americans more than a century ago. Perhaps the realization that refrigeration might also have harmed our health will provide the necessary motivation to reconsider our relationship with it.

At the end of my conversation with economist Lee Craig, I asked whether, if he considered the impact of refrigeration beyond just its initial decades, he'd still conclude that it was a boon to human health. "No, I would not," he replied decisively. "The nutrition issues and the environmental issues and energy issues—all of that would make me re-think it."

7.

THE END
OF COLD

I. The Future of Refrigeration

At one o'clock in the morning, several hours before the boats launch, François Habiyambere, a wholesale fish dealer in Rubavu, in northwest Rwanda, sets out to gather ice. In the whole country, there is just one machine that makes the kind of light, snowy flakes of ice needed to cool the tilapia that, at this hour, are still swimming through the dreams of the fish farmers who supply Habiyambere's business. Flake ice, with its soft edges and fluffy texture, swaddles seafood like a blanket, hugging without crushing its delicate flesh. The flake-ice machine was bought secondhand a few years ago from a Nile perch–processing plant in Uganda. A towering, rusted contraption, it sits behind a gas station on the main road into the southeastern market town of Rusizi, on the border with the Democratic Republic of the Congo. Its daily output would almost fill a typical restaurant dumpster, which is considerably less than the amount required by the five fishmongers who use it.

"The first one who comes gets enough," Habiyambere told me when I accompanied him one day in May. "The rest do not." He said this in a tone of quiet resignation. The machine is five and a half hours' drive south of where he lives, which is why his workday begins in the middle of the night. He rides in one of the country's few refrigerated trucks, driven by a solid, handsome twenty-eight-year-old named Jean de Dieu

Umugenga and laden with spring onions and carrots bound for market. The route is twisty and Umugenga swings around the hairpin bends with panache, shifting in his seat with each gear change, while twangy inanga music plays on the radio.

Sometime after 3:00 a.m., cyclists start to appear. All over rural Rwanda, sinewy young men set out from their homes on heavy steel single-speed bikes that are almost invisible beneath comically oversize loads: bunches of green bananas strapped together onto cargo racks; sacks of tomatoes piled two or three high; dozens of live chickens stacked in pyramids of beaks and feathers; bundles of cassava leaves so massive that, in the predawn light, it looks as though shrubbery is rolling along the side of the road. Over the next four or five hours, as the heat of the day sets in, gradually wilting the cassava leaves and softening the tomatoes, these men will cover hundreds of miles, carrying food from the countryside to sell in markets in the capital, Kigali.

Rwanda is known as le pays des mille collines, land of a thousand hills, but there must be at least ten thousand, their lush, green terraced slopes rising steeply out of a sea of early-morning mist that fills the valleys below. The cyclists coast down each hill, then dismount to push their bikes up the next. When they reach a paved road, some of them may manage to catch a ride hanging on to the back of Umugenga's truck.

Around half past five, as the first flush of dawn appears, members of the Rulindo vegetable cooperative, a few hours northwest of Kigali, head into the fields. Rwandans are notoriously neat, I am told, and the countryside is packed with postage stamp–size plots, like hobbit gardens, hugging the hillside contours in orderly terraces. Chili-pepper bushes and green-bean vines grow in uniform rows; the fertile red soil of the valley floor is pristine and weed-free; every square inch is meticulously cultivated.

By this time, Habiyambere, Umugenga, and I have driven 140 miles down the entire eastern shoreline of Lake Kivu, where the fishing industry of this landlocked country is based. Its waters are dotted with rocky islands and traditional wooden canoes fishing for sambaza, a silvery, sardine-like fish usually eaten deep-fried, with a beer. The canoes travel lashed together in groups of three, their nets attached to long eucalyptus poles that project from the prows and the sterns like insect antennae. On arrival in Rusizi, Habiyambere and Umugenga stop first at the market to unload the vegetables, which will be sold to Congolese traders. Then they head to the ice machine, where, after painstakingly cleaning the truck's interior, they shovel in a small mound of precious flake ice. By 6:45 a.m., they are parked in the shade down at the dock, dozing as they wait for the fishermen to land.

Farther north, closer to the Ugandan border, Charlotte Mukandamage is wiping down the udder of a heifer that she keeps in a wooden stall behind her mud-brick home. Squatting on a plastic jerrican, Mukandamage coaxes a gallon and a half of warm, frothy milk out of the cow and into a small metal pail. Then she carefully picks her way down a steep and slippery mud path carved into the hillside, heading for a concrete marker with a picture of a cow painted on it, where a small crowd has assembled to await the milk collector.

When I tagged along with Mukandamage one morning, we were joined by a half dozen others, including an elderly man in a fedora toting a large pink plastic bucket and a skinny seven-year-old hauling a yellow tin pail nearly half her size. The morning sun was glittering on the tin roofs of nearby homes, and wisps of smoke from woodstoves mingled with mist rising off the hills. Soon a balding man wearing black gum boots came into view: Pierre Bizimana, a farmer and a part-time milk collector. He pushed a bike, over which were slung two battered steel cans, each capable of carrying a little more than thirteen gallons. For the next two hours, in the gathering humidity, Bizimana, his assistant, and I trudged uphill from one station to another, picking up a gallon here and a half gallon there from a few dozen farmers. Then we headed to the nearby town of Gicumbi, where there is a milk-collection center with an industrial chiller.

By 9:30 a.m., Bizimana is heading home to tend to his own cow and a small plot on which he grows sorghum, corn, and beans. Hundreds of miles away, François Habiyambere and Jean de Dieu Umugenga have embarked on the drive back north with a truck full of fresh fish for the Rubavu market. Some of the sweaty cyclists are already making their return journeys too, often with a passenger perched on the cargo rack where the cassava or the chickens had been. And the Rulindo farmers are back from their fields bearing crates of freshly picked peppers and beans. The next morning, the harvest will be loaded onto a RwandAir flight bound for the United Kingdom, where it will be sold in supermarkets. In the meantime, the crates are stacked in a solar-powered cold-storage room, which, at sixty-five degrees, is about twenty degrees warmer than it should be.

I visited Rwanda because it is where the future of refrigeration is being built—and where the urgency and enormous stakes involved in that transformation are hardest to ignore. According to the most recent estimate, 2.76 billion tons of food are wasted globally every year, or 40 percent of everything that is grown worldwide. At least a third of this could be saved by refrigeration. In a country like Rwanda, where fewer than one in five infants and toddlers eat what the World Health Organization classifies as the minimum acceptable diet, such wastage is a matter of life and death.

In countries with a well-developed cold chain, like the United States, food waste mostly takes place at the consumer level, in homes and at restaurants. In Rwanda, as in much of the developing world, the lack of a cold chain means that between a third and a half of everything that I saw being harvested would be lost long before it ever gets that far. Rwanda is also one of the poorest countries in the world: the gross per-capita income is currently $2.28 a day; more than a third of children under five are stunted from malnutrition; and diarrheal outbreaks are so common that they're estimated to reduce the country's GDP by up to 5 percent.

In Kigali, I met the world's first professor of cold economy, Toby Peters, from the University of Birmingham. When I told him about my journeys alongside Rwanda's slowly broiling milk, fish, meat, and vegetables, he defined the problem in systemic terms. "There is no cold chain in Rwanda," he said. "It just doesn't exist." Preventing food loss requires more than a functioning refrigerated warehouse or truck: food has to get cold and remain that way all the way along the chain.

Today in the United States, a green bean grown in, say, Wisconsin will likely have spent no more than two hours, and often much less, at temperatures above forty-five degrees on its way to your fork. As soon as it is harvested, it is rushed to a packhouse to have its "field heat" removed: it is either run through a flume of cold water, known as a hydrocooler, or put in a forced-air chiller, where a gigantic fan pushes refrigerated air through stacked pallets of beans. These processes "precool" the bean, lowering its internal temperature from more than eighty degrees down to the low forties in just a couple of hours. After that, a bean can happily hang out in cold storage facilities, travel in refrigerated trucks, and sit on chilled supermarket shelves for up to four weeks without losing its snap.

In Rwanda, even if the tepid cold-storage room I saw in Rulindo had been running at the correct temperature, down in the low forties, the benefits would have been marginal without the rest of the cold chain in place. In a forty-degree storage room, a bean takes about ten hours to reach the same temperature that precooling achieves in just two. In the whole of Rwanda there is only one forced-air chiller for precooling produce. It's at a government export facility near the airport in Kigali and is almost never used, because it costs too much to run.

For the green bean, the difference between being chilled in two hours and in ten is absolute. Like all fruits and vegetables, a bean cut off from the support of its parent plant will start to consume itself, and the hotter the temperature, the faster it does so. A cold-storage room, con-firmed postharvest expert Natalia Falagán, is of little use without pre-cooling. "And then farmers will say temperature-controlled rooms don't work," Falagán lamented. "No! It's that the fruit you put in there is al-ready mush." The consequences are even more dire for unrefrigerated milk and flake ice–deprived fish. On average, 35 percent of the milk painstakingly gathered on bicycles by people like Pierre Bizimana is suf-ficiently spoiled by the time it reaches the country's dairy-collection cen-ters that it fails quality-control tests and is rejected outright. Unsold, un-iced fish is typically off-loaded to Congolese traders for pennies on the dollar at the end of the day.

Even in China, where the cold chain is under construction, such losses are common. When I visited the wholesale market that supplies 70 per-cent of Beijing's vegetables, lines of trucks were parked just outside, blan-kets and tarpaulins swaddling their trailers. The webs of silvery duct tape holding this makeshift insulation in place glittered with light reflected from the neon sign above the market's entrance gate. I watched as a woman carefully excavated individual, naked stalks of broccoli from a truck packed solid with ice and hay. Her husband, a middle-aged farmer, re-moved his earmuffs to tell me that he expects to have to throw away a quarter of the truckload—more when the weather is warm—as the ice melts and the vegetables rot faster than they can be sold.

Such losses were normal in the United States, too, in the early days of refrigeration. As Arthur Barto Adams, the USDA economist charged with chronicling the fate of the nation's perishables on their way to mar-ket, found in 1916, somewhere between 30 and 40 percent of all the food grown in America was "hauled to the dump-pile" before it even reached a grocery store, icebox, or dinner table. "That this decay of perishables works a great hardship upon both producers and consumers is too evi-dent to require much discussion," Adams observed.

In Rwanda, where farmers are surviving on less than a couple of dol-lars a day, the effect of these losses is crippling. For sub-Saharan Africa as a whole, they are estimated to add up to hundreds of billions of dollars each year. Worldwide, the amount of food that goes uneaten due to a lack of refrigeration represents a lost harvest of sufficient abundance to feed more than 950 million people annually—which is substantially more than the 828 million people that the World Food Programme estimates are currently facing hunger.

Since 2015, when the United Nations issued a call to halve per-capita global food loss and waste by 2030, NGOs, overseas-development agencies, and philanthropic foundations have rushed to fund refrigeration projects in the developing world. A dawning realization that without tackling food loss, it was going to be impossible to achieve the United Nations' 2030 Sustainable Development Goals—numbers one, two, and three of which are "no poverty," "zero hunger," and "good health and well-being"—spurred the launch of a new UN-backed initiative: Global Cooling for All. (I was a keynote speaker at its inaugural meeting in September 2017.) "For poor farmers in developing countries, cooling is key to unlocking a better life, through reduced food losses and increased incomes made possible by cold chains that transport crops to higher paying markets," announced the group's first report, titled *Chilling Prospects.* "Every year, millions of people die due to the absence of cooling that could help address hunger and malnutrition."

The group is cochaired by Rwanda, which President Paul Kagame pledged to transform into a high-income country by 2050; recently his government has come to the conclusion that this cannot be achieved without refrigeration. In 2018, Rwanda announced a National Cooling Strategy, the first in sub-Saharan Africa, and in 2020, it launched a program known as the Africa Centre of Excellence for Sustainable Cooling and Cold Chain, or ACES.

A collaboration between the Rwandan and UK governments and the UN Environment Programme, ACES is designed to harness expertise from within Africa and beyond. Toby Peters at the University of Birmingham is a cofounder, along with Cranfield University's Natalia Falagán.

Several other British universities are involved, as is the University of Rwanda, in Kigali, where the new institution has its campus. The ACES mission is wide-ranging and encompasses research, training, and business incubation, as well as the design and certification of cooling systems. Once construction is complete, its campus will boast the country's first advanced laboratory for studying food preservation and a demonstration hub for the latest refrigeration technology.

Among people involved in international development, Rwanda is considered a good place to do business. There is little corruption; Kagame, though an autocrat, is credited with promoting governmental accountability and transparency. And the country's small size—it is not much larger than Vermont—makes it an ideal testing ground for initiatives that, if successful, can then be deployed across sub-Saharan Africa. ACES has plans to expand from its Kigali hub with spokes across the continent, and the team is also working with the southern Indian state of Telangana to build a similar center there. It is intended to be the smithy in which the links of a new, truly global cold chain will be forged.

B ecause the ACES team was assembled during the COVID-19 era, many of its members had not met in person until May 2022, when Rwanda hosted a UN-sponsored forum on sustainable energy, which showcased ACES, among other initiatives. When Kagame gave an opening address to the forum's delegates—an international assortment of politicians, civil servants, aid workers, entrepreneurs, and academics—ACES served as his example of Africa's potential for ensuring sustainable, equitable development globally, much to the team's delight. "I was in the room, and I felt like jumping out of my chair," said an ebullient Juliet Kabera, the ranking Rwandan member of the initiative, who also heads up the country's Environment Management Authority.

ACES was to host an open day for delegates at its new campus at the culmination of the forum. The weekend before, I accompanied the team on a tour of Rwanda's existing refrigeration infrastructure. Because of the pandemic, some of the Europeans were making their first visit to a country they had been studying for three years. Our first stop was a pair of cold-storage rooms built with European Union funding in 2019, thirty miles south of Kigali on the road to Tanzania. A member of a local farming cooperative walked us over to a low-slung brick structure. Inside, the first things that caught my eye were cobwebs lining the walls. One of the rooms was not functioning, our guide said; the other contained two lonely crates of chili peppers, and the cooling seemed to have been switched on purely in honor of our visit. The spotlessly clean floor certainly did not suggest frequent use. It was also made of wood, a poor choice of material because it is hard to sanitize, so any squashed produce lingers, providing a perfect substrate for fungi and bacteria to grow. Judith Evans, another ACES cofounder and one of the world's leading refrigeration experts, quietly pointed out other design flaws, including the lack of an air curtain at the door, as well as dozens of nails driven through the walls, which would allow heat to bypass the insulation.

"I'm freaking out about this," Falagán whispered as the farmer described how the room worked. "There's no humidity control, no fans for air circulation!" While the team quizzed the unfortunate farmer, I stepped outside and wandered around the corner to see other members of the cooperative loading crates of chilis that had been stored outdoors, under an open-walled shade structure, into the back of a pickup truck. Later, Issa Nkurunziza, a Kigali-based cold-chain expert with the UN Environment Programme, told me that the farmers had confessed to him that the refrigeration unit was simply too expensive to run.

Cold storage alone, without training and a viable business model, risks becoming a white elephant. "People don't understand how to use it," Evans told me. "It's generally not well maintained or serviced." The World Bank, which has funded ten cold-storage rooms in Rwanda in the past few years, has estimated that at least 96 percent of nearby farmers don't use them at all. Such largesse can also trigger unintended consequences. Catherine Kilelu, a food-security researcher in Kenya who is leading the development of an ACES-backed cooling hub there, told me that in one remote community, there was some evidence that the quality of children's diets diminished after the Bill & Melinda Gates Foundation helped fund chilling plants as part of a larger investment in commercializing the country's dairy industry. Previously, Kilelu explained, the yield from evening milking sessions was consumed at home rather than being taken to market. Once a dairy farmer was able to keep this milk salable overnight, however, that source of nutrition disappeared. "You might think, *Well, if they make more money, they can spend that on feeding their kids*, but that's not necessarily the case," she said. "People use it to repair their roofs or buy smartphones or other things they need."

Later on, we visited a packhouse run by Rwanda's National Agricultural Export Development Board. The facility, built in 2017 with World Bank assistance, was much better equipped than anything else we'd seen in the country, but it was stuffed to capacity, with plastic crates full of vegetables stacked twelve high to the ceiling. "Right now, it's just big enough, but with the production plan we have, in six months it will not be," Innocent Mwalimu, a soft-spoken cold-chain specialist, said as he showed us around. As Rwanda emerged from COVID-19, it faced a spiraling balance-of-payments deficit. In response, the government set a target to double the country's perishable exports by 2025. By way of stimulus, companies that use the packhouse are charged less than seven cents per kilo exported, effectively subsidizing the cold chain for agribusiness entrepreneurs. Similar models have been pioneered successfully in Kenya—to the extent that, recently, fruit, vegetable, and cut-flower exports overtook the traditional mainstays of tea, coffee, and tourism to become the largest source of overseas revenue for the Kenyan government.

The downside is that the benefits of this kind of cold-chain investment are not distributed equally. As in the US, refrigeration tends both to require and to enable scale. In Kenya, one study found that three quarters of the country's fruit and vegetable exports are sourced from just seven large, mostly white-owned farms—because they have the capital and the resources to implement stringent international food-safety standards and are perceived as easier to work with and audit. Even companies specifically founded with a mission to install off-grid, affordable cooling systems to reduce postharvest losses and support rural communities have found it challenging to work with Kenya's smallholder farmers. "From an economics point of view, you're forced into bigger systems to make it work," Julian Mitchell, the CEO of one such company, Inspira-Farms, told me. "And that excludes the poorest of the poor"—the farmers who grow more than 90 percent of Kenya's fruit and vegetables, who are left losing half of everything they harvest.

The primary difficulty, as Selçuk Tanatar, the principal operations officer at the World Bank's International Finance Corporation, explained to me, is that operating a cold chain costs the same, if not more, in Nairobi as it does in New York City: five to fifteen cents per kilo of produce. This means that refrigeration adds about 1 percent to the cost of a tomato in the developed world but about 30 percent to its cost in the developing world. "Nobody is going to pay that," Tanatar said. Selling refrigerated food locally isn't feasible; the only financially viable way to build a cold chain is to work with farmers who grow fruits and vegetables that the developed world wants—blueberries, mangoes, French beans. "But then it doesn't really help the local people with food security," Tanatar continued. "You're just getting cheaper and better products to the developed market."

In Rwanda, six million people—nearly half the population—are small-scale farmers, tending an average of less than an acre and a half of land. A solution that does not work for them is not much of a solution at all: a trickle-down cold chain in which the rich grow richer, the poor become poorer by comparison, and, all the while, the former colonists enjoy cheap superfood smoothies.

———

In March 2021, a small, peculiar-looking truck began transporting fruit and vegetables from fields to markets in western Rwanda. From the front, the truck resembles a tank, wider and squatter than I'd expected, and oddly square. It looks the way a truck designed by IKEA might look, and in a sense, that's what it is. The cab is made of lightweight wood-composite panels that can be shipped in flat packs and assembled in a day, without any special tools. Named the OX, the truck was developed in England specifically for emerging markets. It's about half the weight of a standard pickup but able to carry double the load. The windshield and the skid plate meet at a snub-nosed angle, which means that its tires hit steep slopes before the bumper does and that it can ford streams that are up to thirty-five inches deep—both essential attributes for negotiating Rwanda's many severely rutted unpaved roads.

Francine Uwamahoro, OX's managing director for Rwanda, introduced me to a woman with short, orange-dyed hair named Louise Umutoni, saying that she was the company's best driver. "New customers are surprised," Umutoni said. "They don't believe their truck driver is a woman." She took me for a ride as she made her rounds of local farmers. Rwandan roads make for a bone-jarring experience that several drivers described to me as "an African massage." As we drove, Umutoni fielded customer calls on her mobile. The demand for OX trucks is so high that the company currently has to turn down eight in ten requests for transportation.

OX's global managing director, Simon Davis, who left Jaguar Land Rover to take the job, told me that, as innovative as the truck's design is, the secret to its success is the company's business model: the cargo equivalent of a bus service. Most prospective customers can't afford to buy a truck, but they can afford to rent space in a truck operated by OX. "We built our first business model around fifty dollars a day in revenue, total," Davis said. "On our best day so far, we've earned two hundred twenty dollars from a single truck."

———

Umutoni's first customer of the morning was a woman waiting by the side of the road with several baskets of green bananas that she wanted us to take to the nearest city, twelve miles away. She told me that, though OX's rates are higher than those of the men with bicycles, the increased cost is more than covered by the additional income she can make by getting more produce to market faster. Her only complaint about OX was that sometimes when she called there wasn't any space left in the truck; she wanted to start selling to Congolese traders and expand her business further, but first she had to be sure that transportation would be available.

Almost as soon as the first OX truck started rolling around Rwanda, the company began thinking about the next iteration. It sought feedback from drivers like Umutoni. One thing she asked for was better visibility. In rural Rwanda, the roadside is a busy place: goats graze, women sell fruit and vegetables, and children run back and forth, kicking footballs made from inflated condoms wrapped in banana leaves. The new model—still at the prototype stage—is, Davis said, "a bit like driving a conservatory." More important, OX 2.0 is an electric vehicle—the current model runs on diesel—and, as an optional extra, it will be available with a solar-powered refrigeration unit. It thus goes some way toward meeting the need that Innocent Mwalimu and Selçuk Tanatar had pointed out to me: a cold chain with lower operating expenses. OX can power its new truck for less than half the cost of the first-generation diesel prototype.

"For me, having given up on the cold chain, these technologies that can get the operating expenses down—they mean that it's going to maybe be a different story now," Tanatar told me. He said that part of the value of ACES will be in providing a venue to showcase innovations like this to Rwandan farming cooperatives, entrepreneurs, and trainee technicians. Assembling these innovative technologies is Judith Evans's responsibility. She told me that dozens of low-cost, robust solutions are in development or already exist, for all stages of the cold chain. The challenge is matching them to the needs and resources of different communities, crops, and contexts.

"Solar-powered cooling might work perfectly during part of the year, but how do you supplement it during the rainy season?" she said. With colleagues in China, Evans recently designed a shipping pallet with built-in water tubes: these can be frozen solid overnight, then used to keep the food stacked on them chilled for up to three days in an insulated room or truck, like a thermal battery. Combining frosty pallets with a solar-powered OX truck or cold-storage room can help make a cold chain more flexible and more reliable, as well as cheaper to operate. "It's got to be sustainable to be successful," said Evans. "We can't just supply equipment that we use in Europe and America and assume it's going to work."

The ACES campus currently consists of several single-story brick buildings set around a central lawn filled with mauve-flowered jacaranda trees. These will serve as classrooms for teaching future refrigeration technicians. Qualified technicians are in such short supply that when the flake-ice machine I saw in Rusizi breaks down, a mechanic has to be summoned from Uganda to repair it. At the northern edge of the twelve-acre site are a handful of cottages. Some are to be office space for refrigeration companies, both local start-ups and established international corporations; others will provide student housing and a day-care center, intended to encourage female students to train as technicians and entrepreneurs. To the west, land has been set aside for the next phase in ACES's development: a smart farm, to study how preharvest treatments affect postharvest quality and to test novel field precooling equipment.

Rwanda is full of would-be food, agribusiness, and technology entrepreneurs. Africa's "youth bulge" means that young Rwandans are continually warned that there will probably not be jobs waiting for them upon graduation and that they should be prepared to create their own. It seemed as though on any street corner in Kigali one could encounter someone like Donatien Iranshubije, a confident and prepossessing twenty-one-year-old wearing a crisp button-down shirt accessorized with a thin gold chain. Iranshubije cofounded a start-up that offers next-day delivery of fresh fruit and vegetables from rural farming cooperatives to two dozen Kigali families. At the moment, he told me, the company gets around the need for refrigeration by using motorbike couriers to move the food fast. (Rwanda's medical system employs a similar tactic, relying on drones to deliver blood to rural hospitals that would take too long to reach by road.) But as Iranshubije's business expands, he expects to invest in cold storage; for him, as for thousands of others, refrigeration is a prerequisite for growth. The challenge for ACES is to ensure that the urgent need for cold chains in countries like Rwanda is met in a sustainable fashion.

Cold chains present a double bind; both their absence and their presence have huge ecological costs. The UN Food and Agriculture Organization estimates that if global food waste were a country, its greenhouse gas emissions would be the third largest in the world, right behind China and the US. This figure is calculated by adding up the emissions accrued while growing the food (clearing land, applying fertilizer, methane released by rice paddies or cattle, and so on) and doesn't even take into account the increase in greenhouse gas emissions that result when forests are cut down to make way for fields, let alone the vast quantities of water used to irrigate these doomed crops, which, globally, is estimated at nearly a quarter of all freshwater consumption.

Meanwhile, the chemicals and the fossil fuel energy used to refrigerate food already account for more than 2 percent of global emissions.* As countries like Rwanda refrigerate, that is increasing rapidly. Toby Peters, the ACES cofounder, has done the calculations: if every country were to

---

*In total, cooling—which includes air-conditioning as well as refrigeration—accounts for 10 percent of global emissions. As climate change causes more extreme heat events, global demand for air-conditioning is also skyrocketing.

have a cold chain similar to the ones the developed world relies on, these emissions would increase fivefold, at which point they would match the footprint of food waste. "That's just today," he added. "It's not accounting for population growth and a hotter planet."

Refrigeration contributes to rising greenhouse gas levels in two main ways. Generating the power to run cooling equipment, whether it be electricity for warehouses or diesel fuel for trucks, already accounts for more than 8 percent of global electricity usage. (Cold-storage companies are currently the third highest industrial consumers of energy.) Using renewable sources to generate that power would help, but solar-, wind-, geothermal-, and hydro-power generation are growing much too slowly to keep pace with demand.

Refrigeration equipment can also be made more energy efficient—up to a point. While I was in China, I visited the research and development center of Emerson Climate Technologies, one of the largest manufacturers of refrigeration systems in the world. Emerson distributes the compressors, valves, and flow controls that cool many of the country's new dumpling freezers and yogurt display cases. My guide, Clyde Verhoff, vice president of engineering for Emerson's Asia division, is the kind of refrigeration nerd who claims that he can predict the hum that a new component will make before it is even built. A few years ago, Verhoff told me, Emerson helped the Spanish supermarket chain DIA consume 25 percent less electricity in its Shanghai stores by designing an energy-efficient, automated control system. We were standing in a depopulated, football field–sized room containing over a hundred reinforced wire cages. It looked like a dog pound for particularly aggressive breeds, except that instead of an abandoned Rottweiler, each cage housed a new compressor prototype, whirring away to refrigerate the future. Nonetheless, Verhoff shook his head when I asked him whether those kinds of savings could continue. "We're really hitting the limit of what can be done," he yelled over the jet engine–level roar.

The other problem is the refrigerants themselves: the chemicals that are evaporated and condensed by compressors in order to remove heat and thus produce cold. Some of that refrigerant leaks into the atmosphere as a gas—either a little (roughly 2 percent a year from the most up-to-date domestic refrigerators) or a lot (a third, on average, from small delivery trucks). Different refrigeration systems use different refrigerants, some of which, like ammonia, have a negligible global-warming impact. Others, like the hydrochlorofluorocarbons and hydrofluorocarbons (HCFCs and HFCs) that are popular in the developing world and that Kipp Bradford and I used to build our own fridge, are known as super-greenhouse gases because they are thousands of times more warming than CO_2. Project Drawdown, the climate change mitigation project founded by environmentalist Paul Hawken, lists "refrigerant management" as the number one solution to global warming, in terms of potential impact. "We were very surprised," Hawken said when he published the list in 2017. "We didn't see that one coming."

Ironically, HCFCs and HFCs were supposed to be the planet-saving replacement for their predecessors, CFCs, or chlorofluorocarbons. This group of chemicals was first commercialized in the 1930s by Thomas Midgley Jr. as the first nonflammable and nontoxic refrigerants. Midgley, a gifted but notoriously eccentric chemist, worked for General Motors, which manufactured Frigidaire appliances. At the time, these relied on sulfur dioxide as a refrigerant, which causes coughing, burning eyes, and a sore throat if inhaled. The introduction of Midgley's first new CFC refrigerant, which GM called Freon and whose safety Midgley memorably demonstrated by inhaling a lungful, then using it to blow out a candle, helped the fledgling domestic refrigerator industry take off. Unfortunately, half a century later, it emerged that CFCs were extremely harmful, specifically to the layer of atmospheric ozone that shields Earth from damaging levels of the sun's UV radiation.* By the time the world's

---

*Midgley, who was also responsible for adding lead to gasoline, has been described as a "one-man environmental disaster," though, to be fair, he has stiff competition from the inventors of other products of the twentieth century's embrace of "better living through chemistry," from plastics to DDT.

nations came together to agree to eliminate their use in the 1987 Montreal Protocol, CFCs had created a hole in the ozone layer that is still decades away from healing.*

With CFCs banned, refrigerator manufacturers turned to their chemical cousins, HCFCs and HFCs, which turned out to have different but equally disastrous long-term consequences. These too have recently begun to be phased out under the terms of the Kigali Amendment, named for the Rwandan capital where this addendum to the Montreal Protocol was negotiated. Assuming it is successfully enforced—a robust underground market in banned refrigerants has already emerged—the Kigali Amendment has the potential to prevent as much as 0.5 degrees Celsius of global warming. But substantial challenges still lie ahead. Today's replacement refrigerants are usually more expensive, sometimes less efficient, and, because they're often both flammable and toxic, require advanced training to use. Some can't be substituted directly, because they have different operating parameters that require differently designed components. Capturing and safely incinerating the retired HFCs is also expensive and complicated—California's Air Resources Board estimates that, at the moment, when a household fridge reaches the end of its life, more than three-quarters of its refrigerant is lost to the atmosphere, in spite of EPA regulations.

The tagline for ASHRAE, the American Society of Heating, Refrigerating, and Air-Conditioning Engineers, is "Shaping Tomorrow's Built Environment Today." Refrigerating food for a global population of nine billion using today's technology would reshape tomorrow's planetary environment—in the most disastrous manner. Put simply, climate change–induced crop failures and drought will ensure that there is no harvest to store unless we change how we cool our food. "I think it's going to be a disaster," agreed Gu Zhong, managing director of Emerson's China-based rival, DunAn Artificial Environment Equipment Company. "The

---

*It's easy to forget how much of an existential threat this presented. Without the ozone layer to filter it, the sun's ultraviolet radiation would sterilize the surface of the Earth, wiping out all life.

only solution is that we have to go for a new solution. But I don't know what that solution is yet."

In development literature, much has been made of Africa's ability to "leapfrog" richer countries. In Rwanda, a country in which a national network of telephone cables was never laid, cell phones became central to daily life far more quickly than in the US. The same is true for mobile banking and electronic payments. The hope, then, is that Rwanda and its neighbors can do something similar with refrigeration, bypassing inefficient and polluting technologies in favor of more sustainable solutions.

Reinventing the fridge is an enterprise with a distinguished history. Einstein himself filed a patent on an ultraefficient refrigerator with no moving parts back in 1930—although as of 2016 Oxford University engineers were still tinkering with his design. Among the issues that plagued early iterations: it apparently "howled like a jackal."

Judith Evans walked me through some of the novel zero-emissions refrigerators of the future, many of which began to be developed in the 1970s and '80s, spurred by the newly discovered hole in the ozone layer. A couple have since been prototyped, including a thermoacoustic device that uses sound waves to compress helium and cool an ice cream freezer—the result of an unlikely collaboration between Ben & Jerry's, the US Navy, and Pennsylvania State University. Other researchers have created working shopping cart–sized magnetic fridges, which take advantage of the fact that certain metals heat up when they're magnetized and cool down when you take the magnet away. None have been commercialized or seem likely to compete with the dominant century-old vapor-compression system anytime soon. "Cooling is the Cinderella of the energy debate," Toby Peters lamented. "Less than a quarter of one percent of all engineering R&D is spent on cooling."

Even when the technology is proven, change is slow. Evans told me that she has talked to European supermarkets about the potential to use her pallet-style "ice batteries" in their delivery vehicles. "They're interested, but they're very nervous about it," she said. Such reluctance is unsurprising. The system in place already "works," so why take the risk? If these kinds of simple and sustainable solutions can be successfully piloted in developing nations, Evans said, it may lead the way for their adoption by supposedly developed ones.

The more time I spent with the ACES team, the more acutely I felt both their excitement and their anxiety about Rwanda's unbuilt cold chain: get it right, and enter a promised land of food security, prosperity, and sustainability; fail, and wave goodbye to a livable planet while accelerating inequality and exacerbating hunger. "These are the kind of problems that hadn't really even been recognized as problems or challenges before—they were just consequences," Philip Greening, another member of the ACES team, told me. Greening is currently constructing a computer model of Rwanda—a digital twin in which all the possible variants for preserving and moving its food can be implemented, priced, and evaluated—in order to answer such pressing and essential questions as: Where should cooling hubs be placed to be most useful for the communities that need them the most? What will happen if, as currently planned, slaughterhouses are built in rural areas, so that the live chickens I saw, transported on bicycles and slaughtered at home, are replaced by carcasses that need to be moved, stored, and sold under refrigeration? How will exporting 10 percent more fresh produce affect a farming family's nutritional and economic status? Is it worth improving the road network before investing in farm-level precooling facilities?

The use of computer modeling to make such decisions is new and has limitations. Inevitably, there will be simplifications, and some data are likely to be unobtainable. And, of course, humans remain unpredictable. During the COVID-19 pandemic, Greening and Peters, realizing the importance of the cold chain in delivering vaccines, worked with the Bangladeshi government to figure out the most effective possible allocation of the country's refrigerated assets. But Bangladesh's actual vaccination campaign departed significantly from the model's recommendations, as Greening ruefully explained. "In the end, the challenge wasn't so much 'Can we get the vaccine to the right places?' as 'Can we get people to want to be vaccinated?'" The acceptance of refrigerated food may face similar obstacles: just like their counterparts in the United States over a century ago, many Rwandan consumers today believe that refrigerated food isn't fresh. "Traders who sell the rejects from the National Agricultural Export Development Board packhouse on the local market even have to put them in the sun for a while, so that they don't feel cold," Alice Mukamugema, an analyst at the country's Ministry of Agriculture, told me.

Still, for Rwandans, for the ACES team, and for our shared future, not trying is not an option. The population of sub-Saharan Africa is projected to nearly double, reaching more than two billion by midcentury. "Just building the infrastructure—cold chain included—needed to support a billion more people in the next thirty-five years is going to be the biggest construction boom in the history of the world," said Michael Murphy, the founding principal of MASS Design Group, the architecture firm that is working with ACES on the design of its campus. "Unless there are systems that are environmentally conscious, that are socially conscious, that are thinking about the unintended consequences and the failures of the way we've built over the last fifty years, we're *all* facing disaster."

Late one afternoon, I had an appointment to see Christian Benimana, the co–executive director of MASS Design's Kigali office. I'd been riding in cars and trucks all week, so I decided to walk to his office, an hour and a half across town from my hotel. Since the Rwandan genocide, the city's population has exploded, growing from just under 300,000 people in 1994 to more than 1.2 million today, but its streets are surprisingly quiet, lacking the chaotic energy of most cities in the developing world. Kigali is so hilly that all but the poorest people make even short journeys on one of its ubiquitous motorcycle taxis, so for stretches of the walk to Benimana's office I was the only pedestrian.

The lack of bustle on the streets seemed boring at first but gradually became its own source of fascination. The sidewalks were spotless (plastic bags have been banned since 2008), women in high-visibility vests weeded perfectly groomed flower beds and median strips, and there was not a single homeless person to be seen. (The homeless are reportedly moved to what the Rwandan government refers to as "rehabilitation transit centers" but Human Rights Watch calls prisons.) Between anonymous glass office buildings and tidy single-story houses, there were huge expanses of open space: a flock of ibises screeched from an enormous tulip tree; an African spoonbill waded on fuchsia legs along the edges of a muddy river; birds of prey circled above me, riding thermals. Only the smells—sooty diesel fumes and hot bodies crammed together on bicycles and moto-taxis at every intersection—reminded me that I was in a desperately poor country.

Benimana, a reserved but commanding forty-year-old man, told me that in 2007 the Rwandan government announced a visionary master plan to transform Kigali into "an important center of stability and development for the entire continent of Africa." It quickly became clear that the plan was seriously flawed, and there was a public outcry. But the government, rather than pressing on regardless or just giving up, took stock of the complaints and produced a major revision of the plan, which it has since continued to update and implement with considerable success. Some of the results can lack character, Benimana admitted—the city center is a giant roundabout, and its new hotels, malls, and industrial zones are a series of generic boxes—but other aspects are impressive. Wetlands take up a quarter of Kigali's surface area and they are now protected habitats—a distinct improvement on the glorified sewers that the rivers of London and Los Angeles became as those cities urbanized.

"After the genocide, the process of rebuilding was not optional," Benimana said. "And the decision was made early on to set the bar really high—to see whether we can solve some of the structural societal problems that we have, and to become a place that people can learn from."

For Benimana, the ambition of ACES is entirely in keeping with his country's embrace of experimentation and innovation. "We are able to dream things that are beyond what is imaginable, and then act on them," he told me. "Or at least try."

II. The Future May Not Be Refrigerated

"You can lick it," said James Rogers, handing me a two-week-old lime. "We treat the produce with such a small amount of material that you can't see it, you can't taste it, and you can't feel it." Rogers is the CEO and founder of Apeel Sciences, a California start-up that has developed a new way to preserve food.

Apeel's product looks like a brick of whitish flour. Customers—primarily produce farmers and wholesalers—mix it with water and spray it onto fruit and vegetables, where it creates what scientists call a self-forming barrier: a layer of ingredients that, as it dries, assembles itself into a second skin. "James has taken big spoonfuls," Michelle Linn, the company's head of marketing and communications, reassured me.

Refrigeration preserves fruit and vegetables by lowering their metabolism. Produce takes the same number of breaths before dying in the cold; it just takes them more slowly, delaying its inevitable decline. Apeel also works by slowing respiration rates, but it does so using atmospheric adjustment rather than thermal control. Apeel's nanoscale coating is permeable in exactly the right way to slow down the passage of oxygen, carbon dioxide, and water vapor sufficiently that a fruit or vegetable breathes as slowly as possible and remains hydrated for longer. "The optimized little microclimate inside that lime is what gives us the benefit," said Rogers. "It's not the barrier itself—the barrier is just a means to creating the optimal internal atmosphere."

It's the same principle behind the layered films in Jim Lugg's lettuce bags—except that the modified atmosphere is internal to the produce itself, and instead of plastic, Apeel's edible, invisible coating is made using only ingredients extracted from food waste. At the company's Santa Barbara headquarters, Rogers took me into the product-prototyping lab, where hundreds of lemons, limes, avocados, papayas, and red peppers sat on steel racks at room temperature—some coated and bouncy, others naked and afraid. A tray of eight-week-old uncoated lemons were brown and wizened, while their Apeel-enrobed neighbors looked as fresh and plump as if they'd been picked that morning. Immediately below them, uncoated peppers had deflated into a shriveled, wrinkled shadow of their former selves, while the coated ones wouldn't have looked out of place on a grocery store shelf.

At Rogers's urging, I picked up a lime. It was heavy—still juicy—and firm. When I scratched it gently with my fingernail and sniffed it, it smelled tangy, sour, and a little bit floral, as a lime should. Still, I hesitated for a moment before licking it, fearing a waxy note. But there was nothing—no flavor or texture that I could detect beyond the smooth, slightly oily, and lightly dimpled skin of a regular lime.

Rogers first moved to California to watch paint dry. Researchers at the University of California, Santa Barbara, had recently discovered that if they coated two panels with a paint that was capable of harvesting solar energy, then covered one so that the paint dried more slowly, the slow-drying panel would be twice as efficient as the faster-drying one. "Nothing had changed compositionally," said Rogers. "You're baking with the same ingredients, but you get a totally different outcome."

Rogers thought that understanding what was happening to the molecular structure of the paint as it dried would help explain the difference—and point the way to creating even more efficient solar coatings. Practically, what that turned out to involve was a lot of time spent painting panels on campus in Santa Barbara, where he'd enrolled in a PhD program, then driving five and a half hours up to the Bay Area to observe it drying at the atomic scale under the synchrotron light source at the Lawrence Berkeley National Laboratory. Before one of those commutes, he hap-

pened to have read an article about the challenges of feeding a growing global population. With that thought in his head and the Salinas
Valley's lush green ribbons of lettuce blurring together on either side of
the highway, he started mulling the mismatch between food supply and
demand.

"I thought to myself, *Okay, well, other animals must deal with this too,
so how do they do it?*" he said. He was allocated the night shift on the
light source, so after watching his paint dry through the small hours,
he'd spend all day lying in bed reading about squirrel nut-caching strategies, the epic annual migration of millions of wildebeests to find graz-
ing, and the extraordinary body-fat reserves that bears lay down before
hibernation. The human solution, he realized, was to take an abundant
perishable asset—food—and convert it into a less perishable asset. In
early human history, our ancestors might have done this by sharing excess meat from the hunt in a feast, in the expectation that such largesse
would be reciprocated at a later date—anthropologists call this "social
storage"—or by fermenting fruit and grains to form storable liquid calories as alcohol. Today refrigeration served the same end.

Once Rogers had come to the conclusion that the problem was perishability, he began to wonder whether there might be a better solution
to it. This was exactly the same question Britain's chemists had pondered
back in the 1800s, when industrialization kicked urbanization into high
gear, disrupting traditional food supply chains. Back then, as we saw, inventors were experimenting with all sorts of preservation methods—
coatings, antiseptic injections, fumigation, compression, desiccation, and
more—and cooling was commonly seen as the least promising of the
bunch. Rogers was a coatings guy by training—his solar-paint research
had made him an expert in the niche field of thin-film polymer physics—
and once he realized that fresh produce spoils because it respires oxygen
and loses water, his mind immediately went to stainless steel.

"Most people don't think about this, but steel is perishable," explained
Rogers. It reacts with oxygen and water to rust. The decay takes place

over a much longer time frame than it does in an asparagus but has equally destructive long-term results. In the 1800s and early 1900s, metallurgists gradually figured out that if you included small quantities of metals like molybdenum, chromium, and nickel in steel, those sacrificial atoms would react with oxygen to form a thin barrier that physically prevented oxygen from reaching the surface of the steel—in other words, they invented stainless steel. Rogers wondered whether you could do the same thing with food.

Back in Santa Barbara, he told his surfing buddies about the idea. "They were like, 'Yeah, sounds like a good idea, bro, but we don't want to eat your chemicals,'" he said. At first, Rogers argued with them, pointing out that food is made up of chemicals—but then he realized that those molecules could be used to make his barrier. "Then we'd be using food to preserve food," he continued. "Philosophically, how could you argue with that?"

Rogers spent the next couple of years finishing his dissertation and trying to talk himself out of pursuing his idea. "I called my mom and told her, and she said, 'Sweetie, that sounds really nice, but you don't know anything about fresh produce.'" Still, he found himself at the library, reading books and papers on what is happening at a cellular and metabolic level in plant matter as it ages and decays. He entered a venture competition at the university and won. "I kept coming back to all the new opportunities that would be created if you could solve this," he said. "Finally, I was like, *I don't even have a girlfriend. If there's a time to do this, it's now.*"

In 2012, with a small grant from the Gates Foundation and some seed money from private investors, the newly minted Dr. Rogers founded Apeel. At the time, he had no idea what the coating would be—he was just sure it could be made. "It took me two years before I even had a hypothesis that the right solution was lipids," he said. As Rogers walked me through the company's new, much larger headquarters, I began to understand the scope of the task that faced him.

In one lab, a team of scientists was breaking down the rusty red flakes left after a local cannery pressed tomatoes for sauce, in order to mine it for barrier molecules. "We are constantly on the hunt for new food waste that we could use as a feedstock," explained Rogers. "Avocado seeds are really awesome for us, because they're full of fat to feed a baby plant."

Before Rogers could even start combining those molecules to build a barrier, several questions had to be answered. First of all, in order for Apeel's coating to qualify as food in the eyes of the FDA, it has to be made exclusively from materials that are found in high concentrations in the types of fruits and vegetables we eat every day—so Rogers had to figure out what those molecules are, in order to set the parameters of his palette. Then, for each new fruit and vegetable that Apeel hopes to coat, the company's scientists have to establish the nutrient levels and flavor volatiles of a perfectly ripe specimen, as well as the precise rate of respiration and the oxygen–to–carbon dioxide ratio that are most conducive to the expression and maintenance of those qualities. All of these experiments have to be performed afresh for different varieties of those fruits and vegetables—Hass, Bacon, and Fuerte avocados, for example—because they each metabolize according to their own, subtly different rhythm.

Apeel works backward from an understanding of what atmospheric blend keeps an avocado so chill that it's barely metabolizing, to designing the barrier structure that will create those conditions, to finding the molecules that will arrange themselves to form that structure. "The molecules don't react with anything, they don't combine with anything, they're not creating any new chemistry," Rogers explained. "Apart, they do nothing—but as they dry together, they assemble into more than the sum of their parts."

To test how its formulations hold up, Apeel starts by sending its own instruments out to accompany produce along the supply chain. Using that data, it programs an environmental chamber to subject Apeel-coated fruit and vegetables to the exact conditions they will encounter in the real world. To make spraying Apeel seamless for fruit growers, the company has invested in a 3D-printing lab where it can design and test the hoses, nozzles, and blowers needed to apply and dry the coating correctly. For supermarkets, it built a time-lapse photography studio where coated and uncoated radishes and tomatoes sit side by side under professional lighting and a camera on rails glides around taking overhead shots of each pair every hour. "When we first started talking to retailers, we said, 'Hey, we've got a product that gives you better-tasting, longer-lasting, more nutritious produce,'" said Rogers. "They said, 'Okay, but what does it look like?'" By running the produce portraits through various algorithmic analyses, he can now tell produce buyers exactly how much longer a lemon coated with Apeel will remain bright yellow and blemish free.

The company's first formulation was for the finger lime, an Australian fruit that looks a little like an oversize cornichon but is filled with tiny, caviar-like citrus bubbles. It's grown by a neighboring farmer as a specialty crop for restaurants and, without an Apeel coating, has an extremely short shelf life. When I visited, Apeel avocados were two to three weeks away from their first appearance in stores. Apeel-coated lemons, cucumbers, mangoes, and organic apples have since joined them on supermarket shelves, and asparagus, tomatoes, blueberries, and peppers are all on their way—the company told me that it has developed different formulations for more than three dozen fruits and vegetables already. For avocados, its coating means that not only does the fruit last more than twice as long as it would otherwise, but the narrow window in which it is perfect—ripe and creamy but still green—is extended by four or five days. What's more, because the fruit is relaxed, it is more resistant to other stresses, like mold, reducing the need for chemical fungicides.

Today Apeel is used as a supplement to refrigeration, not a replacement. "That's the way that we do things today, because the dogma in the fresh-produce industry is the cold chain, the cold chain, the cold chain, and did I mention the cold chain?" said Rogers. In the lab, Apeel has begun to see evidence that it will be able to quadruple shelf life—at which point it will have effectively matched refrigeration's powers to slow produce decay. In other words, we'd get the same result from a coating made from food waste that requires very little energy for its one-time application as from the entire power-hungry and labor-intensive system we've built to keep our food cold. "That's our North Star," Rogers told me. "If you can do that, think about what it means for people who don't have access to that infrastructure."

The implications of supplying a different solution to the problem of preservation are enormous and wide-ranging—just as the effects of refrigeration have been. "I spend an embarrassingly large amount of my free time working through the permutations of how this plays out," Rogers confessed. In the short term, Apeel is already working to connect smallholders in Latin America and Africa with overseas markets. Demand for avocados, particularly in China, is increasing at such a rate that supply can barely keep up, and the fruit grows well in many of the poorest parts of the world. Apeel could also bring more diversity to our plates by making the most fragile fruits—finger limes, for example—robust enough to survive the commercial supply chain. Freed from the requirement to select for shelf life, plant breeders could focus on improving nutrition, flavor, and the ability to withstand the higher temperatures, increased pest pressures, soil salinity, and drought that accompany climate change.

On a more abstract level, Apeel could redefine freshness once again. "To me, freshness means higher concentrations of the molecules that the fruit produced when it was on the plant," Rogers said. Where once freshness had temporal and spatial constraints—it meant something that was harvested recently and nearby—it might one day become a physiological status that can be defined according to levels of specific chemicals. Maybe a fruit is really only as old as it feels on the inside.*

---

*When, as a joke, I asked Rogers whether Apeel worked as an antiaging treatment for humans, he said that the oxidation and water loss that the coating prevents in fruit are the same things

Beyond that, however, it's easy to speculate and hard to predict. If there's one thing that the story of refrigeration should teach us, it's that changing how we relate to something as fundamental as food has ripple effects extending beyond anything the technology's inventors could have possibly imagined. "I honestly don't know how this plays out in the long term," said Rogers. "What I do know is that there's not going to be a long term if we don't solve some of these short-term problems."

I n 1931, in her monthly feature "Art in Everyday Living" for the popular *Golden Book Magazine*, journalist Leonora Baxter speculated on what the world might look like without refrigeration. "If the stupendous system of food preservation and transportation which supports us were interfered with, even for a short time, our present daily existence would become unworkable," she wrote. "Cities with thousands of inhabitants would fade away," she continued, conjuring up a desperate future in which humans were reduced to beasts in their struggle to survive without a technology that had barely existed thirty years earlier. In short, she concluded, "it is not extravagant to say that our present form of civilization is dependent upon refrigeration."

A couple of years later, Benito Mussolini, the journalist turned Fascist dictator, wrote to the editor of the Italian *Dizionario moderno*, suggesting the addition of some twenty new words to the forthcoming edition. One of these "Mussolini-isms" was *frigoriferare*—literally, to refrigerator-ate—meaning to silence, to sideline, or to not take into account. The Italian friend who pointed this out to me demonstrated its usage in a sample sentence: Il Duce refrigeratored dissent. I might propose another: over the past century, the refrigerator has effectively refrigeratored its critics, refrigeratored its alternatives, and refrigeratored its costs.

I am not antifridge. I have a French-door refrigerator—apparently a

that contribute to wrinkles, so . . . potentially? "We'd rather be the company that solved the food problem and *then* went into cosmetics," he told me, laughing.

very datable one—and I have no desire to give up the ability to enjoy a cold beer and to store butter and cheese through the Angeleno summer. Indeed, unless you are prepared to also grow and slaughter or forage and hunt all your own food, living a fridge-free lifestyle in the developed world is little more than a gesture, albeit an instructive one: you're still being sustained by the cold chain, even if you've banished its frosty fetters from your home.

If history is any guide, it seems unwise to bet against the fridge. When the first and, until recently, the only popular book on refrigeration was published in 1953, its author, historian Oscar Anderson, suggested that this still-new technology would likely soon be improved upon and replaced—perhaps, he suggested, by freeze-drying. Seven decades later, we are more dependent on mechanical cooling than Anderson could have ever imagined.

Since the early days of the ice trade, refrigeration has changed everything about how and what we eat—reshaping trade, transportation, politics, and economics in the process. It has redesigned not only the contents of our plates but also our bodies, our homes, our cities, our landscape, and the global atmosphere. And it has flipped our initial suspicion, or "frigoriphobia," of its space- and time-warping powers into full-blown codependence. Many of those dramatic changes are for the better: only a masochist would wish for a return to the days of swill milk, no bananas, and semipermanent food poisoning. Fridges have freed women from daily shopping and made fresh food both affordable and available year-round. For some farmers, refrigeration has offered a route to riches—or a way to escape the land.

Still, a wholehearted appreciation of those benefits shouldn't prevent us from counting the cold chain's costs and considering its alternatives—especially as the rest of the world starts to refrigerate. Some of its disadvantages are inherent to the technology. It takes a lot of energy to remove heat, which has environmental consequences but also tends to create socioeconomic ones, as smaller farmers growing a diversity of foods for local consumption are less likely to benefit than large landholders cultivating export crops at scale. The resulting concentration and intensification have their own ecological impacts.

Other outcomes are enabled by the fridge rather than fundamental to it. Food waste is used as a justification for refrigeration, but refrigeration also seems to encourage waste. The United States has the world's most developed cold chain, yet it wastes almost as much of its food, by percentage, as Rwanda does for lack of one. Researchers have found that household food waste in China's urban areas has already begun to grow rapidly. Refrigeration tends to shift where the waste takes place, as opposed to eliminating it. Increased wealth typically leads to increased consumption of meat and dairy—a dietary transition that wouldn't be possible without the reduced prices and reliable supply enabled by refrigeration. Given the carbon footprint of livestock, and cattle in particular, if expanding the global cold chain leads more people to adopt an American-style diet, we're going to need a bigger planet. What's more, although raising animals at industrial scale involves cruelties both small and large, those take place out of sight thanks to refrigeration, enabling meat eaters to ignore the consequences of their consumption. "Forgetfulness was among the least noticed and most important of its byproducts," as William Cronon wrote, describing the impact of Swift's refrigerated railcar.

In short, our food system is frostbitten: it has been injured by its exposure to cold. Part of the reason for that is that refrigeration was implemented, for the most part, in order to optimize markets rather than human and environmental health. Many Americans in the early years of the twentieth century recognized the commercial motivation behind its adoption, which helped fuel their suspicions of this novel technology. Cold was a key ingredient in the construction of a food system that prioritizes convenience, abundance, and profit over public health or planetary boundaries. Understanding its role in this system is important, not for the purpose of assigning blame but rather for its potential to help us reimagine and redesign it. To put it another way, refrigeration is essential to the food system we have, but a refrigerated food system doesn't have to look like this. We shouldn't conflate the goal with the tool.

In an essay reflecting on the adoption and effects of a different preservation technology—pasteurization—historian Elting Morison noted the tendency of machines that are invented to satisfy a particular human need or desire to extend their influence beyond that remit, establishing their own logic. "The system of ideas, energy, and machinery we have created to serve some essential human needs, it now appears, may, if not sufficiently tended, shrink human beings to the restricted set of needs the system was designed to satisfy," he warned.

Part of the problem is that, as psychologist Abraham Maslow famously put it, if the tool you have is a hammer, you tend to see every problem as a nail. Ever since we gained the ability to manufacture cold, refrigeration has become the default answer to the problem of preservation. If we unlearn the association between freshness and fridges, cooling could become just one of the ways we store and move food—used more sparingly, more thoughtfully, and much more sustainably. Not putting all our eggs in a single chilled basket would also bolster the resilience of our food system, whose underlying fragility has been recently laid bare by empty supermarket shelves during COVID-19 and Brexit.

Meanwhile, if you bear in mind that refrigeration tends toward scale, you can put in place measures to mitigate unwanted effects, from French regulations on supermarket size to the ACES approach of creating community cooling hubs in order to empower Rwandan smallholders to remain independent. "You have to aggregate at some point in the system," explained Toby Peters. "But if you do it with a community-owned packhouse, you can do it without throwing farmers off their land and while allowing as much of the value to remain with them as possible."

This is not a new concept: in the early decades of the twentieth century, Polly Pennington was also a big proponent of community refrigeration hubs, but they failed to catch on in the United States. In a less individualistic culture, and with the American example in mind—where 3 percent of farms produce nearly half of all the country's food and enjoy an income that is more than four times the average earned by the other 97 percent—it's possible that countries in the developing world might choose differently. It's even possible that their example could help those of us who live in the developed world appreciate that the particular contours of our refrigerated food system are contingent rather than inevitable—that we could imagine it otherwise.

In the developed world, reducing our dependence on refrigeration might also allow us to rebuild our relationship with food. The perpetual winter of known and steady temperature that underlies the permanent global summertime of the supermarket's cornucopia has divorced us from food production and severed our synchrony with the seasons. Natalia Falagán told me she had been working on another project, called Rurban Revolution, which promotes urban agriculture not as a way to feed the city but rather as a way to reconnect city dwellers with the value of food.

"We've been working on lettuce," she said. "It's amazing how little people know about how it grows, when it grows, how long it takes to grow." Her hypothesis—which, she told me, has proven correct thus far—is that if people have the experience of tending to a fruit or vegetable crop themselves, at however small a scale, they will eat more of it and waste less.

Reconnected to the cycle of the natural world, we might see through the cold chain's illusory promise of eternal abundance. And we might once again treat food—our most intimate connection to Earth—with the respect it deserves.

Epilogue

MELTDOWN

The Norwegian archipelago of Svalbard lies in the Arctic Ocean, closer to the North Pole than to the Arctic Circle. It is the northernmost point reachable by scheduled passenger plane—farther north from Oslo than Oslo is from Rome; far enough north that in summer the sun never sets, and in winter polar night prevails. When I visited in October, the sun rose at about 10:00 a.m., briefly illuminating the gray, pebble-strewn valleys and steep, glacier-covered mountains that encircle the colorful houses of the island chain's main settlement, Longyearbyen. Just a few hours later it set, plunging the landscape back into shades of violet blue, occasionally illuminated by the greenish glow of the aurora borealis.

Having begun as a whaling port and evolved into a coal-mining town, Longyearbyen is now a hub for tourists hoping to catch a last glimpse of climate change's poster child, the polar bear. Under the terms of the 1920 treaty that recognized Norwegian sovereignty over the archipelago, citizens of any nation are allowed to live and work in Svalbard, and the archipelago's second-largest town, Barentsburg, is almost entirely inhabited by ethnic Russians. Another Russian settlement, Pyramiden, was abandoned in 1998; its crumbling apartment blocks, gymnasium, and Lenin bust (the northernmost in the world) are a forlorn monument to Soviet aesthetics.

Most of Svalbard's 2,900 inhabitants now work in government administration, research, or tourism, although a few coal mines are still operational. No one is allowed to be born in Svalbard—pregnant women fly to the mainland to give birth—and although people can die there, burial has been banned since the 1950s, when scientists exhumed the bodies of those who died during the Spanish flu outbreak and found that the permafrost had preserved their tissues so well that they still contained live viral material. No fruits, vegetables, or grains grow there, and, although locally caught reindeer, seal, grouse, and whale are all found on local restaurant menus, the bulk of the island's food supply is brought by container ship or on one of the two daily flights from Oslo.

Nonetheless, this Arctic desert is home to more crop diversity than anywhere else on Earth. A couple of miles outside of Longyearbyen, a steel doorway embedded in a snow-covered mountain leads into the Svalbard Global Seed Vault: refrigeration's great promise to preserve the future of food.

The so-called doomsday vault is not open to the public—in fact, it's rarely open at all. Since the first seeds were deposited in February 2008, that massive metal door has swung open only a handful of times a year, mostly to receive new accessions. I was lucky enough to be invited to tag along with Åsmund Asdal, the Seed Vault's coordinator, on one of those occasions, to see how a subterranean freezer has become the backup system for the world's crops.

The idea—to store a reserve copy of each of the different seeds preserved by all of the world's agricultural gene banks—first emerged in 2003. The commercial crop varieties grown in fields today contain only a tiny fraction of the genetic diversity of their species. Different genes, as well as mutations of each individual gene, called alleles, carry the code for different potential traits, from pest resistance to particular flavor notes and from blossom timing to drought tolerance. If the seeds from old or forgotten varieties and wild relatives aren't saved in gene banks, plant breeders have nothing to draw on to develop new and improved crops—or to save existing ones from fatal threats.

When a scientist like Harry Klee wants to develop a new tomato, he starts by searching through gene banks to find varieties that have the traits he wants, drawing on that tried-and-tested DNA as his source material rather than trying to synthesize a new gene from scratch and insert it. But having access to that diversity isn't just nice; it's necessary.

In the late 1960s, a virus carried by a bug called the brown planthopper started spreading across Asia, shriveling rice on the stem. In some regions, entire harvests were lost. Indonesia, one of the worst-affected countries, was estimated to have suffered more than $100 million in damage alone, without taking into account the money spent on pest control. As farmers faced ruin and countries struggled to feed their populations, one of the world's most essential crops was facing an uncertain future. In response, scientists screened nearly 26,000 varieties of rice stored in gene banks across the region before finding a traditional variety from central India that carried a gene necessary for resistance to the virus. They successfully bred it with domesticated rice, averting disaster. Much of the rice grown in Asia today is the offspring of that cross.

In the 1990s, agronomist Cary Fowler surveyed the world's gene banks for the UN Food and Agriculture Organization. In his report, he documented the destruction of seeds during conflicts in Burundi, Afghanistan, and Iraq, due to flooding and fire in the Philippines, and following equipment malfunctions in Italy and Madagascar. Each time seeds were lost, unique genes—DNA that had evolved once over thousands of years and might never evolve again—were lost with them. Given how essential the contents of agricultural gene banks are to humanity's continued ability to feed itself, Fowler was horrified at the vulnerabilities he saw.

In response, he came up with the concept of creating a Noah's ark for seeds: the Svalbard Global Seed Vault. It is intended to ensure that, if disaster strikes a particular gene bank, all is not lost. A duplicate copy of its seeds, lodged deep inside the Arctic permafrost, will ensure that future generations still have access to all the genetic resources millennia of evolution can offer, in order to breed crops that can withstand whatever challenges emerge in the decades to come.

Its location was chosen, Fowler wrote, because it was "virtually immune to the kinds of problems we had witnessed elsewhere." Svalbard is distant from the front lines of conflict, out of range of hurricanes and tornadoes, and unaffected by earthquakes. Burying the vault in a former mine shaft meant that a thick layer of frozen soil and rock should preserve the seed samples, even without power.

Just before my visit, the doomsday vault had been called upon for the first time. War in Syria had cut off access to a gene bank that holds more than 140,000 varieties of wheat, fava beans, lentils, chickpeas, and barley—many of them extinct in the wild. Fortunately, the gene bank's staff had deposited 80 percent of its holdings at Svalbard before they were forced to flee. As they set up a new headquarters in Lebanon, they withdrew their seeds to reestablish the collection. (Once the stock is replenished by being grown and harvested in greenhouses nearby, backup copies will be deposited in the Svalbard vault again.) No one knows what has become of the old gene bank. From Lebanon, the staff are in touch with their former neighbors, who report that "sometimes, the lights are on." Among the team's current projects: developing a variety of wheat that uses genes from one of the seeds retrieved from Svalbard to help fight a fly infestation currently decimating fields in Washington State.

Thirty boxes of seeds were due to be deposited during my visit in 2016. Some had arrived on my flight, others on a shipping container the week before. Our small party—a couple of other Norwegian staff members, an Estonian photographer, a Japanese film crew, a Brazilian agronomist, and two Swedish journalists—had accompanied Åsmund Asdal to the airport to watch him pass the boxes through the X-ray machine for oversize baggage at security. The boxes aren't opened after they depart the donating institution, but they are scanned to make sure they contain only seeds.

There were a couple of blue plastic tubs holding cowpeas, mung beans, and Bambara groundnuts from the International Institute of

Tropical Agriculture in Ibadan, Nigeria, a half dozen more labeled "BEANS" from Colombia, a couple of green ones containing barley from Okayama University in Japan, and some bulging, Saran-wrapped cardboard boxes from Burundi. "These are our first deposits from Singapore," said Asdal, lifting up a Perspex shoebox crammed with foil envelopes and wrapped with a gray ribbon and putting it onto the belt. Asdal told me he was from the Telemark region of Norway, where his relatives used to harvest ice for the queen of England from a nearby pond. "They said that if she could read *The Times* through the ice, the quality was approved," he said. "The harvesting stopped in the 1960s, when I was six—everyone had refrigerators by then."

Standing next to me to watch the seed scanning was Bente Næverdal, who lives in Svalbard and manages the vault for Statsbygg, the government institution that operates state-owned buildings. The responsibility didn't bother her, she told me. "Svalbard is a safe place. I'm not afraid anything would happen."

Not much goes on at the vault day to day, she explained. The heavy steel door at the entrance remains shut; the keys are locked in a safe in her office. She can monitor the temperature and gas levels in the seed chamber from her desk, as well as any activity from the motion, fire, and smoke detectors and security cameras that protect it. There are not many moving parts—just a transformer to power the pumps that remove snowmelt that sometimes seeps into the entrance tunnel and the refrigeration equipment that brings the seed chamber down to minus 0.4 degrees, the temperature at which seeds can remain viable for decades at least, and likely centuries. "It's the most important refrigerator in the world," Næverdal told me. "It doesn't have to do anything fancy—it just has to work."

With the boxes scanned, we made plans to meet at the vault the next day for the deposit to take place. Asdal warned one of the Swedish journalists that her gel pen would freeze in the seed chamber and urged us to dress as warmly as we could. "One guy spent a half hour in there in shorts, but he was German," he said. "It will be a very cold day."

Clad in my RefrigiWear jacket and a pair of borrowed ski pants, I arrived at 9 o'clock the next morning. In the ultramarine light before dawn, the vault's entrance rose like a shard from the mountainside. The door was open a sliver, glowing yellow. The night before, it had rained, and the mountain above and fjord below were invisible behind a blanket of thick gray fog. All I could see as I peeked inside was a portable construction light hanging on a ladder, buckets surrounding its base.

Asdal broke the news: water had entered under the door and shorted the transformer, knocking out the pumps and the refrigeration system for a couple of hours. The "flood" was an inch and a half deep at most, and it had reached only about 230 feet into the 400-foot-long tunnel before freezing solid, but the combination of water and electricity made it too dangerous for us to enter. After planning for months and traveling for days, this was the closest I would get to the Svalbard Global Seed Vault freezer, described by its founder Cary Fowler as offering "'fail-safe' protection for one of the important natural resources on Earth."

Asdal was quick to reassure us all that the seeds were fine and the temperature had remained steady inside the chamber. Even so, he and his colleagues were visibly shaken. "It is the thing every gene bank manager is afraid of—losing refrigeration," said Gustavo Heiden, the Brazilian scientist who coordinates the rescue of crop wild relatives for his own country's gene bank. "I emailed my colleagues and said, 'Look, it can happen here too, even at the doomsday vault.'" "It never rains like that on Svalbard," added Næverdal. "You are seeing climate change." "This is supposed to last for eternity," said Asdal, shaking his head.

That October was the warmest and wettest in Svalbard's history. In March 2019, the Norwegian Meteorological Institute announced the Arctic archipelago had experienced one hundred consecutive months of above-average temperatures. No other place in the world is heating up faster: the permafrost itself has become unstable.

———

Back at my hotel, pondering what to do with my suddenly free day, I picked up a book of Viking folklore. In old Norse mythology, I read, the coming of Ragnarök, the twilight of the gods, is preceded by endless winter. The apocalypse is heralded by a creeping cold that expands to encircle the globe, holding the world fast in its remorseless grip. "First, there is a winter called the Fimbul-winter, when snow drives from all quarters, the frosts are so severe, the winds so keen and piercing, that there is no joy in the sun," says Odin, the god of poets. Then wolves swallow the sun and moon, floods overwhelm the land, and Earth sinks into the sea. "Fire rages, heat blazes, and high flames play 'gainst heaven itself," Odin continues. By the end, the gods are dead.

With power over space, time, and the seasons—a power that rests on the creation of an all-consuming artificial winter—we are almost as gods. As the new Arctic we have built for our food plays its part in melting the old Arctic faster than anyone could have imagined, we had better muster the collective wisdom and action to avoid their fate.

Acknowledgments

My adventures in the artificial cryosphere began in early 2010, when I emailed Matt Coolidge at the Center for Land Use Interpretation to see whether he'd be interested in collaborating on a project to document the US coldscape. Our research, which was supported by the Graham Foundation for Advanced Studies in the Fine Arts and conducted in collaboration with Ben Loescher, Steve Rowell, Sarah Simons, Aurora Tang, and Marissa Looby, culminated in an exhibition—*Perishable: An Exploration of the Refrigerated Landscape of America*—on display at CLUI's Los Angeles headquarters in spring 2013. Some early research on this topic was also conducted at Columbia University's Graduate School of Architecture, Planning and Preservation alongside Kubi Ackerman and Eugenia Manwelyan, with support from the President's Global Innovation Fund.

My journey continued overseas when Michael Pollan and Malia Wollan invited me to join the inaugural class of fellows for the University of California, Berkeley 11th Hour Food and Farming Journalism Fellowship. With their support, and the invaluable assistance of guest editors Jack Hitt and Alan Burdick, I was able to report on China's refrigeration boom for *The New York Times Magazine*. The resulting story benefited greatly from the translation and reporting assistance provided by Owen Guo and the editorial attention of Hugo Lindgren and Bill Wasik, as well as fact-checking by Karen Fragala-Smith. Parts of that story appear in this book, alongside previously unpublished reporting from China. Michael Pollan has continued to be a supporter of my writing, on this topic and others, and his encouragement, assistance, and advice have been invaluable.

Sina Najafi at *Cabinet* and Craig Cannon and Tim Hwang, authors of *The Container Guide*, were early enthusiasts, and I am grateful to them for the opportunity to share parts of my research in their publications. My reporting in Rwanda was commissioned by *The New Yorker*, where the resulting story was expertly edited by the incomparable Leo Carey, fact-checked by Natalie Meade, and copyedited by Andrew Boynton.

I owe a huge thanks to the Knight Science Journalism Program at MIT, which funded my research into the "fridge diet" with a 2020–2021 Project Fellowship. I'm grateful to program staff Deborah Blum, Ashley Smart, and Bettina Urcuioli, as well as to my fellow Fellows, for all their help and companionship. One of them, Lindsay Gellman, is not only an accomplished journalist in her own right, but also fact-checked this book, tirelessly tracking down every last detail. I can't thank her enough. (Of course, any remaining errors are entirely my responsibility.)

This book was also supported by a generous grant from the Alfred P. Sloan Foundation for the Public Understanding of Science & Technology. By funding editorial, scientific, and fact-checking review, in addition to providing a budget for research materials, travel, and an authorial stipend, the Sloan Foundation helped make this book much stronger. I owe particular gratitude to vice president and program director Doron Weber for his ongoing investment in my explorations at the intersection of food and science.

I am lucky to have encountered several librarians, archaeologists, and archivists who have been beyond generous with their assistance. Thanks to Megan and Rick Prelinger of the Prelinger Library; Gretchen Rings, museum librarian and head of library collections at the Field Museum; Diane Wunsch, special collections librarian at the USDA National Agricultural Library; Clayton Harper and Josh Smith at the American Merchant Marine Museum; Sarah Wilmot, outreach curator and science historian at the John Innes Centre; Frank Bowles at Cambridge University Library; Katherine Fox at the Baker Library at Harvard Business School; and Joe Bagley, city archaeologist at Boston Landmarks Commission; as well as staff at the National Archives at Kew, the Minnesota Historical Society in Saint Paul, and the British Library. In addition, Rebecca Federman of the New York Public Library went above and beyond to help me access online resources; Sci-Hub was (and thankfully still is) an invaluable resource for those whose work depends on consulting research journals but who are not blessed with academic library access; and, when all else failed, the extraordinarily generous Irwin Goldman at the University of Wisconsin–Madison downloaded papers for me.

Over the past decade, dozens and dozens of people have taken the time to meet and talk all things cold with me, answering my questions, sending me references, sharing their insights, and showing me around fascinating corners of the artificial cryosphere. In addition to the individuals you have already met within the pages of this book, for whose generosity and enthusiasm I am endlessly grateful, I owe thanks to the following: Daniel Cooke, Javier Cortés, Chris Mckeon, and Peter Yee at Americold; Paula Glover, Cody Hughes, and Scotty DePriest of Refrigi-Wear; Nolwenn Robert-Jourdren at the International Institute of Refrigeration; Corey Rosenbusch, Bsrat Mezghebe, and Megan Costello at the Global Cold Chain Alliance; Laura Oleniacz and Amanda Padbury at North Carolina State University; Malcolm Tucker at the London Canal Museum; Christian James of the Grimsby Institute of Further and Higher Education; Angel Medina Vaya and Daniel Simms at Cranfield University; Kevin Senseny at Citrosuco North America, Inc.; Jon Miles at NewCold; Michelle Linn at Apeel; Bryan Armbrust, Jared Oren, Charles Smith, and Kerry Larsen at the US Army Corps of Engineers Cold Regions Research and Engineering Laboratory; Wayt Gibbs and his colleagues at Global Good and Intellectual Ventures; Kota Homma at Snow Shop Kobiyama; Tom Rodriguez and Thom L. Thomas at Los Angeles Cold Storage; Paul Singh at University of California, Davis; Sujoy Roy at Lawrence Berkeley National Laboratory; Matt Poese at Pennsylvania State University; Tim Ray at Maritime International Cold Storage in New Bedford; Kevin Davis at Rebound Technologies; author Steve Silberman and beekeeper William Parrett; Mike Tipton at the University of Portsmouth; Deana R. Jones of the USDA Agricultural Research Service; Martin J. Loessner at ETH Zurich; Benjamin Chapman at North Carolina State University; and Sam Bompas of Bompas & Parr.

In China, above and beyond the names already mentioned, I also owe thanks to Ralph Bean, Josephine Lau, Dechen Pemba, Tim McLellan, Douglas MacDonald, Mark Dunson, Jim Harkness, and many more. In Rwanda, I am grateful to Amy Maxmen, Andrew Blum, Ayubu Kasasa, Sam Peters, Leyla Sayin, and Issy MacFarlane. Luigi Guarino, Jeremy Cherfas, and Brian Lainoff helped secure my invitation to visit the Svalbard Global Seed Vault—my thanks are due also to them.

Alexis Madrigal shared enthusiasm, ideas, and his thoughts on the location of Croker Global Foods in Wolfe's *A Man in Full*; Martin Dixon at Sub Brit helped me reach Sylvia Beamon; Kevin Slavin introduced me to Kipp Bradford; Darien Williams pointed me to Miss Chiquita and her catchy jingle; Federico Sanna introduced me to Mussolini's neologisms and translated the relevant article for me; Jim Webb and Melissa Forbes hosted me while I visited the USDA National Agricultural Library; Jenny Raymond and Jake Barton were my constant cheerleaders and also kindly hosted me while I visited Master Purveyors, D'Arrigo New York, and the American Merchant Marine Museum; my brother, Mat, is always game to answer questions on classical matters; and my parents generously loaned me accommodation for a much-needed writing retreat.

I've been thinking and talking about refrigeration for a decade now, so the list of friends, acquaintances, and strangers who have not only put up with my monologues on the topic but also shared suggestions, thoughts, and encouragement is long. I'm grateful to them all, but particularly to Ellie Robins and Lizzie Prestel for accompanying me every step of the way, and to Wayne Chambliss for not only alerting me to ethylene's connection to the Oracle of Delphi but also dedicating hours he didn't have to reading the manuscript, twice. His comments and suggestions have improved it immensely. Cynthia Graber, my cohost and partner in making *Gastropod*, has been endlessly patient and supportive throughout.

Helen Thorpe with the Lighthouse Writers Workshop offered invaluable advice during the writing process and made transformative editorial suggestions on an early draft. My agent Nathaniel Jacks was a source of support throughout; my gratitude also to Lyndsey Blessing and the rest of the team at InkWell Management.

The team at Penguin Press—Casey Denis, Victoria Lopez, Hilary Roberts, Alicia Cooper, Darren Haggar, Lauren Lauzon, and Juli Kiyan—is the best. I can't thank them enough. Above all, I'm grateful for the privilege of working with Ann Godoff, who has an entirely justified reputation as one of the industry's most brilliant and incisive editors. Her guidance in shaping and shepherding this book was invaluable.

Finally, there aren't enough words to thank my husband, Geoff Manaugh, who not only came up with the title for this book, found and suggested sources I would never have come across on my own, accompanied me on visits to dozens of refrigerated warehouses despite hating the cold, and read and edited every word I've written on this topic, but also gave me the inspiration and encouragement I needed to become a writer in the first place. I love you.

Selected Sources

In the following pages, I've provided a list of the sources on which I drew for each chapter, as a guide to the experiences, visits, interviews, books, articles, documentaries, podcasts, exhibitions, and online resources that supplied me with facts or informed my thinking.

When providing contemporary dollar equivalents for historic values, I have relied on MeasuringWorth.com.

Throughout, I drew on the encyclopedic Roger Thévenot, *A History of Refrigeration throughout the World* (Paris: International Institute of Refrigeration, 1979). Two books that were also invaluable throughout and that shaped my thinking on this topic are Susanne Freidberg, *Fresh: A Perishable History* (Cambridge, MA: Belknap Press, 2010), and Carolyn Steel, *Hungry City: How Food Shapes Our Lives* (London: Chatto & Windus, 2008).

1. Welcome to the Artificial Cryosphere

Much of this chapter is based on my experience working at Americold facilities in Southern California in March 2018. For this chapter, I also conducted interviews with Adam Feiges, former principal at Cloverleaf Cold Storage (January 2013); Toby Peters, professor of cold economy at the University of Birmingham (numerous conversations since 2014); Denis P. Blondin, assistant professor in the Faculty of Medicine and Health Sciences at the Université de Sherbrooke (January 2023); and Ryan Silberman, co-CEO, Scotty DePriest, vice president of operations, and Cody Hughes, director of quality, at RefrigiWear (January 2022). I visited Shawn Deaton, operations director at the Textile Protection and Comfort Center at North Carolina State University, in February 2022. Mike Tipton, professor of human and applied physiology at the Extreme Environments Laboratory, School of Sport, Health & Exercise Science at the University of Portsmouth, also provided insights.

I drew on the following books: Bill Streevor, *Cold: Adventures in the World's Frozen Places* (New York: Back Bay Books, 2010); Paul Theroux, *The Mosquito Coast* (Boston: Houghton Mifflin, 1982); Varlam Shalamov, *Kolyma Stories* (New York: New York Review of Books, 2018); Ken Parsons, *Human Thermal Environments: The Effects of Hot, Moderate, and Cold Environments on Human Health, Comfort, and Performance* (Boca Raton, FL: CRC Press, 2007); Ingvar Holmér and Kalev Kuklane, eds., *Problems with Cold Work: Proceedings from an International Symposium Held in Stockholm, Sweden, Grand Hôtel Saltsjöbaden, November 16–20, 1997* (Solna, Sweden: Arbetslivsinstitutet, 1998); and Tom Wolfe, *A Man in Full* (New York: Farrar, Straus and Giroux, 1998).

Royal Society, "Royal Society Names Refrigeration Most Significant Invention in the History of Food and Drink," news release, September 13, 2012.

To make sense of my experience of working in the cold, I also consulted the following: Medical Record, "Effects of Intense Cold upon the Mind," *Ice and Refrigeration*, July 1895, p. 30; Ingvar Holmér, "Effects of Work in Cold Stores on Man," *Scandinavian Journal of Work, Environment & Health* 5 (1979): 195–204; Yutaka Tochihara, "Work in Artificial Cold Environments," *Journal of Physiological Anthropology and Applied Human Science* 24, no. 1 (2005): 73–76; Marika Falla et al., "The Effect of Cold Exposure on Cognitive Performance in Healthy Adults: A Systematic Review," *International Journal of Environmental Research and Public Health* 18 (September 2012): 9725; The Editors, "Cold Stress at Work: Preventive Research," *Industrial Health* 47 (2009): 205–6; Hannu Anttonen et al., "Safety at Work in Cold Environments and Prevention of Cold Stress," *Industrial Health* 47 (2009): 254–61; Tiina M. Mäkinen et al., "Health Problems in Cold Work," *Industrial Health* 47 (2009): 207–20; Hein A. M. Daanen et al., "Manual Performance Deterioration in the Cold Estimated Using the Wind Chill Equivalent Temperature," *Industrial Health* 47 (2009): 262–70; Hein A. M. Daanen et al., "Human Whole Body Cold Adaptation," *Temperature* 3, no. 1 (2016): 104–18; Benjamin S. Bleier et al., "Cold Exposure Impairs Extracellular Vesicle Swarm–Mediated Nasal Antiviral Immunity," *Journal of Allergy and Clinical Immunology* 151, no. 2 (February 2023): 509–25; Massachusetts Eye and Ear Infirmary, "Scientists Uncover Biological Explanation behind Why Upper Respiratory Infections Are More Common in Colder Temperatures," news release, December 6, 2022; John M. Grady, "Metabolic Asymmetry and the Global Diversity of Marine Predators," *Science* 363, no. 6425 (January 2019); Ed Yong, "Why Whales, Seals, and Penguins Like Their Food Cold," *Atlantic*, January 24, 2019; William Park, "Why Some People Can Deal with the Cold," *BBC Future*, March 10, 2021; Beth Roars, "You Can Hear Cold," bethroars.com, January 5, 2021; Tim Adams, "Ice Baths and Snow Meditation: Can Cold Therapy Make You Stronger?" *Observer*, May 7, 2017; Tim Lewis, "The Big Chill: The Health Benefits of Swimming in Ice Water," *Guardian*, December 23, 2018; and David Robson, "The Big Idea: How Keeping Warm Wards Off Loneliness," *Guardian*, November 28, 2022.

On the history of the hoodie, I referenced Nazanin Shahnavaz, "The Secret History of the Hoodie," *i-D*, May 23, 2016; Denis Wilson, "A Look under the Hoodie," *New York Times*, December 23, 2006; "Tales of the Hood: The Epic Story of Hoodie Fashion," Epic Hoodie Fashions, November 9, 2018; and "History of the Hoodie," Triple Crown Products, April 19, 2021.

On the challenges posed by ice cream's airiness, see U. K. Dubey and C. H. White, "Ice Cream Shrinkage: A Problem for the Ice Cream Industry," *Journal of Dairy Science* 80, no. 12 (1997): 3439–44.

For data on US and global refrigerated warehouse capacity, I consulted Victoria Salin, "2020 Global Cold Storage Capacity Report," Global Cold Chain Alliance (August 2020). To arrive at a figure for the proportion of refrigerated foods in the average America diet, I drew on US food-availability data from the USDA, calculated the percentage that each food category makes up in the average diet, and then added up those foods that typically pass through the cold chain at some point on their way from farm to fork. The result is an approximation, but an informed one.

2. The Conquest of Cold

Throughout this chapter and in several subsequent chapters, I drew on Carroll Gantz, *Refrigeration: A History* (Jefferson, NC: McFarland, 2015); Oscar Edward Anderson Jr., *Refrigeration in America* (Princeton, NJ: Princeton University Press, 1953); Jonathan Rees, *Refrigeration Nation: A History of Ice, Appliances, and Enterprise in America* (Baltimore: Johns Hopkins University Press, 2013); Barry Donaldson and Bernard Nagengast, *Heat & Cold, Mastering the Great Indoors* (Atlanta: ASHRAE, 1995); and Helen Peavitt, *Refrigerator: The Story of Cool in the Kitchen* (New York: Reaktion Books, 2017).

I. Stop the Rot

The beginning of this chapter was based on my experience building a refrigerator with Kipp Bradford, who was senior research scientist at the MIT Media Lab at the time and went on to found Gradient, where he was chief technology officer. It was informed by input from Judith Evans, director of refrigeration developments and testing and professor at London South Bank University.

After Life: The Strange Science of Decay was directed by Dani Carlaw and Fred Hepburn and aired in December 2011 on BBC Four.

On the struggle to understand cold, I referenced the following: Streevor, *Cold*; John Aubrey, *"Brief Lives," Chiefly of Contemporaries, Set Down by John Aubrey, between the Years 1669 & 1696*, vol. 1, ed. Andrew Clark (Project Gutenberg, 2014); Robert Boyle, "New Experiments and Observations Touching on Cold; or, an Experimental History of Cold Begun," in *The Works of the Honourable Robert Boyle*, vol. 2 (London: A. Millar, 1744); René Descartes, "Meditations on First Philosophy: Meditations III" (1641), trans. Elizabeth S. Haldane, in *The Philosophical Works of Descartes* (Cambridge: Cambridge University Press, 1911).

On food rot and preservation, I drew on Jack A. Gilbert and Josh D. Neufeld, "Life in a World without Microbes," *PLOS Biology* 12, no. 12 (December 2014): e1002020; Sue Shephard, *Pickled, Potted, and Canned: How the Art and Science of Food Preserving Changed the World* (New York: Simon & Schuster, 2000); Maguelonne Toussaint-Samat, *A History of Food* (Hoboken, NJ: Wiley-Blackwell, 2008); Massimo Montanari, *Food Is Culture* (New York: Columbia University Press, 2004); and Clifton Fadiman, *Any Number Can Play* (New York: Avon, 1957).

II. The Ice Harvest

While writing this section, I harvested ice at the Thompson Ice House in February 2014 and returned to enjoy ice cream frozen using our harvest in July 2014. I also drew extensively on Gavin Weightman, *The Frozen Water Trade: A True Story* (New York: Hachette, 2004), and I consulted the Tudor Ice Company Collection at the Baker Library, Harvard Business School, Boston, MA. I spoke with Monica Smith, professor in UCLA's Department of Anthropology and author of *Cities: The First 6,000 Years*, in March 2022.

Although I was in touch with Sylvia Beamon, to my regret we weren't able to meet before her death in 2021. To describe her research and the early history of ice storage, I drew on the following: Sylvia P. Beamon, "Icehouse Reflections: Thirty Years On," *Subterranea* 56 (April 2021): 70–76; Sylvia P. Beamon and Lisa G. Donel, "An Investigation of Royston Cave," *Proceedings of the Cambridge Antiquarian Society* 68 (1978): 47–58; Bob Trubshaw, "Royston Cave and Related Caverns," *Mercian Mysteries*, no. 18, February 1994; Daniel Stables, "A Secret Site for the Knights Templar?" *BBC Travel*, January 1, 2023; Sylvia P. Beamon and Susan Roaf, *The Ice-houses of Britain* (Abingdon, UK: Routledge, 1990); and "Sylvia Beamon: The Life of Royston Academic, Archaeologist and Activist," *Royston Crow*, December 20, 2021.

I also referenced the following: Nicholas St. Fleur, "What Was Kept in This Stone Age Meat Locker? Bone Marrow," *New York Times*, October 9, 2019; Ruth Blasco et al., "Bone Marrow Storage and Delayed Consumption at Middle Pleistocene Qesem Cave, Israel (420 to 200 ka)," *Science Advances* 5, no. 10 (October 2019); Elizabeth David, *Harvest of the Cold Months: The Social History of Ice and Ices* (London: Michael Joseph, 1994); Tim Buxbaum, *Icehouses* (Oxford: Shire, 1992); Sarah Mytton Maury, *An Englishwoman in America* (London: Thomas Richardson and Son, 1848); "The Confectioners Have Been Able to Lay In a Store of Ice to Freeze Their Creams in Summer," *Times*, January 21, 1822, p. 3; Ezra Pound, *Shih-ching: The Classic Anthology Defined by Confucius* (Cambridge, MA: Harvard University Press, 1954); and Susanne Freidberg, "The Triumph of the Egg," *Comparative Studies in Society and History* 50, no. 2 (2008): 400–423.

III. A Machine to Produce Cold

To tell the story of the Columbian Exposition, the "Greatest Refrigerator on Earth," and the tragedy of July 11, 1893, I relied on the following sources: Hubert Howe Bancroft, *The Book of the Fair: An Historical and*

Descriptive Presentation of the World's Science, Art, and Industry, as Viewed through the Columbian Exposition at Chicago in 1893, vol. 1 (New York: Bounty Books, 1893); "Extra: Flame and Death," *Evening World*, July 10, 1893, p. 1; "From the World's Fair," *Clearwater Echo*, April 28, 1893, p. 2; "Over a Score Dead," *Champaign Daily Gazette*, July 12, 1893, p. 1; "Down in the Flames," *Marion Star*, July 11, 1893, p. 1; "Girt by Flames High in Air," *New York Times*, July 11, 1893, p. 1; "Down in the Flames," *Piqua Daily Call*, July 11, 1893, p. 1; "Baptism of Fire," *Union County Journal*, July 13, 1893, p. 1; and Eric Westervelt, "Greatness Is Not a Given: 'America the Beautiful' Asks How We Can Do Better," *All Things Considered*, National Public Radio, April 4, 2019.

On the history of mechanical cold, in addition to Thévenot, Gantz, Anderson, Rees, Donaldson and Nagengast, and Peavitt, as mentioned above, I consulted the following: Joseph Black and William Cullen, *Experiments upon Magnesia Alba, Quick-lime and Other Alcaline Substances; to Which Is Annexed, an Essay on the Cold Produced by Evaporating Fluids, and Some Other Means of Producing Cold* (Edinburgh: W. Creech, 1777); Hamish MacPherson, "William Cullen: Time This Scottish Inventor Was Brought In from the Cold," *National*, April 10, 2022; Benjamin Franklin, *Experiments and Observations on Electricity* (London: E. Cave, 1769), 363–68; Andy Pearson, "The Birth of the Refrigeration Industry in London: 1850–1900" (presented before the Institute of Refrigeration, December 5, 2018); Nigel Isaacs, "Sydney's First Ice" (2011), *Dictionary of Sydney*, State Library, New South Wales; "Harrison, James (1816–1893)," L. G. Bruce-Wallace, *Australian Dictionary of Biography*, vol. 1 (1966); Tim Lee, "James Harrison Invented Australia's First Ice-Making Machine, but Is Now Forgotten," *Landline*, ABC, March 31, 2022; J. E. Siebel, "Refrigeration in Its Relation to the Fermenting (Brewing) Industry of the United States," *The Western Brewer: And Journal of the Barley, Malt and Hop Trades* 33 (December 1908): 666–67; Martin Stack, "A Concise History of America's Brewing Industry," *EH.Net Encyclopedia* (July 2003); and Susan K. Appel, "Artificial Refrigeration and the Architecture of 19th-Century American Breweries," *IA: The Journal of the Society for Industrial Archeology* 16, no. 1 (1990): 21–38. I also referenced Robert Browning's poem "Andrea del Sarto," published in the collection *Men and Women*, in 1855.

3. The Way of All Flesh

As I wrote this chapter, I found the following books on the history of meat invaluable: William Cronon, *Nature's Metropolis: Chicago and the Great West* (New York: W. W. Norton, 1991); Chris Otter, *Diet for a Large Planet: Industrial Britain, Food Systems, and World Ecology* (Chicago: University of Chicago Press, 2020); Betty Fussell, *Raising Steaks: The Life and Times of American Beef* (San Diego: Harcourt, 2008); Maureen Ogle, *In Meat We Trust: An Unexpected History of Carnivore America* (Boston: Houghton Mifflin Harcourt, 2013); and Joshua Specht, *Red Meat Republic: A Hoof-to-Table History of How Beef Changed America* (Princeton, NJ: Princeton University Press, 2019).

I. Where's the Beef?

The story of Gustavus Swift is largely based on Louis F. Swift Jr. and Arthur Van Vlissingen, *The Yankee of the Yards: The Biography of Gustavus Franklin Swift* (Chicago & New York: A. W. Shaw, 1927). Throughout this section, I drew on Thévenot, *A History of Refrigeration*, and James Troubridge Critchell and Joseph Raymond, *A History of the Frozen Meat Trade* (London: Constable, 1912), which begins with the stirring line "Much that is extremely interesting, and a little that is romantic, is to be found in the history of the industry to which this book is devoted." On Swift's success, I also consulted Gary D. Libecap, "The Rise of the Chicago Packers and the Origins of Meat Inspection and Antitrust" (Working Paper No. 29, National Bureau of Economic Research, September 1991); and Mary Yeager Kujovich, "The Refrigerator Car and the Growth of the American Dressed Beef Industry," *Business History Review* 46 (1970): 460–82.

To understand how cities were supplied with meat, I consulted Upton Sinclair, *The Jungle* (New York: Doubleday, Page, 1906); Diane Purkiss, *English Food: A Social History of England Told through the Food on Its Tables* (Glasgow: William Collins, 2022); George Dodd, *The Food of London: A Sketch of the Chief Varieties, Sources of Supply, Probable Quantities, Modes of Arrival, Processes of Manufacture, Suspected Adulteration, and Machinery of Distribution* (London: Longman, Brown, Green, and Longmans, 1856); Steel, *Hungry City*; Catherine McNeur, *Taming Manhattan: Environmental Battles in the Antebellum City* (Cambridge, MA: Harvard University Press, 2014); Julius Friedrich Sachse, *The Wayside Inns on the Lancaster Roadside between Philadelphia and Lancaster* (United States: Press of the New Era Printing Company, 1912); Richard Perren, "The North American Beef and Cattle Trade with Great Britain, 1870–1914," *Economic History Review* 24, no. 3 (August 1971): 430–44; Helen Watkins, "Fridge Space: Journeys of the Domestic Refrigerator" (PhD diss., University of British Columbia, 2008); and Charles Dickens, *Great Expectations* (1861; Project Gutenberg, 1998).

My references for Britain's meat famine include *Substances Used as Food: As Exemplified in the Great Exhibition* (London: Society for Promoting Christian Knowledge, 1854); Wentworth Lascelles Scott, "On the Supply of Animal Food to Britain, and the Means Proposed for Increasing It," *Journal of the Society of the Arts*, no. 796 (February 21, 1868); Kenneth J. Carpenter, "The History of Enthusiasm for Protein" (paper presented at the History of Nutrition symposium given by the American Institute of Nutrition at the 69th Annual Meeting of the Federation of American Societies for Experimental Biology, Anaheim, CA, April 21–26, 1985); Mark R. Finlay, "Quackery and Cookery: Justus von

Liebig's Extract of Meat and the Theory of Nutrition in the Victorian Age," *Bulletin of the History of Medicine* 66, no. 3 (1992): 404–18; "Pressed Beef and Desiccated Beef-Juice," *Lancet* 98, no. 2498 (July 15, 1871): 105; Rebecca J. H. Woods, "The Shape of Meat: Preserving Animal Flesh in Victorian Britain," *Osiris* 35 (2020).

To tell the story of Charles Tellier, I also referenced "Too Late to Aid Tellier," *New York Times*, October 20, 1913; "Fund for Charles Tellier," *Cold Storage and Ice Trade Journal* 44 (October 1912): 68; and Robert Bruner, Sarah Costa, and Sean Carr, "Le Frigorifique: Charles Tellier and the Creation of the Cold Chain," Darden Case No. UVA-ENT-0232 (December 20, 2022). For Paris's contemporary urine-diversion scheme, see Nicola Twilley, "Waste Not, Want Not," *New Standard*, 2024.

Archaeologist Louise Cooke pointed me toward the impact of refrigeration on the landscape of the North York Moors. My account also drew on Alastair J. Durie, "Game Shooting: An Elite Sport c. 1870–1980," *Sport in History* 28, no. 3 (September 2008): 431–49; Stephen Croft and Dr. Louise Cooke, "This Exploited Land: The Trailblazing Story of Ironstone and Railways in the North York Moors," Landscape Conservation Action Plan, October 2015; and Richard Christopher Chiverrell, "Moorland Vegetation History and Climate Change on the North York Moors during the Last 2000 Years" (PhD diss., University College of Ripon and York St. John, 1998).

Additional sources included David McWilliams, "Floating Fridges Changed History," *Irish Independent*, May 6, 2015; Ian Arthur, "Shipboard Refrigeration and the Beginnings of the Frozen Meat Trade," *Journal of the Royal Australian Historical Society* 92, no. 1 (June 1, 2006); and Alexi Giannoulias, "57. Photo of the Reversal of the Chicago River (1900)," *100 Most Valuable Documents at the Illinois State Archives: The Online Exhibit*.

II. Better Living through Chemistry

Much of this section is drawn from the Mary E. Pennington Papers, Special Collections, US Department of Agriculture National Agricultural Library, Beltsville, MD.

For my description of the cold storage banquet, I consulted "The National Convention: A Big Success," *Egg Reporter*, November 6, 1911, pp. 22–65; "Cold Storage Luncheon," *Bulletin of the American Warehousemen's Association* 12 (1911): 325; "Poultry, Butter and Egg Men Meet," *Ice and Refrigeration* 41 (November 1911): 177–83; "Chicago Cold Storage Banquet," *San Antonio Express*, October 20, 1911, p. 6; "Try It On San Antonio," *San Antonio Express*, November 6, 1911; "The Great Cold Storage Banquet," *National Provisioner*, September 27, 1913, p. 123; and "The Abuse of Cold Storage," *Inter Ocean* (Chicago), October 30, 1911, p. 6.

On ptomaine poisoning, I referenced "Achievements in Public Health, 1900–1999: Control of Infectious Diseases," *Morbidity and Mortality Weekly Report* 48, no. 29 (July 30, 1999): 621–29; "Leading Causes of Death," National Center for Health Statistics, CDC; Nykole Nevol, "Deadly Summers: Infant and Child Deaths in 19th Century Rochester, New York," *GREAT Day Posters* 87 (2021); Stanford T. Shulman, "The History of Pediatric Infectious Diseases," *Pediatric Research* 55 (2004): 163–76; Sara Josephine Baker, *Fighting for Life* (New York: Macmillan, 1939); Edward Geist, "When Ice Cream Was Poisonous," *Bulletin of the History of Medicine* 86, no. 3 (Fall 2012): 333–60; Bill Bynum, "Discarded Diagnoses: Ptomaine Poisoning," *Lancet* 357 (March 31, 2001): 1050; and Anne Hardy, "A Short History of Food Poisoning in Britain," *Social History of Medicine* 12, no. 2 (1999): 293–311.

On freshness and food fear, in addition to my conversation with Susanne Freidberg, I drew on the following: Freidberg, *Fresh*; H. P. Lovecraft, "Cool Air," *Tales of Magic and Mystery*, March 1928; Susanne Freidberg, "The Triumph of the Egg," *Comparative Studies in Society and History* 50, no. 2 (2008): 400–423; Betsey Dexter Dyer and Jonathan Brumberg-Kraus, "Cultures on Ice: Refrigeration and the Americanization of Immigrants in the First Half of the Twentieth Century," in *Food and Material Culture: Proceedings of the Oxford Symposium on Food and Cookery 2013* (Devon, UK: Prospect, 2014); Deborah Blum, *The Poison Squad: One Chemist's Single-Minded Crusade for Food Safety at the Turn of the Twentieth Century* (New York: Penguin Press, 2018); and Weldon B. Heyburn, Chairman, *Report of Committee and Hearings Held before the Senate Committee on Manufactures Relative to Foods Held in Cold Storage*, 61st Cong., 3rd sess., SR 1272 (Washington, DC: Government Printing Office, 1911).

To tell the story of Polly Pennington, in addition to her papers, I consulted Barbara Heggie, "Profiles: Ice Woman," *New Yorker*, September 6, 1941, pp. 23–30; Anne Pierce, "Rescuing the Perishables: The Story of a Remarkable Woman Who Has Done a Remarkable Work in Food Conservation," *Field Illustrated* 32, no. 1 (January 1925): 16–48; Lisa Mae Robinson, "Regulating What We Eat: Mary Engle Pennington and the Food Research Laboratory," *Agricultural History* 64, no. 2 (Spring 1990): 143–53; "A Woman's Work for Pure Storage Food," *Oregon Sunday Journal*, March 20, 1910, p. 73; and Elizabeth D. Schafer, "Pennington, Mary Engle (1872–1952)," Encyclopedia.com.

III. When Muscle Becomes Meat

Much of this section is drawn from my visits to Master Purveyors in 2011 and 2023, as well as from Sam Solasz and Judy Katz, *Angel of the Ghetto: One Man's Triumph over Heartbreaking Tragedy* (New York: New Voices Press, 2020). I also met with meat scientists Stephen James and his son Christian James of the Food Refrigeration and Process Engineering Research Centre at the Grimsby Institute in April 2019; their book, S. J. James and C. James, *Meat Refrigeration* (Cambridge: Woodhead, 2002), was an invaluable resource. Keith E. Belk, professor and holder of the Monfort Endowed Chair at Colorado State University, was extremely helpful in reviewing the biochemistry of meat aging and the endogenous enzyme systems involved in this process.

Eleanor Parker, *Winters in the World: A Journey Through the Anglo-Saxon Year* (London: Reaktion Books, 2022) is an entirely delightful book and also my source for *Blotmonad*. Benjamin Franklin was far and away the best founding father, and his letters can be found in *The Papers of Benjamin Franklin*, published online by a team of scholars at Yale University.

While writing this section, I also consulted the following: Shane Hamilton, *Trucking Country: The Road to America's Wal-Mart Economy* (Princeton, NJ: Princeton University Press, 2008); Robert G. Cassens, *Meat Preservation: Preventing Losses and Assuring Safety* (Trumbull, CT: Food & Nutrition Press, 1994); Fidel Toldra, ed., *Lawrie's Meat Science*, 8th ed. (Cambridge: Woodhead, 2017); Doree Lewak, "This New Yorker Is Alive Because of His Butchering Skills," *New York Post*, June 12, 2016; C. L. Davey et al., "Carcass Electrical Stimulation to Prevent Cold Shortening Toughness in Beef," *New Zealand Journal of Agricultural Research* 19, no. 1 (1976): 13–18; Elisabeth Huff Lonergan et al., "Review: Biochemistry of Postmortem Muscle: Lessons on Mechanisms of Meat Tenderization," *Meat Science* 86 (2010): 184–95; Linda M. Samuelsson, "Effects of Dry-Aging on Meat Quality Attributes and Metabolite Profiles of Beef Loins," *Meat Science* 111 (2016): 168–76; Dashmaa Dashdorj et al., "Review: Dry Aging of Beef," *Journal of Animal Science and Technology* 19 (May 2016); Jeff. W. Savell, "Dry-Aging of Beef: Executive Summary," Center for Research and Knowledge Management, National Cattlemen's Beef Association, 2008; G. C. Smith et al., "Postharvest Practices for Enhancing Beef Tenderness," Center for Research and Knowledge Management, National Cattlemen's Beef Association, 2008; Fred E. Deatherage, "Early Investigations on the Acceleration of

Post-mortem Tenderization of Meat by Electrical Stimulation," Ohio State University; D. M. Stiffler et al., "Electrical Stimulation: Purpose, Application and Results," Bulletin B-1375, Texas Agricultural Extension Service, 1982; David Goodsell, "Molecule of the Month: Calcium Pump," *PDB-101*, March 2004; The Kudos Science Trust, "Dr. Carrick Devine—Agricultural Science: Lifetime Achievement," October 31, 2016, YouTube video, 3:26, youtube.com/watch?v=hwh7TeCM5hY; and Tender Beef, "HVES Animation—Short Technical and Economic Explanation," HVES, March 10, 2020, https://hves.eu/en/video-en/.

To paint a picture of the scale of refrigeration-enabled meat consumption, I drew on the following: Vaclav Smil, "Eating Meat: Evolution, Patterns, and Consequences," *Population and Development Review* 28, no. 4 (December 2002); Carys E. Bennett et al., "The Broiler Chicken as a Signal of a Human Reconfigured Biosphere," *Royal Society Open Science* 5, no. 180 (December 12, 2018); Hannah Ritchie, "Wild Mammals Make Up Only a Few Percent of the World's Mammals," *Our World in Data*, December 15, 2022; Hannah Ritchie, "How Much of the World's Land Would We Need in Order to Feed the Global Population with the Average Diet of a Given Country?" *Our World in Data*, October 3, 2017; World Wildlife Federation, "The Amazon in Crisis: Forest Loss Threatens the Region and the Planet," November 8, 2022; Liz Kimbrough, "How Close Is the Amazon Tipping Point?" *Mongabay*, September 20, 2022; Paul Josephson, "The Ocean's Hot Dog: The Development of the Fish Stick," *Technology and Culture* 49, no. 1 (January 2008): 41–61; and Alister Doyle, "Ocean Fish Numbers Cut in Half Since 1970," *Scientific American*, September 16, 2015.

Finally: Francis Bacon, *Sylva Sylvarum: Or, a Natural History in Ten Centuries* (London: Bennet Griffin, 1683).

4. Inside the Time Machine

I. Sleeping Beauties

This section draws on my visits and conversations with Natalia Falagán, senior lecturer in food science and technology at Cranfield University; as well as interviews with Kate Evans, professor in the Department of Horticulture at Washington State University, in September 2022, and Jim Lugg, president of J. Lugg & Associates, in August 2022. It also draws extensively from the archives of the Low Temperature Research Station, which are held as part of the records of the Department of Scientific and Industrial Research: Food Investigation Board, 1916–1960, at the National Archives, Kew, as well as the Low Temperature Research Station and related papers, 1917–1989, at Cambridge University Library's Department of Manuscripts and University Archives.

The following books were a helpful resource in the writing of this section: Andrew Deener, *The Problem with Feeding Cities: The Social Transformation of Infrastructure, Abundance, and Inequality in America* (Chicago: University of Chicago Press, 2020); Catherine Price, *Vitamania: How Vitamins Revolutionized the Way We Think about Food* (New York: Penguin Books, 2016); Joan Morgan and Alison Richards, *The New Book of Apples: The Definitive Guide to Over 2,000 Varieties* (London: Ebury Press, 2002); and Arthur B. Adams, *Marketing Perishable Farm Products* (New York: Longmans, Green, 1916).

On supplying cities with fruit and vegetables, I consulted Eunice Fuller Barnard, "In Food, Also, a New Fashion Is Here," *New York Times Magazine*, May 4, 1930, and "Britain on the Brink of Starvation: Unrestricted Submarine Warfare," *Historic England*, February 1, 2017.

To tell the story of Kidd and West's work at the LTRS, I referenced the following: H. B. S. Montgomery and A. F. Posnette, "Franklin Kidd, 12 October 1890–7 May 1974," *Biographical Memoirs of Fellows of the Royal Society* 21 (November 1975): 406–30; P. D. Sell, "Cyril West,

1887–1986," *Watsonia* 16 (1987): 361–63; F. Kidd, "Food Research under the Department of Scientific and Industrial Research," *Proceedings of the Royal Society of London. Series A, Mathematical and Physical Sciences* 205, no. 1083 (March 7, 1951): 467–83; Franklin Kidd, Cyril West, and M. N. Kidd, *Gas Storage of Fruit: The Use of Artificial Atmospheres of Regulated Composition, Either Alone or in Conjunction with Refrigeration, for the Purpose of Preserving Fresh Fruit during Overseas Transport or in Land Stores,* Department of Scientific Research, Food Investigation, Special Report No. 30 (London: HMO Stationery Office, 1927); and Franklin Kidd and Cyril West, "Gas-Storage of Fruit IV: Cox's Orange Pippin Apples," *Journal of Pomology and Horticultural Science* 14, no. 3 (1937): 276–94; "'Gas' Storage of Apples," *Times* (London), March 24, 1930, p. 20; and Sir William Hardy, "Some Recent Developments in Low Temperature Research," *Journal of the Society of Chemical Industry* 52, no. 3 (January 20, 1933): 45–49.

On the history of controlled atmosphere storage, I also drew on Stefanie Glinski, "The Ancient Method That Keeps Afghanistan's Grapes Fresh All Winter," *Atlas Obscura*, March 25, 2021; J. É. Bérard, "Mémoire sur la maturation des fruits," *Annales de chimie et de physique* 16 (1821); "An Old Fruit House," *Ice and Refrigeration*, July 1895, pp. 23–25; "Fruit Shipments in Carbonic Acid Gas," Official Report of the Eighteenth Fruit Growers' Convention (Sacramento, CA: 1895), pp. 148–50; Dana G. Dalrymple, "The Development of an Agricultural Technology: Controlled-Atmosphere Storage of Fruit," *Technology and Culture* 10, no. 1 (January 1969): 35–48; and John M. Love, "Robert Smock and the Diffusion of Controlled Atmosphere Technology in the US Apple Industry, 1940–60" (Cornell Agricultural Economics Staff Paper No. 88-20, August 1988).

On the practice of controlled-atmosphere storage, I consulted the following: D. Bishop, "Controlled Atmosphere Storage," in *Cold and Chilled Storage Technology,* ed. Clive Dellino (Boston: Springer, 1990), 53–92; Kate Prengaman, "The Challenges of Storing Organic Apples," *Good*

Fruit Grower, January 24, 2018; "Controlled Atmosphere Storage: A Technical Information Bulletin of the Northwest Horticultural Council," June 2, 2023; Kate Prengaman, "How Low Can You Go," *Good Fruit Grower,* May 1, 2017, pp. 42–45; Steven Morris, "Fruit Farm Manager Jailed over Deaths of Men Who 'Scuba Dived' for Apples," *Guardian,* July 1, 2015; S. J. Kay, *Postharvest Physiology of Perishable Plant Products* (New York: Van Nostrand Reinhold, 1991); Miguel Espino-Díaz et al., "Biochemistry of Apple Aroma: A Review," *Food Technology & Biotechnology* 54, no. 1 (December 2016): 375–97; John K. Fellman et al., "Relationship of Harvest Maturity to Flavor Regeneration after CA Storage of 'Delicious' Apples," *Postharvest Biology and Technology* 27 (2003): 39–51; Fritz K. Bangerth et al., "Physiological Impacts of Fruit Ripening and Storage Conditions on Aroma Volatile Formation in Apple and Strawberry Fruit: A Review," *HortScience* 47, no. 1 (2012): 4–10; and James J. Nagle, "Use Widens for Cooling System," *New York Times,* April 6, 1966, p. 55.

On the difficulties of storing one of my favorite apples, see Adriano Arriel Saqet, "Storage of 'Cox Orange Pippin' Apple Severely Affected by Watercore," *Erwerbs-Obstbau* 62 (September 2020): 391–98; E. T. Chittenden et al., "Bitter Pit in Cox's Orange Apples," *New Zealand Journal of Agricultural Research* 12, no. 1 (1969): 240–47; and D. S. Johnson, "Investigating the Cause of Diffuse Browning Disorder in CA-Stored Cox's Orange Pippin Apples," *ISHS Acta horticulturae* 857: IX International Controlled Atmosphere Research Conference (April 2010).

On the Cosmic Crisp, I drew on the following: Brooke Jarvis, "The Launch," *California Sunday Magazine,* July 18, 2019; and M. Sharon Baker, "The Next Big Apple Variety Was Bred for Deliciousness in Washington," *Seattle Business Magazine,* November 24, 2017.

On lettuce, I consulted the following: John Steinbeck, *East of Eden* (New York: Viking Press, 1952); Gabriella M. Petrick, "The Arbiters of Taste: Producers, Consumers and the Industrialization of Taste in America, 1900–1960" (PhD diss., University of Delaware, 2006); Gabriella M.

Petrick, "'Like Ribbons of Green and Gold': Industrializing Lettuce and the Quest for Quality in the Salinas Valley, 1920–1965," *Agricultural History* 80, no. 3 (2006): 269–95; "Microbiological Surveillance Sampling: FY21 Sample Collection and Analysis of Lettuce Grown in Salinas Valley, CA," US Food & Drug Administration, September 8, 2022; Paul F. Griffin and C. Langdon White, "Lettuce Industry of the Salinas Valley," *Scientific Monthly* 81, no. 2 (August 1955): 77–84; "Green Gold," *The Fridge Light*, CBC, October 24, 2017; Craig Claiborne, "News of Food: Salads," *New York Times*, December 8, 1958, p. 41; Ed Dinger, "Fresh Express Inc.," Encyclopedia.com; Dave Stidolph, "Vacuum Cooling, Once Radical Idea, Reigns Supreme," *Watsonville Register-Pajaronian*, December 13, 1954, p. 11; Robert H. Kieckhefer interviewed by Arthur J. McCourt, Weyerhaeuser Company Historical Archives, February 20, 1976; Harland Padfield and William E. Martin, *Farmers, Workers, and Machines* (Phoenix: University of Arizona Press, 1965); Molly Oleson, "Jim Lugg: Founder of the Modern Salad," *Breakthroughs*, Fall 2017; James Lugg et al., "Establishing Supply Chain for an Innovation: The Case of Prepackaged Salad," *ARE Update* 20, no. 6 (2017): 5–8; "A System for Abundance" in *FOOD: Transforming the American Table* (exhibition at National Museum of American History, November 2012); and Joe Mathews, "Salinas and Yuma Are 500 Miles Apart—but Agribusiness Is Growing Them Closer," *Zocalo Public Square*, October 22, 2018.

II. Oracular Bananas

In addition to the papers of the Low Temperature Research Station, cited above, this section draws on my visits to Banana Distributors of New York, Inc., as it was known when Paul Rosenblatt was in charge (October 2011), and D'Arrigo New York Wholesale Delivery & Banana Facility, as it is called today, where my host was Gabriela D'Arrigo (February 2023). It also draws on my conversations with Jim Lentz, president and owner of Thermal Technologies (September 2022); John Hale, adjunct professor at the University of Louisville (April 2022); Henry Spiller, former director of the Central Ohio Poison Center and the Kentucky Regional Poison Center (April 2022); and Steve Barnard, founder and CEO of Mission Produce (August 2022).

On ethylene and plants, see Peter Smith, "The Peas That Smelled the Leaky Pipe," *Smithsonian*, June 1, 2012; Arthur F. Sievers and Rodney H. True, *A Preliminary Study of the Forced Curing of Lemons as Practiced in California* (Washington, DC: Government Printing Office, 1912); R. B. Harvey, "Artificial Ripening of Fruits and Vegetables" (University of Minnesota Agricultural Experiment Station Bulletin 247, October 1928); R. Gane, "Production of Ethylene by Some Ripening Fruits," *Nature* 134, no. 1008 (December 29, 1934); Michael J. Haydon et al., "Sucrose and Ethylene Signaling Interact to Modulate the Circadian Clock," *Plant Physiology* 175, no. 2 (October 2017): 947–58; Hirokazu Ueda et al., "Plant Communication: Mediated by Individual or Blended VOCs?," *Plant Signaling & Behavior* 7, no. 2 (February 2012): 222–26; Anja K. Meents and Axel Mithöfer, "Plant–Plant Communication: Is There a Role for Volatile Damage-Associated Molecular Patterns?," *Frontiers in Plant Science* 11 (October 15, 2020); Dominique Van Der Straeten, "Ethylene in Vegetative Development: A Tale with a Riddle," *New Phytologist* 194, no. 4 (March 2012): 895–909; Jan Kępczyński, "Gas-Priming as a Novel Simple Method of Seed Treatment with Ethylene, Hydrogen Cyanide or Nitric Oxide," *Acta physiologiae plantarum* 43, no. 117 (July 2021); Renata Bogatek and Agnieszka Gniazdowska, "Ethylene in Seed Development, Dormancy and Germination," *Annual Plant Reviews* 44 (April 2018): 189–218; Adrien Fernet, "Ethylene to Delay the Germination of Onions," *FreshPlaza*, February 21, 2019; Alain Soler et al., "Forcing in Pineapples: What Is New?" *Newsletter of the Pineapple Working Group, International Society for Horticultural Science*, 2006, pp. 27–31; Mandy Kendrick, "The Origin of Fruit Ripening,"

Scientific American, August 17, 2009; and Robert Nicholas Spengler, "Origins of the Apple: The Role of Megafaunal Mutualism in the Domestication of Malus and Rosaceous Trees," *Frontiers in Plant Science* 10, no. 617 (May 27, 2019): 1–18.

On ethylene and people: A. B. Luckhardt and J. B. Carter, "The Physiologic Effects of Ethylene," *Journal of the American Medical Association* 80, no. 11 (March 17, 1923): 765–70; Jelle de Boer et al., "New Evidence of the Geological Origins of the Ancient Delphic Oracle (Greece)," *Geology* 29 (2001): 707–10; Henry A. Spiller, John R. Hale, and Jelle Z. De Boer, "The Delphic Oracle: A Multidisciplinary Defense of the Gaseous Vent Theory," *Journal of Toxicology: Clinical Toxicology* 40, no. 2 (2002): 289–96; John R. Hale et al., "Questioning the Delphic Oracle," *Scientific American* 289, no. 2 (August 2003): 66–73; Doomberg, "Science by Press Release," *Substack*, September 16, 2022; Alexander H. Tullo, "The Search for Greener Ethylene," *Chemical & Engineering News*, March 15, 2021; Stanley P. Burg and Ellen A. Burg, "Role of Ethylene in Fruit Ripening," *Plant Physiology* 37, no. 2 (March 1962): 179–89; and "2022 NIHF Inductee Sylvia Blankenship: The Horticultural Hero," National Inventors Hall of Fame, 2022.

Regarding bananas, I drew on the following sources: Dan Koeppel, *Banana: The Fate of the Fruit That Changed the World* (New York: Hudson Street Press, 2008); Muhammad Siddiq, ed., *Handbook of Banana Production, Postharvest Science, Processing Technology, and Nutrition* (Hoboken, NJ: Wiley, 2020); Frederick Upham Adams, *Conquest of the Tropics: The Story of the Creative Enterprises Conducted by the United Fruit Company* (New York: Doubleday, Page, 1914); Robert Baden-Powell, *Scouting for Boys* (London: H. Cox, 1908); "The Banana as the Basis of a New Industry," *Scientific American* 80, no. 9 (March 4, 1899); Francis X. Clines, "First Banana: A Welcome to a New Land," *New York Times*, July 31, 1994, p. 33; Matt Blitz, "The Origin of the 'Slipping on a Banana Peel' Comedy Gag," *Today I Found Out*, November 29, 2013; L.

Williams, "Refrigeration and the S. S. Venus," *Unifruitco* 4 (April 1929): 528–32; John Soluri, "Accounting for Taste: Export Bananas, Mass Markets, and Panama Disease," *Environmental History* 7, no. 3 (July 2002): 386–410; Roderick Abbott, "A Socio-economic History of the International Banana Trade, 1870–1930" (EUI Working Paper RSCAS 2009/22, European University Institute, May 2009); "Tariff Readjustment—1929," Hearings before the Committee on Ways and Means, House of Representatives, 70th Cong., 2nd sess., no. 37 (Washington, DC: Government Printing Office, 1929); Ulrike Praeger et al., "Effect of Storage Climate on Green-Life Duration of Bananas," 5th International Workshop, Cold Chain Management, Bonn, Germany, June 2013; Daniel Krieger, "Mozart's Growing Influence on Food," *Japan Times*, November 25, 2010; "Explosion in Pittsburgh's Produce District Last Week," *Chicago Packer*, December 26, 1936; Kathleen Donahoe, "The Pittsburgh Banana Company Explosion," 2015, Archives & Special Collections, University of Pittsburgh; Michael Le Page, "Bananas Threatened by Devastating Fungus Given Temporary Resistance," *New Scientist*, September 21, 2022; and T.W., "Where Did Banana Republics Get Their Name?," *Economist*, November 21, 2013.

On avocados, I referenced the following: Charles Oilman Henry, "How to Benefit from 'Pre-Ripe'," California Avocado Society 1984 Yearbook 68, pp. 37–41; and Brook Larmer, "How the Avocado Became the Fruit of Global Trade," *New York Times*, March 27, 2018.

III. Trading Futures

This section is informed by my conversations with Irwin Goldman, professor in the Department of Horticulture at the University of Wisconsin–Madison (September 2022), and Shawn Hackett, president and CEO of Hackett Financial Advisors (August 2022), as well as my visit to Citrosuco North America, Inc., in March 2013.

SELECTED SOURCES

I drew on the following books: John McPhee, *Oranges* (New York: Farrar, Straus and Giroux, 1975); Alissa Hamilton, *Squeezed: What You Don't Know about Orange Juice* (New Haven: Yale University Press, 2010); Tetra Pak, *The Orange Book: A Unique Guide to Orange Juice Production* (2017); and Emily Lambert, *The Futures: The Rise of the Speculator and the Origins of the World's Biggest Markets* (New York: Basic Books, 2010).

I also referenced the following: Shane Hamilton, "Cold Capitalism: The Political Ecology of Frozen Concentrated Orange Juice" (Working Paper No. 35, Program in Science, Technology, and Society, Massachusetts Institute of Technology, February 2003); Thirteenth Annual Citrus Statistical Survey, Agricultural Marketing Service, Florida Department of Agriculture, March 1961; "From the Concentrate Revolution to the Decline of Florida Citrus" in the online exhibition *Bittersweet: The Rise and Fall of the Citrus Industry in Florida*, Florida Memory, State Library and Archives of Florida; Angelico Law, "Brazil's Foothold in the Orange Juice Industry," October 5, 2018; "The 100-Year Journey of the UF/IFAS Citrus Research and Education Center," UF/IFAS Communications, November 2017; United States International Trade Commission, "In the Matter of Certain Orange Juice from Brazil," Investigation No. 731-TA-1089 (Washington, DC: Heritage Reporting, 2012); Duane D. Stanford, "Coke Has a Secret Formula for Orange Juice, Too," *Bloomberg Businessweek*, February 4–10, 2013; "How Orange Juice Was Built," *Proof*, January 2021; Josephine Peterson, "Port of Wilmington Worker Critically Injured after Falling 50 Feet in Juice Storage Tank," *News Journal*, January 9, 2019; "Crowning Achievements in Bulk Aseptic Storage," *Food Engineering*, October 3, 2007; "Tropicana Products, Inc.," Reference for Business, Company History Index; and Marisa L. Zansler, "Overview of Recent OJ Retail Sales Trends and Florida Processor Movement," FDOC Live Webinar Series, April 30, 2020.

5. A Third Pole

I. Meet the Thermo King

Most of this section is drawn from the Frederick Jones papers, Manuscripts Collection, Minnesota Historical Society, Saint Paul, MN. I also consulted Kathleen Peippo, "Thermo King Corporation History," in *International Directory of Company Histories*, ed. Tina Grant (Detroit, MI: St. James Press, 1996), 505–507.

I also drew on the following books: F. Scott Fitzgerald, *The Great Gatsby* (New York: Charles Scribner's Sons, 1925); Hamilton, *Trucking Country*; Warren Belasco and Roger Horowitz, eds., *Food Chains: From Farmyard to Shopping Cart* (Philadelphia: University of Pennsylvania Press, 2008); Mark Kurlansky, *Birdseye: The Adventures of a Curious Man* (New York: Anchor Books, 2013); Tom Philpott, *Perilous Bounty: The Looming Collapse of American Farming and How We Can Prevent It* (New York: Bloomsbury, 2020); Michael Pollan, *The Omnivore's Dilemma: A Natural History of Four Meals* (New York: Penguin Press, 2006); Eric Schlosser, *Fast Food Nation: The Dark Side of the All American Meal* (New York: Houghton Mifflin, 2001); Sarah Murray, *Moveable Feasts: From Ancient Rome to the 21st Century, the Incredible Journeys of the Food We Eat* (New York: St. Martin's Press, 2007); Robert D. Heap, *Guide to Refrigerated Transport* (Paris: International Institute of Refrigeration, 2010); and Sasha Issenberg, *The Sushi Economy: Globalization and the Making of a Modern Delicacy* (New York: Avery, 2008).

On the extended impact of refrigerated trucking, I referenced the following articles: Vera J. Banks and Judith Z. Kalbacher, "The Changing US Farm Population," *Rural Development Perspectives*, March 1980, pp. 43–46; Eli Tan, "A French-fry Boomtown Emerges as a Climate Winner—as Long as It Has Water," *Washington Post*, August 21, 2023; Steve Kay, "Tyson's Beef Battleship," *Meat & Poultry*, November 28,

2016; Shane Hamilton, "The Economies and Conveniences of Modern-Day Living: Frozen Foods and Mass Marketing, 1945–1965," *Business History Review* 77 (Spring 2003): 33–60; Tara Garnett, "Food Refrigeration: What Is the Contribution to Greenhouse Gas Emissions and How Might Emissions Be Reduced?" (Food Climate Research Network working paper, April 2007); Kovie Biakolo, "A Brief History of the TV Dinner," *Smithsonian*, November 2020; Reuters, "How Four Big Companies Control the U.S. Beef Industry," June 17, 2021; Georgina Gustin, "Air Pollution from Raising Livestock Accounts for Most of the 16,000 US Deaths Each Year Tied to Food Production, Study Finds," *Inside Climate News*, May 11, 2021; Donald Carr, "Manure from Unregulated Factory Farms Fuels Lake Erie's Toxic Algae Blooms," Environmental Working Group, April 9, 2019; and Burke W. Griggs, Matthew R. Sanderson, and Jacob A. Miller-Klugesherz, "Farmers Are Depleting the Ogallala Aquifer Because the Government Pays Them to Do It," *Trends* (American Bar Association), February 27, 2022.

On refrigerated air transportation, I consulted the following: Jack Kinyon, "Air Transport Command–Airlift During WWII," Air Mobility Command Museum; Mathew Noblett, "The Transatlantic Relationship: The Berlin Airlift," Academy for Cultural Diplomacy; Arthur Veysey, "Jets Make Any Day a Day for Strawberries," *Chicago Tribune*, April 20, 1969; "Perishable Logistics: Cold Chain on a Plane," *Inbound Logistics*, January 2014; Anna King, "Sweet Northwest Cherries Get a New First-Class Direct Flight to Asia," *NWNews*, June 12, 2015; Anna King, "Tariffed Northwest Cherry Growers Don't Have Much Time to Sort Out Marketing Strategy," *Northwest Public Broadcasting*, April 9, 2018; David Parkinson, "When Fish Began to Fly," *Globe and Mail*, July 12, 2007; Rojda Akdag, "How Bluefin Tuna Became a Top Cargo in Air Freight," *More Than Shipping*, July 27, 2015; Scott Mall, "Flashback Friday: The History of Air Freight," *FreightWaves*, May 24, 2019; and *Sushi: Global Catch*, directed by Mark Hall (Kino Lorber, 2011).

SELECTED SOURCES

II. Reefer Madness

This section relied on several conversations with Barbara Pratt at her offices in Florham Park, New Jersey, at her family farm, Wilkens Fruit & Fir Farm, Yorktown Heights, New York, and over the phone, as well as the Mobile Research Laboratory records from the Sea Land, Inc., Papers at the American Merchant Marine Museum, Kings Point, New York. I also drew on my conversation with Steve Barnard.

I found the following books helpful to understand the story of container and reefer shipping: Alexander Klose, *The Container Principle: How a Box Changes the Way We Think* (Cambridge, MA: MIT Press, 2015); Marc Levinson, *The Box: How the Shipping Container Made the World Smaller and the World Economy Bigger* (Princeton, NJ: Princeton University Press, 2006); and Thomas Taro Lennerfors and Peter Birch, *Snow in the Tropics: A History of the Independent Reefer Operators* (Boston: Brill, 2019).

On containerization, I also referenced the following: "Malcolm Purcell McLean, Pioneer of Container Ships, Died on May 25th, Aged 87," *Economist*, May 31, 2001; Betty Joyce Nash, "The Voyage to Containerization," *Economic History*, 2012, pp. 39–42; Brian J. Cudahy, "The Containership Revolution: Malcom McLean's 1956 Innovation Goes Global," *TR News*, no. 246 (September–October 2006); Sarah Murray, "World Would Be Lost without Big Metal Box," *Weekender*, April 29–30, 2006; "Tankers to Carry 2-Way Pay Loads," *New York Times*, April 27, 1936, p. 54; Judah Levine, "The History of the Shipping Container," Freightos, April 24, 2016; Jean-Paul Rodrigue, Theo Notteboom, and Athanasios Pallis, "The Changing Geography of Seaports," in *Port Economics, Management and Policy* (New York: Routledge, 2022); Leah Brooks et al., "The Local Impact of Containerization" (working paper, Division of Research and Statistics, Board of Governors of the Federal Reserve System, April 28, 2021); Anna Nagurney, "Global Shortage of Shipping Containers Highlights Their Importance in Getting Goods to Amazon

Warehouses, Store Shelves and Your Door in Time for Christmas," *Conversation*, September 21, 2021; "Banana-Shipping Invention to Cut Hundreds of Longshoreman Jobs," *South Florida Sun Sentinel*, May 8, 1989; Haylle Sok, "50 Years of Container Refrigeration Innovation," *Global Trade Magazine*, December 27, 2017; "Drewry: Shipping Line Share of Reefer Market to Hit 85% by 2021," *Container Management*, September 27, 2017; "Tuna Travels," Maersk, January 21, 2020; and John Churchill, "The Queen of Cool," Maersk Stories, May 10, 2013.

On the impact of reefer shipping, I consulted the following: Joanna Blythman, "Strange Fruit," *Guardian*, September 6, 2002; David Karp, "Most of America's Fruit Is Now Imported. Is That a Bad Thing?" *New York Times*, March 13, 2018; Biing-Hwan Lin and Rosanna Mentzer Morrison, "A Closer Look at Declining Fruit and Vegetable Consumption Using Linked Data Sources," *Amber Waves*, July 5, 2016; Cheryl Schweizer, "Washington Apple Growers Facing Some Exporting Challenges," *KREM2*, January 17, 2023; Joseph Leitmann-Santa Cruz and Ambassador Robert Pastorino, "Accession of Chile to NAFTA: Benefits for Chile and the United States," Institute for Agriculture & Trade Policy, May 14, 2000; M. Jahi Chappell et al., "Food Sovereignty: An Alternative Paradigm for Poverty Reduction and Biodiversity Conservation in Latin America," *F1000 Research* 2 (November 2013); Peter Williams and Warwick E. Murray, "Behind the 'Miracle': Non-traditional Agro-Exports and Water Stress in Marginalised Areas of Ica, Peru," *Bulletin of Latin American Research* (2018); Ayesha Tandon, "'Food Miles' Have Larger Climate Impact Than Thought, Study Suggests," *CarbonBrief*, June 20, 2022; Molly Leavens, "Do Food Miles Really Matter?" Harvard University Sustainability, March 7, 2017; Seung Hee Lee et al., "Adults Meeting Fruit and Vegetable Intake Recommendations—United States, 2019," *MMWR* 71 (2022): 1–9; and US Food Imports data set, Economic Research Service, USDA.

III. Building a New Arctic

This section is largely based on my visits to Springfield Underground and Carthage Underground in 2017; my visit to NewCold Wakefield in 2019; my tour of the Port of Wilmington, Delaware, with Vered Nohi in 2013; my visit with Boston's city archaeologist, Joseph Bagley, at his lab in 2014; and my extensive travel and reporting in China in January 2014, including visits to Chen Zemin and Sanquan in Zhengzhou, to Dai Jianjun and Longjing Caotang, and with more than a dozen other people and places involved in China's cold chain. It also draws on my conversations with Matt Coolidge and the CLUI team, as well as my interviews with Adam Feiges and Marc Wulfraat (February 2023).

I also consulted the following books: Hamilton, *Trucking Country*; Erna Risch, *The Quartermaster Corps: Organization, Supply, and Services*, vol. 1 (Washington, DC: Center of Military History, US Army, 1995); and David L. Manuel, *Men and Machines: The Brambles Story* (Sydney: Ure Smith, 1970).

Additional references include "The Ozarks," *Conspiracy Theory with Jesse Ventura*, season 2, episode 4, November 26, 2012; "Boston's Cold Corner," *Ice and Refrigeration* 9, no. 6 (1895): 375–95; and Rick LeBlanc, "Another Sneak Attack, War Heralded Pallet in Industry," *Pallet Enterprise*, May 3, 2002.

For the story of Vernon, California, I consulted Charles F. McElwee, "The Supply-Chain Empire," *City Journal*, Winter 2022; Sam Allen and Hector Becerra, "Complex Financial Deals and Energy Projects Cost Vernon Millions," *Los Angeles Times*, August 13, 2011; Adam Nagourney, "Plan Would Erase All-Business Town," *New York Times*, March 1, 2011; and Hadley Meares, "Vernon: The Implausible History of an Industrial Wasteland," *Curbed LA*, May 19, 2017.

On cold chain expansion, see Rich Lachowsky et al., "Sold on Cold: Temperature-Controlled Development Pipeline Reaches New Record,"

Newmark Research, March 2023; Jeff Berman, "CBRE Report Presents a Bright Future for Cold Storage Warehouse Growth," *Logistics Management*, June 12, 2019; Spencer Brewer, "'We're Off to the Races': Barber Partners and Bain Capital Announce Nationwide Cold Storage Joint Venture," *Dallas Business Journal*, May 3, 2022; and Salin, "2020 Global Cold Storage Capacity Report."

In reporting on China, many of the documents and articles I consulted were originally in Chinese and were translated for me by Owen Guo. I also found the following useful: Peter Grant, "Cold Snap: Developers Pour Money into Cold Storage in China," *Wall Street Journal*, January 3, 2017; Nicola Davison, "China's Growing Appetite for Pork Creates New Pollution," *China Dialogue*, January 10, 2013; "Hog Heaven: China Builds Pig Hotels for Better Biosecurity," *Bloomberg*, August 2, 2021; Yijing Zhou, "The Food Retail Revolution in China and Its Association with Diet and Health," *Food Policy* 55 (August 1, 2015): 92–100; Jennifer Timmons, "Chicken Feet Are a Big Deal," *Delmarva Farmer*, February 2, 2018; Wang Jun et al., "China Food Manufacturing Annual Report," USDA Global Agricultural Information Network, February 2013; Peter Cohan, "How Did Wal-Mart Crack Open China?," *Forbes*, May 18, 2012; W. Wang et al. "China's Food Production and Cold Chain Logistics"; Joanna Bonarriva et al., "China Agricultural Trade: Competitive Conditions and Effects on U.S. Exports," US International Trade Commission investigation no. 332-518, publication 4219, March 2011.

6. *The Tip of the Iceberg*

I. Cold Case

This section draws on my meeting with Jon Steinberg in February 2019, as well as on conversations with Beko's Justin Reinke; Mary Kay Bolger, retired senior product development manager at Whirlpool Corporation;

and William Kwon, senior product manager at LG Electronics for refrigeration and home appliances (all in February 2023).

Throughout this section, I consulted, as always, the encyclopedic Thévenot, *A History of Refrigeration*, as well as Anderson, *Refrigeration in America*; Gantz, *Refrigeration*; Rees, *Refrigeration Nation*; and Peavitt, *Refrigerator*; as well as Watkins, "Fridge Space." I also drew on Jonathan Rees, *Refrigerator* (New York: Bloomsbury Academic, 2015).

My sources for America's fridge-peeping obsession include the following: LG Electronics USA, "Behind Closed Doors: Survey Takes a Fresh Look at What Your Fridge Says About You," news release, June 28, 2017; JP Mangalindan, "For Some Strange Reason, Periscope Users Are Obsessed with Your Fridge," *Mashable*, March 28, 2015; Ellissa Bain, "What Is the TikTok Fridge Challenge? Social Media's Most Pointless Trend," *HITC*, April 23, 2020; Michael D. Shear, Katie Rogers, and Annie Karni, "Beneath Joe Biden's Folksy Demeanor, a Short Fuse and an Obsession with Details," *New York Times*, May 14, 2021; Joe Taysom, "David Bowie Used to Store His Urine in the Fridge to Stop Witches Stealing It," *Far Out Magazine*, August 24, 2020; and John Keefe, "Quiz: Can You Tell a 'Trump' Fridge from a 'Biden' Fridge?," *New York Times*, October 27, 2020.

Data on American fridge ownership can be found in "The Effect of Income on Appliances in U.S. Households," US Energy Information Administration, November 21, 2001.

Ruth Schwartz Cowan's essay "How the Refrigerator Got Its Hum" can be found in *The Social Shaping of Technology*, ed. Donald MacKenzie and Judy Wajcman (Milton Keynes, UK, and Philadelphia: Open University Press, 1985), 208–18. For data on fridge adoption, see H. Laurence Miller, "The Demand for Refrigerators: A Statistical Study," *Review of Economics and Statistics* 42, no. 2 (1960): 196–202. On the refrigerator's hum, see *The Velvet Underground*, directed by Todd Haynes (Apple TV+, 2021), and Micah Loewinger, "Mysteries of Sound," *On the Media*, August 25, 2023.

Robert and Helen Lynd's fascinating study was published as *Middletown: A Study in American Culture* by Robert S. Lynd and Helen M. Lynd (New York: Harcourt, Brace, 1929).

To explore female liberation (or lack thereof) through refrigeration, I turned to the following: US Census Bureau, "Chapter D. Labor Force, Wages, and Working Conditions" (Series D 1-238), in *Historical Statistics of the United States, 1789–1945*; Jeremy Greenwood, *Evolving Households: The Imprint of Technology on Life* (Cambridge, MA: MIT Press, 2019); Jeremy Greenwood et al., "Engines of Liberation," *Review of Economic Studies* 72 (2005): 109–33; Jeremy Greenwood et al., "Technology and the Changing Family" (National Bureau of Economic Research Working Paper 17735, January 2012); and Ruth Schwartz Cowan, *More Work for Mother: The Ironies of Household Technology from the Open Hearth to the Microwave* (New York: Basic Books, 1983).

To understand the impact of refrigeration on kitchens and cities, I also drew on Sarah Archer, *The Midcentury Kitchen: America's Favorite Room, from Workspace to Dreamscape* (New York: Countryman Press, 2019); Steel, *Hungry City*; and Ian Chodikoff, "Sites and Scenes," *Canadian Architect*, November 1, 2007.

Earnest Elmo Calkins's essay helping pave the descent into our current era of disposable consumer capitalism was published as "What Consumer Engineering Really Is" in *Consumer Engineering: A New Technique for Prosperity*, ed. Roy Sheldon and Egmont Arens (New York: Harper, 1932). See also Stan Goldblatt, "America's Oldest Fridge Still Keeping Cool," *New York Post*, July 5, 2013; Bernard A. Nagengast and Randy Schrecengost, *Adventures in Heat and Cold: Men and Women Who Made Your Lives Better* (Atlanta: ASHRAE, 2020).

II. Freshness Guaranteed

Jeanne E. Arnold et al., *Life at Home in the 21st Century: 32 Families Open Their Doors* (Los Angeles: Cotsen Institute of Archaeology Press,

2012), was published as the culmination of a nine-year interdisciplinary research project led by UCLA's Center on Everyday Lives of Families. It's fascinating, horrifying, and as one of its authors, Anthony P. Graesch, a newly married, child-free graduate student when the study began, told a reporter, "the very purest form of birth control devised."

I interviewed Martha Stewart for *Kinfolk* magazine, issue 25, September 5, 2017.

On the relationship between refrigeration and consumer food waste, I drew on Tara Garnett, "Food Refrigeration: What Is the Contribution to Greenhouse Gas Emissions and How Might Emissions Be Reduced?" (Food Climate Research Network working paper, April 2007); Jonathan Bloom, *American Wasteland: How America Throws Away Nearly Half of Its Food (and What We Can Do about It)* (Boston: Da Capo Lifelong Books, 2010); William Rathje and Cullen Murphy, *Rubbish! The Archaeology of Garbage* (Phoenix: University of Arizona Press, 2001); Jeff Harrison, "William L. Rathje: 1945–2012," *University of Arizona News*, June 5, 2012; and the USDA's Food Waste FAQs.

For the story of WhiteWave paying dairy slotting fees, see William Shurtleff and Akiko Aoyagi, *History of White Wave (1977–2022): America's Most Creative and Successful Soyfoods Maker* (Lafayette, CA: Soyinfo Center, 2022); Sam Fromartz, "Starting a Business: This Entrepreneur Didn't Cry over Spilt Soy Milk," *St. Louis Post-Dispatch*, August 11, 2003; and Bethany Mclean, "Profile in Persistence: In 1977 Steve Demos Had an Idea to Sell Soy-Based Foods to Health-Conscious Americans," *CNN Money*, May 1, 2001.

According to Bill Yenne, writing in *Great American Beers: Twelve Brands That Become Icons* (Saint Paul, MN: MBI, 2004), the earliest date-labeled food or beverage was Lucky Lager, brewed by the General Brewing Corporation in San Francisco. In advertising materials, it was identified as both "one of the world's really fine beers" and "the original age-dated beer."

On sell-by and best-before dates, I consulted Steve Lawrence, "Should

All Foods Have 'Spoil Dates'?," *Chicago Tribune*, October 6, 1977; Judy Hevrdejs, "Because It's Dated Is It Also Fresh?," *Chicago Tribune*, January 18, 1979; Richard Milne, "Arbiters of Waste: Date Labels, the Consumer and Knowing Good, Safe Food," *Sociological Review* 60, no. 2 supp. (2013): 84–101; The Editors, "Read This Before June 11," *Bloomberg*, June 10, 2015; Casey Williams, "Al Capone's Brother May Have Invented Date Labels for Milk," *HuffPost*, August 3, 2016; Gigen Mammoser, "Al Capone and the Short, Confusing History of Expiration Dates," *Vice*, December 17, 2016; Debasmita Patra et al., "Evaluation of Global Research Trends in the Area of Food Waste Due to Date Labeling Using a Scientometrics Approach," *Food Control* 115 (2020); and "When Did 'Best Before' Dates Begin?," *Spectator*, August 6, 2022. I also drew on the following online resources: Lynne Olver, "Shelf-Life Dating (USA)" in "The Food Timeline"; and "Food Product Dating," USDA Food Safety and Inspection Service.

Jihyun Ryou's work can be found at savefoodfromthefridge.com.

On the contentious issue of whether to vaccinate chickens or refrigerate eggs, I consulted the following: William Neuman, "U.S. Rejected Hen Vaccine Despite British Success," *New York Times*, August 24, 2010; and D. R. Jones et al., "Impact of Egg Handling and Conditions during Extended Storage on Egg Quality," *Poultry Science* 97, no. 2 (2018): 716–23.

On "smart" freshness-detection technology, I drew on the following: Tamar Haspel, "This Groundbreaking Technology Will Soon Let Us See Exactly What's in Our Food," *Washington Post*, March 26, 2016; Margi Murphy, "A Very Cool Invention: This Grundig Fridge Can Tell When Food Is Going Off by Detecting Its Pongy Odour," *Sun*, September 4, 2017; and Michael Wolf, "Is Amazon Considering Making a Smart Fridge? Probably Not (but Maybe)," *Spoon*, November 3, 2017.

III. The Taste of Cold

Parts of this section were drawn from my conversations and correspondence with Harry Klee, emeritus professor in the Horticultural Sciences Department at the University of Florida in 2022 and 2023. I also spoke to Tassos Stassopoulos, managing partner and CIO of Trinetra, in February 2023 and visited Dr. Wang Guoli, director of the National Engineering Research Center for Agricultural Product Logistics, in Jinan, China, in January 2014. I spoke to historian of alcohol David Wondrich for an episode of *Gastropod*: Nicola Twilley and Cynthia Graber, "The Cocktail Hour," *Gastropod*, May 26, 2015.

Waldo Jaquith, "On the Impracticality of a Cheeseburger," waldo.ja quith.org, December 3, 2011.

On California common or steam beer, I consulted Jeff Alworth, "The Making of a Classic: Anchor Steam," *Beervana*, May 6, 2020; and Ian Steadman, "'Kim Jong-Ale': North Korea's Surprising Microbrewery Culture Explored," *Wired* UK, April 29, 2013.

On the historical enjoyment of ice cream and other frozen delights, I consulted David, *Harvest*; Jeri Quinzio, *Of Sugar and Snow: A History of Ice Cream Making* (Berkeley: University of California Press, 2009); and Ivan Day, *Ice Cream: A History* (London: Shire, 2011).

On the science of brain freeze, see American Physiological Society, "Changes in Brain's Blood Flow Could Cause 'Brain Freeze'" (press release), *ScienceDaily*, April 12, 2012; and Melissa Mary Blatt et al., "Cerebral Vascular Blood Flow Changes during 'Brain Freeze,'" *Federation of American Societies for Experimental Biology Journal* 26, no. S1 (April 2012).

On cold and taste, I drew on the following: R. Eccles et al., "Cold Pleasure. Why We Like Ice Drinks, Ice-Lollies and Ice Cream," *Appetite* 71 (2013): 357–60; Ann-Marie Torregrossa et al., "Water Restriction and Fluid Temperature Alter Preference for Water and Sucrose Solutions," *Chemical Senses* 37, no. 3 (March 2012): 279–92; and K. Talavera et al.,

"Heat Activation of TRPM5 Underlies Thermal Sensitivity of Sweet Taste," *Nature* 438 (2005): 1022–25. The Coca-Cola anecdote is drawn from Mark Prendergast, *For God, Country and Coca-Cola: The Definitive History of the Great American Soft Drink and the Company That Makes It* (New York: Basic Books, 2013). *Cooking with Cold* was published by the Kelvinator Sales Corporation of Detroit in 1932.

I also referenced Harold McGee, *On Food and Cooking: The Science and Lore of the Kitchen* (New York: Scribner, 2004); Allie Rowbottom, *JELL-O Girls: A Family History* (New York: Little, Brown, 2018); Helen Veit, "An Economic History of Leftovers," *Atlantic*, October 7, 2015; "Easiest Way to Improve the Flavor of Soups and Stews," *Cook's Illustrated*, January–February 2008; "Got Leftovers? Tips for Safely Savoring Foods a Second Time Around," *Food Technology Magazine*, March 2, 2015; Aaron Hutcherson, "Why Certain Foods Taste Better the Next Day," *Washington Post*, March 3, 2023; Bella Isaacs-Thomas, "The Food Science behind What Makes Leftovers Tasty (or Not)," *PBS NewsHour*, December 23, 2022; Luis Villazon, "Why Does Bolognese, Stew and Curry Taste Better the Next Day?," *BBC Science Focus*, August 24, 2017; "The Abuse of Cold Storage," *Inter Ocean* (Chicago), October 30, 1911, p. 6; Associated Press, "Failing Ice Cellars Signal Changes in Alaska Whaling Towns," November 25, 2019; and Y. Yang et al., "Knowledge of, and Attitudes towards, Live Fish Transport among Aquaculture Industry Stakeholders in China: A Qualitative Study," *Animals* 11, no. 9 (September 13, 2021).

On tomatoes, I referenced Barry Estabrook, *Tomatoland: How Modern Industrial Agriculture Destroyed Our Most Alluring Fruit* (Kansas City, MO: Andrews McMeel, 2012); Alex Philippidis, "Mistakes Shorten First Approved GMO's Shelf Life," *Genetic Engineering & Biotechnology News*, April 12, 2016; Thomas Whiteside, "A Reporter at Large: Tomatoes," *New Yorker*, January 24, 1977; Denise M. Tieman et al., "Identification of Loci Affecting Flavour Volatile Emissions in Tomato Fruits," *Journal* of *Experimental Botany* 54 (2006): 887–96; S. Mathieu et al., "Flavour Compounds in Tomato Fruits: Identification of Loci and Potential Pathways Affecting Volatile Composition," *Journal of Experimental Botany* 60 (2009): 325–37; D. M. Tieman et al., "The Chemical Interactions Underlying Tomato Flavor Preferences," *Current Biology* 22 (2012): 1–5; Linda M. Bartoshuk and H. J. Klee, "Better Fruits and Vegetables through Sensory Analysis," *Current Biology* 23 (2013): 374–78; Bo Zhang et al., "Chilling-Induced Tomato Flavor Loss Is Associated with Altered Volatile Synthesis and Transient Changes in DNA Methylation," *PNAS Biological Sciences* 113, no. 44 (October 17, 2016): 12580–85; and D. Tieman et al., "A Chemical Genetic Roadmap to Improved Tomato Flavor," *Science* 355 (2017): 391–94.

IV. The Fridge Diet

This section draws on my meeting with Jelena Bekvalac at the Museum of London in September 2016, as well as follow-up conversations as the project progressed. Gaynor Western and Jelena Bekvalac documented their project and its findings in illustrated detail in their book, *Manufactured Bodies: The Impact of Industrialisation on London Health* (Oxford: Oxbow Books, 2020). Simon Szreter, professor of history and public policy at University of Cambridge, and Johan Mackenbach, professor of public health at Erasmus University Medical Center in Rotterdam, also provided critical feedback.

It also draws on my conversation with Lee Craig, Alumni Distinguished Professor of Economics at North Carolina State University. The antebellum puzzle first surfaced in Robert W. Fogel et al., "The Economic and Demographic Significance of Secular Changes in Human Stature: The U.S. 1750–1960" (NBER Working Paper, April 1979). Lee Craig's papers on this topic include "The Effect of Mechanical Refrigeration on Nutrition in the United States," with Barry Goodwin and

Thomas Grennes, *Social Science History* 28 (Summer 2004): 325–36, as well as "The Short and the Dead: Nutrition, Mortality, and the 'Antebellum Puzzle' in the United States," with Michael Haines and Thomas Weiss, *Journal of Economic History* 63 (June 2003): 385–416. I also consulted Robert W. Fogel et al., "Secular Changes in American and British Stature and Nutrition," *Journal of Interdisciplinary History* 14, no. 2 (Autumn 1983): 445–81; and John Komlos, "A Three-Decade History of the Antebellum Puzzle: Explaining the Shrinking of the U.S. Population at the Onset of Modern Economic Growth," *Journal of the Historical Society* 12 (December 2012): 395–445.

I spoke with Jonathan Rees, professor of history at Colorado State University, Pueblo, in April 2023. Rees referenced research discussed earlier in this chapter: *Middletown: A Study in American Culture* by Robert S. Lynd and Helen M. Lynd. Additional sources for understanding shifts in diet and nutrition include Lizzie Collingham, *The Biscuit: The History of a Very British Indulgence* (London: Bodley Head, 2020); Otter, *Diet for a Large Planet*; Chris Otter, "The British Nutrition Transition and its Histories," *History Compass* 10, no. 11 (2012): 812–25; Joyce H. Lee et al., "United States Dietary Trends Since 1800: Lack of Association Between Saturated Fatty Acid Consumption and Non-communicable Diseases," *Frontiers in Nutrition* 8 (2021); Roderick Floud et al., *The Changing Body: Health, Nutrition, and Human Development in the Western World since 1700* (Cambridge: Cambridge University Press, 2011); Carrie R. Daniel et al., "Trends in Meat Consumption in the USA," *Public Health Nutrition* 14, no. 4 (April 2011): 575–83; Nina Teicholz, "How Americans Got Red Meat Wrong," *Atlantic*, June 2, 2014; and the US Department of Agriculture Economic Research Service (ERS) Food Availability (per Capita) Data System.

For changes in the nutritional content of fruits and vegetables, I referenced the following: Estabrook, *Tomatoland*; Donald R. Davis et al., "Changes in USDA Food Composition Data for 43 Garden Crops, 1950 to 1999," *Journal of the American College of Nutrition* 23, no. 6 (2004):

SELECTED SOURCES

669–82; Roddy Scheer and Doug Moss, "Dirt Poor: Have Fruits and Vegetables Become Less Nutritious?," *Scientific American*, April 27, 2011; and Rachel Lovell, "How Modern Food Can Regain Its Nutrients," *BBC Future*, 2023. For losses during refrigerated storage, I drew on Joy C. Rickman et al., "Nutritional Comparison of Fresh, Frozen and Canned Fruits and Vegetables. Part 1. Vitamins C and B and Phenolic Compounds," *Journal of the Science of Food and Agriculture* 87 (2007): 930–44; and James Wong, "The Decline and Fall of Broccoli's Nutrients," *Guardian*, April 9, 2017.

On the link between cold storage, reduced salt intake, and gastric cancer, I consulted the following: Christopher P. Howson, Tomohiko Hiyama, and Ernst L. Wynder, "The Decline in Gastric Cancer: Epidemiology of an Unplanned Triumph," *Epidemiologic Reviews* 8, no. 1 (1986): 1–27; David C. Paik et al., "The Epidemiological Enigma of Gastric Cancer Rates in the US: Was Grandmother's Sausage the Cause?," *International Journal of Epidemiology* 30, no. 1 (February 2001): 181–82; David Coggon et al., "Stomach Cancer and Food Storage," *Journal of the National Cancer Institute* 81, no. 15 (August 2, 1989): 1178–82; and Masoud Amiri et al., "The Decline in Stomach Cancer Mortality: Exploration of Future Trends in Seven European Countries," *European Journal of Epidemiology* 26, no. 1 (January 2011): 23–28.

On food poisoning, as part of my reporting in China, I spoke to Mike Moriarty of A.T. Kearney, Inc., and Richard Brubaker, adjunct professor at China Europe International Business School. I also drew on Anne Hardy, "Food Poisoning: An On-going Saga," *History and Policy*, January 13, 2016; A.T. Kearney, Inc., "Food Safety in China," 2007; Cameron J. Reid et al., "A Role for ColV Plasmids in the Evolution of Pathogenic *Escherichia coli* ST58," *Nature Communications* 13, no. 1 (2022); Madeline Drexler, *Emerging Epidemics: The Menace of New Infections* (Washington, DC: Joseph Henry Press, 2002); and CDC/National Center for Health Statistics.

Finally, on the microbiome, I spoke to Justin Sonnenburg in March

365

2023 and consulted Hannah C. Wastyk et al., "Gut-Microbiota-Targeted Diets Modulate Human Immune Status," *Cell* 184, no. 16 (August 5, 2021): 4137–153.e14; Aashish R. Jha et al., "Gut Microbiome Transition across a Lifestyle Gradient in Himalaya," *PLOS Biology* 16, no. 11 (November 2018); and Erica D. Sonnenburg and Justin L. Sonnenburg, "The Ancestral and Industrialized Gut Microbiota and Implications for Human Health," *Nature Reviews Microbiology* 17 (June 2019): 383–90. I also spoke to Fuchsia Dunlop in 2013 and referred to the transcript of the public workshop Fecal Microbiota for Transplantation, held in Bethesda, Maryland, on May 3, 2013, by the US FDA.

7. *The End of Cold*

I. The Future of Refrigeration

This section is mostly based on my reporting in Rwanda for *The New Yorker* in 2022, including extensive travel in Rwanda and a visit to OX's R&D facility in Leamington Spa, as well as conversations with Toby Peters, professor of cold economy and codirector of the Centre for Sustainable Cooling at the University of Birmingham; Brian Holuj, program manager at the United Nations Environment Programme; Juliet Kabera, director general of the Rwanda Environment Management Authority; Judith Evans; Natalia Falagán; Steve Cowperthwaite, head of international stratospheric ozone and fluorinated greenhouse gases at the UK Department for Environment Food and Rural Affairs; Phil Greening, professor at Heriot Watt University and director of the Centre for Sustainable Road Freight and the Centre for Logistics and Sustainability; Issa Nkurunziza, agriculture cold chain expert at UNEP U4E Rwanda Cooling Initiative; Jean Baptiste Ndahetuye, operations research coordinator at ACES; Francine Uwamahoro, Simon Davis, Sam Dargan, Ferdinand Munezero, Louise Umutoni, and Jean de Dieu Umugenga at OX Delivers; Catherine Kilelu, senior research fellow at the African Centre for Technology Studies; Innocent Mwalimu, horticulture quality assurance officer at the National Agricultural Export Development Board; Guillaume Nyagatare, senior lecturer at the University of Rwanda; Paul Imulia of Hortifresh Ltd.; Alice Mukamugema, crop products supply chain and market analyst at MINAGRI; Lisa Kitinoja, founder and president of the Postharvest Education Foundation; Vincent Gasasira, founder of Post-Harvest Plus; Pascal Dukuzamuhoza, district agriculture inspector at MINAGRI; Donatien Iranshubije; François Habiyambere; Julian Mitchell, CEO of Inspira Farms; Michael Murphy, founding principal and executive director, Christian Benimana, co–executive director and senior principal, and Tilly Lenartowicz, director of environmental engineering at MASS Design Group; Selçuk Tanatar, principal operations officer at International Finance Corporation, World Bank Group; Jimmy Washington, director of sustainability and cold chain development at Carrier; Eric Trachtenberg, practice lead and senior director of Millennium Challenge Corporation's Land and Agricultural Economy Practice Group; and Pat Hughes, adviser at Integrated Cold Chain Management. In China, I visited Beijing Xinfadi Agricultural Products Wholesale Market, as well as Emerson's Research & Solutions Center in Suzhou, where I spoke with Clyde Verhoff and Mark Bills.

On Rwanda, I also consulted the following resources: Anne-Michèle Paridaens and Sashrika Jayasinghe, "Rwanda: Comprehensive Food Security Analysis 2018," World Food Programme Vulnerability Analysis and Mapping (Rome: United Nations World Food Programme, 2018); World Bank, "GDP per Capita: Rwanda"; James Noah Ssemanda et al., "Estimates of the Burden of Illnesses Related to Foodborne Pathogens as from the Syndromic Surveillance Data of 2013 in Rwanda," *Microbial Risk Analysis* 9 (2018): 55–63; Harald von Witzke et al., "The Economics of Reducing Food-Borne Diseases in Developing Countries: The Case of Diarrhea in Rwanda," *Agrarwirtschaft* 54, no. 7 (2005): 314–17; Mohamed M. El-Mogy and Lisa Kitinoja, "Review of Best Postharvest Practices for Fresh Market Green Beans" (PEF White Paper 19-01, The

Postharvest Education Foundation, February 2019); Adams, *Marketing Perishable Farm Products*; Ministry of Finance and Economic Planning (MINECOFIN), Republic of Rwanda, "Vision 2050," December 2020; Ministry of Environment, REMA, and United Nations, "Rwanda National Cooling Strategy," 2019; National Agriculture Export Board, "Cold Chain Assessment: Status of Cold Chain Infrastructure in Rwanda," January 2019; University of Birmingham, "Africa's Clean Cooling Centre of Excellence Moves Closer to Boosting Farmers' Livelihoods," news release, December 8, 2020; and "ACES Synthesis Report on Rwandan Agriculture and Vaccine Cold-chain Equipment, Policies, Programmes and Practices."

On the nexus of refrigeration, food waste, hunger, and climate change, I drew on the following: Esben Hegnsholt, "Tackling the 1.6-Billion-Ton Food Loss and Waste Crisis," Boston Consulting Group, August 20, 2018; World Wildlife Federation, "Driven to Waste: Global Food Loss on Farms," July 2021; J. L. Dupont, A. El Ahmar, and J. Guilpart, "The Role of Refrigeration in Worldwide Nutrition (2020), 6th Informatory Note on Refrigeration and Food," International Institute of Refrigeration, March 2020; Sustainable Energy for All, "Chilling Prospects: Providing Sustainable Cooling for All," July 1, 2018; Nadia El-Hage Scialabba, "Food Wastage Footprint: Impacts on Natural Resources, Summary Report," Food and Agriculture Organization of the United Nations, 2013; Nadia El-Hage Scialabba, "Food Wastage Footprint & Climate Change," UN Food & Agriculture Organisation, November 2015; Carbon Trust, Cool Coalition, High-Level Champions, Kigali Cooling Efficiency Program, and Oxford Martin School at the University of Oxford, "Climate Action Pathway: Net-Zero Cooling," December 9, 2020; Toby Peters and Phil Greening, "A Seven-Point Plan to Tackle the World's Biggest Cooling Challenge," University of Birmingham; Yabin Dong et al., "Greenhouse Gas Emissions from Air Conditioning and Refrigeration Service Expansion in Developing Countries," *Annual Review of Environment and Resources* 46 (2021): 59–83; Brent R. Heard and Shelie A. Miller, "Critical Research Needed to Examine the Environmental Impacts of Expanded Refrigeration on the Food System," *Environmental Science and Technology* 50, no. 22 (October 2016): 12060–12071; Stefan Ellerbeck, "IEA: More Than a Third of the World's Electricity Will Come from Renewables in 2025," World Economic Forum, March 16, 2023; Weizhen Tan, "What 'Transition'? Renewable Energy Is Growing, but Overall Energy Demand Is Growing Faster," *CNBC*, November 3, 2021; and Toby Peters and Leyla Sayin, "The Cold Economy" (ADBI Working Paper 1326, Asian Development Bank Institute, Tokyo, 2022).

On refrigerants, I drew on Eric Dean Wilson, *After Cooling: On Freon, Global Warming, and the Terrible Cost of Comfort* (New York: Simon & Schuster, 2021); Paul Hawken, *Drawdown* (New York: Penguin, 2017); Diane Toomey, "Paul Hawken on One Hundred Solutions to the Climate Crisis," *Yale Environment 360*, July 25, 2017; Fred Pearce, "Inventor Hero Was a One-Man Environmental Disaster," *New Scientist*, June 7, 2017; UN Environment Programme, "Rebuilding the Ozone Layer: How the World Came Together for the Ultimate Repair Job," September 15, 2021; Nick Campbell, "Why Does the Illegal Trade in Refrigerant Gases Matter to Europe's Energy Security?," *Politico*, September 8, 2022; Samuel Smith, "Refrigerant Leak Detection and Regulatory Update," Copeland, July 24, 2018; James M. Calm, "The Next Generation of Refrigerants—Historical Review, Considerations, and Outlook," *International Journal of Refrigeration* 31, no. 7 (November 2008): 1123–33; Lisa Tryson and Torben Funder-Kristensen, "Momentum Grows for Low GWP Refrigerants," *Contracting Business*, December 13, 2019; and Christina Theodoridi, "The Unexpectedly Exciting World of Refrigerants," NRDC, August 5, 2021.

To illustrate the efforts to create alternative methods of mechanical cooling, I drew on the following: Alok Jha, "Einstein Fridge Design Can Help Global Cooling," *Guardian*, September 20, 2008; Gene Dannen, "The Einstein-Szilard Refrigerators," *Scientific American*, January 1997;

Keng Wai Chan and Malcolm McCulloch, "The Einstein-Szilard Refrigerator: An Experimental Exploration," *ASHRAE Transactions* 122, no. 1 (2016); Jennifer Ouellette, "Chill, Baby, Chill," *Cocktail Party Physics*, November 30, 2010; Noah Schachtman, "Hear That? The Fridge Is Chilling," *Wired*, January 6, 2003; General Electric, "Not Your Average Fridge Magnet: These High-Tech Magnets Will Keep Your Butter (and Beer) Cold," news release, February 7, 2014; and Andrew Turley, "The Future of Cool," *Chemistry World*, January 12, 2012.

I also referenced the following: Susanne Freidberg, *French Beans and Food Scares: Culture and Commerce in an Anxious Age* (New York: Oxford University Press, 2004); Emmanuel Ntirenganya, "NAEB Upgrades Packing House to Ensure Safety of Horticulture Exports," *New Times*, March 2, 2020; Macharla Kamau, "Horticulture Extends Its Lead as Biggest Foreign Exchange Earner," *Sunday Standard*, May 7, 2022; Reuters, "Kenya Horticulture Export Earnings up 5% Yr/Yr in 2021," February 22, 2022; Kristen Hall-Geisler, "The OX Is a Flat-Pack Truck for the Developing World," *TechCrunch*, September 27, 2016; Human Rights Watch, "'Why Not Call This Place a Prison?' Unlawful Detention and Ill-Treatment in Rwanda's Gikondo Transit Center," September 24, 2015; Caroline Kimeu, "What Tanzania Tells Us about Africa's Population Explosion as the World Hits 8bn People," *Guardian*, November 15, 2022; and Andrew Blum, "Planning Rwanda," *Metropolis*, November 1, 2007.

II. The Future May Not Be Refrigerated

This section is largely based on my visit to Apeel in May 2018. I also drew on the following: Stephanie Strom, "An (Edible) Solution to Extend Produce's Shelf Life," *New York Times*, December 13, 2016; Leonora R. Baxter, "The New Ice Age," *Golden Book Magazine*, no. 73 (January 1931); Matteo Motolese, "Il Duce 'frigoriferò' il dissenso," trans. Federico Sanna, *Il Sole 24 Ore*, February 23, 2020; Anderson Jr., *Refrigeration in America*; Elting E. Morison, *Men, Machines, and Modern Times*, 50th anniversary ed. (Cambridge, MA: MIT Press, 2016); Mary E. Penning-ton, "Influence of Refrigeration on the Food Supply" (read at the 22nd Annual Meeting of the American Warehousemen's Association, Pitts-burgh, December 5–7, 1912); Farming and Farm Income, USDA Eco-nomic Research Service; and Jessica Davies, "Rurban Revolution: Can Ruralising Urban Areas through Greening and Growing Create a Healthy, Sustainable & Resilient Food System?," Lancaster University, Principal Investigator, UK Research and Innovation (March 2019–March 2021).

Epilogue: Meltdown

I visited Svalbard in October 2016. I also drew on the following: Cary Fowler, *Seeds on Ice: Svalbard and the Global Seed Vault* (Westport, CT: Prospects Press, 2016); Lyndsey Matthews, "There's a Remote Norwe-gian Town Where You're Not Allowed to Die," *Men's Health*, March 14, 2018; "Banking against Doomsday," *Economist*, March 10, 2012; "Banks for Bean Counters," *Economist*, September 10, 2015; *Brown Planthopper: Threat to Rice Production in Asia* (Proceedings of a Symposium, Interna-tional Rice Research Institute, Manila, Philippines, 1979); Kazushige Sogawa et al., "Mechanisms of Brown Planthopper Resistance in Mudgo Variety of Rice," *Applied Entomology & Zoology* 5, no. 3 (1970): 145–58; S. P. Singh et al., "Indian Rice Genetic Resources and Their Contribu-tion to World Rice Improvement—a Review," *Agricultural Reviews* 24, no. 4 (2003): 292–97; Helen Sullivan, "A Syrian Seed Bank's Fight to Survive," *New Yorker*, October 19, 2021; Suzanne Goldenberg, "Global Seed Vault Dispatches First Ever Grain Shipment," *Guardian*, October 19, 2015; Mark Sabbatini, "'No Other Place in the World Is Warming Up Faster Than Svalbard': March Will Be 100th Straight Month of Above-Average Svalbard Temperatures, Weather Service Says," *Icepeople*, March 25, 2019; and *The Younger Edda*, trans. Rasmus B. Anderson (Chicago: Scott, Foresman, 1901).

Index

Adams, Arthur Barto, 114, 284
Adams, Frederick Upham, 138, 140, 160
Afghanistan, 116, 317
Africa Centre of Excellence for
    Sustainable Cooling and Cold
    Chain (ACES), 285–88, 291–94,
    298–301, 311
agribusinesses, 169, 209, 288, 292–93.
    See also specific businesses
Agricultural Revolution, 37–38
agriculture
    controlled atmosphere and, 121, 125
    and diet/health, 264–65
    gene banks, 316–18, 320
    industrialization of, 169
    introduction of, 37–40
    prerefrigeration, 244
    urban, 312
    See also farmers
air conditioning, 8, 86, 165–66, 182, 293
air freight, 21, 23, 174–78, 209, 282, 316
Alden, Charles, 70
Amana, 228–30
Amato (Americold supervisor), 4, 14,
    16–17, 21
Amazon.com, 243
Ambrosi, Jimmy, 12–13, 16
American Civil War, 59–60, 263–64
American Linde Refrigerating
    Company, 140
American Society of Heating,
    Refrigerating, and Air-Conditioning
    Engineers (ASHRAE), 296

American Warehousemen's Association, 94
Americold, 3–4, 9–17, 21–23, 194–95,
    198–201
Anheuser-Busch, 20
antebellum puzzle, 263–66
Anthropocene, 106–7
Antle, Bud, 127–28
Apeel Sciences, 301–8
Appert, Nicolas, 38–39
apples, 33, 83, 267
    ethylene and, 131–32, 135–36,
        239–41
    farming/harvesting of, 178, 252
    preservation/storage of, 111–12, 114–24,
        129–32, 156–58
    shipping of, 174, 182, 189
    spray coating for, 306
Argentina, 22, 70, 76–77, 81–82
Armour, 82, 167
Armour, Philip, 90
Asdal, Åsmund, 316, 318–20
atmospheric engineering/modification,
    112, 116–18, 136, 141–46, 158, 239.
    See also controlled atmosphere
Austen, Jane, 48
Australia, 59, 70, 76–82, 197, 240,
    266, 306
avocados, 136, 144–46, 158, 188, 239, 241,
    302, 305–7

Bacon, Francis, 28, 44, 107
Banana Distributors of New York, 136–37,
    141–44

bananas, 111, 252, 268
  ethylene and, 136, 141–44, 190, 252
  history of, 137–39, 142, 158–160
  refrigeration and, 137–41, 142, 156, 158–60, 182
  ripening of, 142–45, 158–59, 190
  in Rwanda, 280, 291
  shipping of, 139–41, 159, 174, 181–82, 190, 199, 268
Bangladesh, 298–99
Barnard, Eunice Fuller, 113
Barnard, Steve, 145–46, 188
Baxter, Leonora, 308
Beamon, Sylvia, 43–49, 51, 53
Beard, James, 251
Becher, J. J., 132
beer, 54, 59, 60, 111, 180, 220, 236, 239, 244–45, 247, 258
Behr, Robert, 153
Beijing, China, 208, 210–11, 220, 284
Beko, 228, 232, 242–43
Bekvalac, Jelena, 260–64, 266
Bell, Henry, 78–79
Bell-Coleman refrigerator, 78–79
Ben & Jerry's, 297
Benimana, Christian, 299–301
Bérard, Jacques Étienne, 116, 122
Bernardez, Juan "Ricky," 97–98
Biden, Joe, 219
Birdseye, Clarence, 171
Bizimana, Pierre, 281–82, 284
"Blind Dates: How to Break the Codes on the Foods You Buy," 237
Blondin, Denis, 16
Boeing, 175
Bolger, Mary Kay, 229–30, 234–35, 241
Booth's Cold Storage (Chicago), 83
Borden, Gail, 70
Boston, MA, 15, 50, 51, 53, 59, 75, 137, 140, 196
Boyle, Robert, 28
Bradford, Kipp, 27–32, 222, 227, 295
brain freeze, 245–46
Brambles, 197
Brazil, 131, 146, 149–50, 258, 318, 320
Breakstone, Myron, 20–21
Breeden, Jack Charles, 170

breweries, 54, 59–61, 244–45
British Association of Refrigeration, 131–32
Brownell, Henry, 84
Bruce Church Inc., 126, 129, 131
Brunsing, Rex, 127–28, 148
Brydone, Patrick, 246
Bubbly Creek sewer (Chicago), 65–66
Burundi, 317, 319
Busch, 61. See also Anheuser-Busch
Busch, Adolphus, 244–45
butchers/butchery, 22, 66, 72–73, 78–80, 96–97, 103–5, 173, 226

cacao beans, 184–85
Calgene, 252
California, 237
  iced railcars and, 169, 211
  refrigerated warehouses in, 3, 198–99
  shipping meat/produce from, 169, 170, 175, 188, 200, 268
  See also Americold; Apeel Sciences; Los Angeles, Salinas Valley (CA), Vernon (CA)
California Air Resources Board, 296
California Avocado Society, 145
Calkins, Earnest Elmo, 227–28, 230
canned food, 37–40
Capone, Al, 236
Carbonic Acid Gas Company (San Jose, CA), 116–17
cargo ships/shipping, 76–79, 174–76, 179–82, 190, 200. See also reefers (refrigerated containers); shipping containers
Carlos (Americold employee), 17, 22–23
Carrier, 182
Carter, J. B., 133
Carter, Jason, 11–12
caves, 59–60
  for cheese, 7, 191–95, 198
  for storing food, 43–46
cellars, 46, 48, 60–61, 116, 224, 238, 244, 250, 267
Centennial Exposition (Philadelphia), 138
Central America, 144, 158–60, 268
Centre for Human Bioarchaeology (London), 261

Cesar (Americold employee), 22–23
CFCs, 295–96
Champion, 20
charqui (jerked beef), 71, 87
Check Their Fridge (blog), 218
cheese, 14, 20, 238, 248, 258
    additional nutrition from, 265, 269
    caves for, 7, 191–95, 198
    food poisoning from, 85
    preservation of, 36–39, 46, 88
    shipping of, 174
cheeseburger experiment, 243–44
"Chemical and Bacteriological Study of
        Fresh Eggs, A" (Pennington), 92
chemical preservatives, 36–37, 70, 88–90,
        136, 270, 303
chemistry, 69–70, 89–91, 112–13, 116, 251
Chen Zemin, 205–9
Chicago, IL, 164, 166, 199
    futures trading in, 154–55
    meat-packers in, 65–66, 79, 167, 170,
        172–73, 211
    refrigerated warehouses in, 83–85
    shipping beef/livestock from, 71–73, 169
    stockyards in, 88–89, 172–73
    See also Cold Storage Banquet of 1911
Chicago River, 65, 82
Chicago World's Fair, 55–57, 75
Chile, 175–76, 182, 188–89, 191, 199
Chilling Prospects, 285
China, 187, 292, 296–97
    avocados and, 146, 307
    chilled desserts of, 245
    farmers in, 209, 213
    fast food in, 259
    food preservation in, 213, 273
    food waste in, 310
    fresh fish in, 213, 250–51
    greenhouse emissions of, 293
    refrigeration in, 207–14, 258–59, 294
    See also cold chain: in China; frozen
        foods: Chinese dumplings; *specific
        cities*
Chiquita, 190, 199
Christie, Agatha, 87
CIA, 159
Citrosuco, 146–47, 151, 153, 199

Claiborne, Craig, 251
climate change, 293–96, 307, 315, 320–21.
        See also global: warming
cliometrics, 263–67
Cloverleaf Cold Storage, 198
Coca-Cola, 150–51, 153, 167, 247, 257
cold
    effects on food/drinks, 8–10, 245–47
    effects on health, 15–17
    effects on humans, 9–11, 14–17, 19–20,
        245–47
    and the food system, 309–12
    preservative power of, 46, 74, 88
    science behind, 246–47
    sensory pleasures of, 244–49
    slows growth of bacteria, 98, 104, 271
    study of, 28
cold chain
    in China, 176, 209, 211–12, 271, 284
    and delivering vaccines, 298–99
    ecological costs of, 293–99, 309
    explanation of, 5–9
    and food distribution, 201–2
    and food waste, 282–83
    fresh produce and, 305–7
    and freshness, 236
    global, 6, 187–88, 205, 286, 310
    investment in, 287–89
    reshaping of, 204
    in Rwanda, 286–92, 298, 310
    in US, 211–12, 282–83, 310
    See also developing countries
cold storage
    experimental years of, 55–62, 88
    fear/suspicion of, 84–89, 106
    history of, 43–49
    studies of, 57–58
    temperature of, 282–83
    underground, 194–95, 250
    See also refrigeration
Cold Storage and Ice Trade Journal, 78
Cold Storage Banquet of 1911, 82–85, 89,
        94, 213, 249
Coleman, J. J., 78–79
Collingham, Lizzie, 267–68
community cooling hubs, 298, 311–12
comparative advantage principle, 128–29

Congolese traders, 281, 284, 291
consumer engineering, 227–30
controlled atmosphere
    bags, 5, 130–31, 301–2
    storage, 118–25, 130, 158, 239
convenience foods, 22, 171–72, 203, 232,
    251, 310
Cooking with Cold (Kelvinator), 247–48
Cook's Illustrated, 248–49
"Cool Air" (Lovecraft), 86–87
Costco, 146, 232
COVID-19, 5, 219, 271, 286–88,
    298–99, 311
Cowan, Ruth Schwartz, 223
Craig, Lee, 263–68, 274–75
Critchell, James, 80–82, 87
crop diversity, 316
Crosley Shelvador refrigerator, 228
cryosphere, artificial, 3, 6–9, 13, 15, 59, 62,
    142, 146, 191
cryosphere, natural, 6, 15
Cuba, 50–51, 54
Cullen, William, 57–58
Cultural Revolution (China), 206

Dai Jianjun, 212–14, 273
dairy, 107, 212, 239, 244
    becomes a staple, 94
    consumption of, 69, 310
    and diets, 67, 113, 205, 258, 268, 270
    and health, 91, 265
    preservation of, 88
    refrigeration of, 9, 12, 14, 198–99,
        221, 254
    See also cheese; ice cream; milk
Dalrymple, Dana, 121, 124
Danone, 3–4
D'Arrigo, Andrew and Stephen, 137
D'Arrigo, Gabriela, 137, 141–44, 239
D'Arrigo New York, 137
date labels, 236–38, 243
dating, 8, 217–20, 233, 257
David, Elizabeth, 47
Davis, Simon, 290–91
de Boer, Jelle Zeilinga, 134–35
De La Vergne company, 61
Deaton, Shawn, 18–20

deforestation, 107–8, 158–59, 293
Demos, Steve, 235–36
Deng Xiaoping, 206, 210
Descartes, René, 28
developing countries, 295, 309–12
    building cold chains in, 6, 8, 285,
        289, 293
    delivering vaccines in, 298–99
    diets of, 205, 282
    lack of cold chain in, 282–83
    refrigeration projects in, 285–86
    See also specific countries
Dickens, Charles, 68, 72, 260
diet
    in 1800s, 66–69, 269
    American, 67–68, 172, 270, 310
    and animal protein, 69, 82, 112–13, 173
    cartographically eclectic, 170–72
    decline in, 264, 288
    in developing world, 205, 282
    of hunter-gatherers, 37–38
    modern Western, 272
    prerefrigeration, 37–39, 267–68
    and refrigerated ships/containers,
        181–82, 186, 188
    refrigeration's impact on, 40, 264–70, 310
    and vitamin-rich foods, 112–13, 126,
        148–49
digital consumers, 205
diseases, 15
    bacterial infections, 270–72, 274
    chronic inflammation, 272–74
    diarrhea, 61, 85, 270–72, 283
    history/study of, 260–66
    of plants/trees, 135, 156, 159
    refrigeration and, 61, 94, 265–66, 274
    related to food, 85, 94, 270–75
    See also food: poisoning
diversity, of food, 160, 169, 225, 267–68,
    309–10, 316
DNA, 254, 317
Dodd, George, 67–68
Dowie, Harry, 84, 88–89
drinks
    chilled by ice, 50, 54, 60–61, 71
    cocktails, 244–45
    mechanical refrigeration for, 60

and temp/taste receptors, 245–47, 272
See also beer; breweries; orange juice
drought, 131, 296, 307, 316
drying food, 35–36, 67, 71, 273
DunAn Artificial Environment
    Equipment Company (China),
    296–97
Dunlop, Fuchsia, 273
DuPont, 20–21

East of Eden (Steinbeck), 124, 126–27
Eastman, Timothy C., 76, 78
Eastman Kodak, 127
economic growth, 209, 264–65
economics, 8, 309
Edison, Thomas, 223
Egg Reporter, The, 83–85
eggs, 37, 67, 92–95, 239–41
Egypt, 16
Eichengreen, Meyer, 83
Einstein, Albert, 297
electric refrigerators, 31, 61, 208,
    220–27
Electrolux-Servel, 222–23
Elizabeth, Queen, 319
Emerson Climate Technologies (China),
    294, 296
energy
    efficiency, 204–5, 297, 307
    in refrigeration, 193, 222, 275,
        293–94, 309
    renewable, 77–78, 294, 302
    sustainable, 286
environmental impact
    of cattle ranching, 107–8
    dietary shifts and, 259
    of freezer trawlers, 107
    of global cold chain, 293–99
    of growing bananas, 158–59
    of industrialization, 266
    of large-scale farming, 174
    of manure lagoons, 211
    and natural ice trade, 61
    of reefer shipping, 189–90
    of refrigeration, 7–8, 275, 293–94, 309–10
    of slaughterhouses, 82
environmental policies, 200, 296

Espinoza, Anthony, 3–4, 9–12, 14, 22
ethylene, 132–36, 141–46, 158, 190,
    239–42, 252, 255
Evans, Judith, 287, 291–92, 297–98
Evans, Kate, 122–24
Express Channel Food Logistics, 210–11

Falagán, Natalia, 120, 242
    as ACES cofounder, 285, 287
    on lettuce/salad, 129, 131, 312
    on refrigerating fruit, 239
    research of, 111–12, 156–57, 312
    on sustainable food system, 158
farmers, 94, 157, 239
    adopt ice-filled trucks, 168–69
    benefit from refrigeration, 54, 77,
        189, 309
    in developing countries, 285, 311–12
    ecological impacts of, 309–10
    and frozen meat trade, 78–81
    in Great Britain, 80–81
    large-scale, 156, 172–74, 189
    of orange trees, 154–56
    small-scale, 172, 289, 309–12
    and smart farms, 292
    and spray preservative, 301, 306
farm-to-table food, 4–7, 283
fast food, 6, 7, 106, 153, 259
FDA, 237–38, 241, 305
Federal Meat Inspection and Pure Food
    and Drug Acts, 89–90, 92
Feiges, Adam, 13, 198–200, 205
fermented foods, 36–37, 39, 213,
    273–74, 303
field refrigeration units, 167
Fineberg, Al, 165–66
fish, 171, 213
    preservation of, 35–36, 177
    preserved in ice, 46, 49, 54, 279–81, 284
    purchased alive, 250–51
    refrigeration and, 181, 243
    Rwanda's industry of, 279–81
    shipping of, 23, 174–78, 190, 251
    "tuna coffins" and, 177–78
fish sticks, 45, 107, 149, 172, 178, 232
fishermen, 43, 49, 54, 71, 89, 107,
    177–78, 213

flake ice, 177, 279–81, 284, 292
flash-frozen food, 171–72
flavor, 135, 213
  and climate change, 307
  of cold-stored apples, 118, 121–24
  improved by cold, 84, 95–106, 248–49
  loss of, 33, 71, 95, 142, 269–70
  of meat, 71, 79
  of prerefrigeration-era foods, 36–37
  restoring it, 150–53
  science behind, 252–55
  and spray preservative, 302, 305–6
Florida, 59, 114, 148–53, 156, 199,
  252–53, 256
Florida State University, 253
Florida Tomato Committee, 253
flowers/flower bulbs, 21, 175, 183, 288
Fogel, Robert, 264
Fogelman, Brian, 146–47, 149–52
food
  available year round, 94, 106, 181, 187,
    225, 268, 309
  laws for, 92, 237, 240
  our relationship with, 224, 241, 308, 312
  production of, 169–71, 174, 224, 312
  rotting/spoilage of, 32–40, 232
  safety of, 85, 87–95, 131, 289
  security in, 288–89, 298
  storage of, 44, 55, 238–41, 244, 250
  supply of, 5, 234, 303
Food Network, 230
food poisoning, 85, 87, 94, 238, 270–71,
  274, 309
Food Research Laboratory, 90
food system, 5, 169, 178–81, 198, 244, 255,
  271, 307, 310–12
food waste, 35, 268, 312
  in developing countries, 282–85, 289, 310
  global, 293–94
  and produce preservation, 302, 304–5, 307
  a refrigeration by-product, 232–35,
    241, 310
  in the US, 232–35, 238, 284, 310
Ford, Henry, 79, 163
Fowler, Cary, 317–18, 320
France, 38–39, 54, 60–61, 76–78, 87, 191,
  226–29, 311

Franklin, Benjamin, 58, 100–101
freeways, 170, 200
freeze-drying, 309
Freezine, 88, 94
Freidberg, Susanne, 86–87
Freon, 295
freshness, 311
  definition of, 152, 307
  and food date labels, 236–38
  limited life span of, 33–35
  of meat, 68–74, 79, 87
  prerefrigeration, 240–41
  preserved foods and, 37–40
  and refrigerated food, 40, 83–88,
    249–51, 299
  and refrigerators, 235–38, 242–43
  science behind, 86, 242
Frigidaire, 222, 227–28, 295
frigoriferare, 308
frigoriphobie, 87, 94, 309
frozen foods, 3
  automated warehouses for, 202–5
  Chinese dumplings, 8, 205–10,
    212–13, 294
  freezer burn and, 207, 232
  introduction of, 171–72
  snow pits for, 45–46
  storage of, 6, 9–11, 14, 20–22, 195, 228
  See also orange juice; *specific types*
Frozen Water Trade, The (Weightman), 52
fruit
  as commodities, 137, 139, 158–59
  harvesting of, 139, 255, 307
  Pratt's studies of, 186
  refrigeration of, 137–42, 146, 241–43
  refrigeration's effects on, 239, 251,
    254–55
  shipping of, 209, 252, 254
  varieties lost, 156
  See also ethylene; produce; *specific types*
Fulton Market (New York City), 54, 196
fungicides, 306
futures market, 154–56, 256–59

Gane, Richard, 135
Garbage Project (University of AZ),
  233–34

Garnett, Tara, 172, 232–33
Gary, Gabe, 194–95
Gates Foundation, 288, 304
GDP, 264–65, 283
General Electric, 218, 222–23, 227–28, 234
General Foods, 171, 184
General Motors, 222, 295
genetic engineering, 157, 252–55
geography of food, 40, 158, 169, 173, 195, 236
germ theory, 61, 264
Germany, 61, 103, 114, 174–75, 227
G.I. Bill, 170
Glasgow & Thunder brewery (Australia), 59
global
    cold chain, 6, 205, 286, 310
    commodities, 137, 146
    economy, 8, 198
    food system, 5, 178, 271, 312
    food waste, 282–85, 293–94
    market, 189
    population growth, 303
    seed vault, 316–20
    supply chains, 180
    trade, 176, 181, 190, 200
    warming, 8, 293–97
Global Cold Chain Alliance, 6
Global Cooling for All, 285
Goldman, Irwin, 157
Goodwin, Barry, 264
Great Britain, 237, 245, 303
    apples in, 122
    automated warehouses in, 202–5
    banana trade of, 139–40
    cold storage research in, 114–15, 123, 131–32
    food preservation in, 37–38, 70–71, 87, 111–12
    ice pit experiment in, 44–46
    icehouses in, 47–49, 62
    Industrial Revolution of, 37–38, 261, 263
    meat shipped to, 78–79, 87
    prevents scurvy, 113
    refrigeration in, 61, 87, 118, 213, 226
    "rot box" experiment in, 32–35
    wartime food anxiety in, 114

See also London, UK; Low Temperature Research Station (LTRS); NewCold Wakefield; Royston Cave (UK)
Great Depression, 168–69, 225
Great Exhibition of 1851 (London), 70–71
Great Expectations (Dickens), 72, 260
Great Pittsburgh Banana Company Explosion, 144
"Greatest Refrigerator on Earth," 55–57
Greece/Greeks, 36, 47, 134–36
greenhouse gases, 8, 293–97
Greening, Philip, 298–99
Grennes, Thomas, 264
Griesemer, Louis, 191–95
Grundig, 243
Guatemala, 158–59, 188
gut microbiome, 7, 270, 272–74

Habiyambere, François, 279–82
Hackett, Shawn, 154–56
Hale, John, 134–35
Hallock, MN, 163–65
Hardy, William B., 131–32, 135
Harrison, James, 59
HarvestFresh fridge, 242
Harvey, R. B., 133
Haussmann, Baron, 76–77
Hawken, Paul, 295
Hawksmoor, Nicholas, 47
HCFCs/HFCs, 295–96
health, 61, 91, 310
    animal protein and, 69, 112–13
    chemical preservatives and, 88–90
    cold's effects on, 15–17
    decline in, 264, 266, 270
    dietary shifts and, 258–59
    and early cold-storage foods, 85, 88–90, 95
    and food preservation, 241
    and frozen foods, 212
    of Londoners, 260–64
    prerefrigeration, 267–68
    refrigeration's impact on, 5, 8, 188, 260–75
    See also food: safety; gut microbiome; nutrients; nutrition; vitamins
Heiden, Gustavo, 320
height, 8, 264–67

Henry, Gil, 145
Henry, O. 159
Hill, Walter, 164
homegrown produce, 188, 225, 243–44,
    251, 256, 268, 312
Honduras, 158–59
Honeywell, 10
Horrisberger, Jim, 150, 153
Hotel Sherman (Chicago), 83–84
household technologies, 226
"How the Refrigerator Got Its Hum"
    (Cowan), 223
hunger, 7, 35, 81, 285, 298
hunter-gatherers, 37–38, 40, 59, 303, 309

ice, mechanically made, 57–62, 125–28,
    244–45, 279–81
ice, natural
    early industry in, 49–55, 309
    harvesting of, 40–43, 47–54, 61, 319
    and income/social class, 256–59
    industrialization of, 49–53, 58, 66
    for preserving food, 71, 76, 78, 88, 92–93
    in refrigerated warehouses, 196
    shipping of, 49–52, 59–61, 66, 73
Ice Age, 47
Ice and Refrigeration, 10
ice cream, 42–43, 244–47
    early history of, 47, 50–54, 71
    food poisoning from, 85
    and future of cooling, 297
    "penny licks," 54, 91
    shipping of, 190
    social status and, 258
    storing of, 14, 193, 203
iceberg lettuce, 125–31
iceboxes, 52, 54, 220–21, 228
iced railcars, 268
    meat shipped in, 71, 73–76, 88, 166,
        211, 264
    natural ice used for, 54, 73–76, 93
    produce shipped in, 125–27, 137, 166, 211
    public health and, 88
    railways' investment in, 169
    Thermo King units for, 170
    vast implications of, 211
icehouses, 40–44, 46–52, 54, 62, 113

Iceman Cometh, The (O'Neill), 221
icemen, 221
Ideal X, 179–80
immigrants, 54, 60, 67–68, 81, 128, 138
Imperial Direct West India Mail Service,
    139–40
income, 208, 231, 257–59, 264–65, 285,
    288, 291, 312
India, 60, 191, 205, 257–58, 286, 317
Industrial Revolution, 37–38, 261, 263, 268
industrialization, 87, 169
    and advent of refrigeration, 274
    of cold, 49, 51, 58
    and food preservation, 36–38
    of food system/supply, 244, 303
    impact on health, 263–64, 266
    of meat processing, 88, 101, 172–73
    of produce, 136, 252, 255, 272
    and urban workers' diets, 113
Instagram, 146, 219
International Association of Refrigerated
    Warehouses, 7
Intercontinental Exchange
    (ICE Futures, NYC), 154
Iowa Beef Packers (IBP), 173
Iranshubije, Donatien, 293
Irish independence, 81, 108
Italy, 47–49, 227, 246, 317

James, Stephen, 98–100, 105
Japan, 16, 143, 176–78, 196, 240, 250,
    270, 319
Japan Airlines (JAL), 176–78
Jaquith, Waldo, 243–44
Jefferson, Thomas, 51, 220
jellied foods, 247–49
Jones, Fred McKinley, 163–70, 182.
    See also Thermo King
*Journal of the American Medical
    Association, The*, 85, 133
juice preservation, 146–53
Jungle, The (Sinclair), 65–66, 88
junk food, 272

Kabera, Juliet, 286
Kagame, Paul, 285–86
Kelvinator, 221–22, 227–28, 247–48

Kennedy, John F., 107
Kenya, 175, 288–89
Kidd, Franklin, 115–23, 130, 132, 135, 156, 158
Kieckhefer, Robert, 126–28
Kieckhefer Box Company, 127
Kigali, Rwanda, 280, 283, 286, 293, 299–300
Kigali Amendment, 296
Kilelu, Catherine, 288
kitchens, 8, 32, 167, 171, 220, 223, 225, 230–31, 239–40
Klee, Harry, 251–56, 316–17
Knights Templar, 43–44
Koeppel, Dan, 159–60
Kraft, 7, 191–94, 198
Kroger, 146, 200

laborers/workers
    disputes, 131, 180
    meat/dairy intake of, 69, 113
    poorly compensated, 105, 173–74, 187
    unionized, 128, 172–73
Latin America, 159–60, 189, 307
LED lighting, 242
leftovers, 218, 221, 247–49, 258
Lentz, Jim, 142–43, 146, 158
lettuce, 5, 125–31, 241–42, 268, 301, 303, 312
LG Electronics, 218–19
Liebig, Justus von, 69–71
Lincoln, Ken, 40–41
Lineage Logistics, 205
Liu Peijun, 208–11
livestock, 6, 91, 271
    environmental impact of, 107, 174, 310
    in Europe/UK, 80–82
    fed "purified diets," 113
    raised in cities, 67–68
    refrigeration's impact on, 79–82, 172–73
    shipping of, 69–76, 87, 182
    slaughtering of, 79–82, 98–101, 105, 172, 244
    in the US, 69–73, 79–82
    See also meat
London, UK
    cold storage in, 87, 118
    food supply of, 49, 67–68
    ice shipped to, 52, 53
    ice wells in, 48
    meat shipped to, 70, 76, 78–80
    polluted rivers in, 300
    skeleton study and, 260–64, 266
    See also Great Exhibition of 1851 (London)
Longjing Caotang, 212
Los Angeles, CA, 198, 200, 217, 220, 231, 300
Lovecraft, H. P., 86–87
Low Temperature Research Station (LTRS), 114–15, 123, 131–32, 135
Luciano, Juan, 137, 141–44
Luckhardt, A. B., 133
Lugg, Jim, 125, 129–31, 301
Lynd, Robert and Helen, 225, 267–68

Mackenzie, John, 80
Madden, Martin B., 84
Maersk, 179, 183, 190
Malden, Mortimer, 20–21
malnutrition, 283, 285
Man in Full, A (Wolfe), 11
Marks & Spencer (UK), 237
MASS Design Group, 299–300
Massachusetts Institute of Technology, 95
Master Purveyors (Bronx), 96–98, 101–6, 173, 239
Maury, Sarah Mytton, 53
Maytag, 229, 241
McGavin, George, 32–35
McLean, Malcom, 179–81
McPhee, John, 148–49
McWilliams, David, 81
meat, 298
    American diet and, 66–67, 170, 173
    America's supply of, 65–68, 72–73
    for the cities, 67–73, 76
    "cold shortening" of, 100–101, 132
    cooled mechanically, 75–79
    dry-aging of, 96, 102–6, 250
    electrical stimulation of, 5, 100–102, 105–6
    improved by cold, 95–106, 248–49
    industrialization of, 88, 101, 172–73

meat *(cont.)*
  lab-grown, 120
  main producers of, 82, 173
  preservation of, 36, 70–71, 74, 87–90
  production/consumption of, 79–82,
    106, 108, 310
  public suspicion of, 87, 94
  refrigeration of, 99–101, 113, 198–99, 243
  regulating temperature/humidity for, 239
  shipped overseas, 75–82, 87, 174, 200, 266
  shipping of, 23, 140, 181–82
  supply chains for, 65–69
  vacuum-packed, 22, 173
  "wet-aging" of, 105–6, 173, 250
  See also ripening: of meat
meat-packers, 65–66, 79, 82, 88, 90, 94, 100,
    105, 197–98 *See also* butchers/
    butchery; Chicago, IL: meat-packers in
mechanical refrigeration
  development/evolution of, 6, 57–58,
    196, 238
  experimental years of, 54–62, 75–79, 249
  fear/suspicion of, 249, 274
  first mobile units for, 166–67
  vast expansion of, 205
Meditations (Descartes), 28
Mellowes, Alfred, 222
Mexico, 146, 183, 185, 205
Michigan, 75, 85, 121, 171
microbes, 34–37, 46, 88, 91, 95
Midgley, Thomas Jr., 295
milk, 3, 5, 9, 35–36, 45–46, 67, 89, 91,
    236–39, 244, 258, 281–84. *See also*
    dairy; Silk soy milk
Millard Refrigerated Services, 200
Minute Maid, 148, 150–53, 155, 172, 255
Mission Produce (CA), 145–46
Missouri. *See* Springfield
    Underground (MO)
USS *Monitor*, 222
Monitor Top refrigerator, 222–25, 227
Montana, 22, 237
Moore, Thomas, 220–21
Mort, Thomas, 76, 78–79
Mosquito Coast, The (Theroux), 8–9
Mukamugema, Alice, 299
Mukandamage, Charlotte, 281

Muncie, IN, 225, 267–68
Murphy, Michael, 299
Museum of London, 261, 266
Mussolini, Benito, 308
Mwalimu, Innocent, 288, 291

Næverdal, Bente, 319–20
Napoleon, 10, 38
National Poultry, Butter and Egg
    Association, 83–84, 88–89
Native Americans, 80, 250, 268
Neljubow, Dimitry, 133
Nelson, Phil, 152
New Deal, 225
New Orleans, LA, 60, 140
New York Board of Health, 114
New York City, 176, 178, 200
  brewing beer in, 60
  cooling by ice in, 54, 61, 126, 196
  and distributing bananas, 136–37,
    141–44
  food-distribution site in, 96
  hot summers of, 60, 85–86
  livestock raised in, 68
  meatpacking businesses in, 96–98, 102–4
  population growth in, 67
  and shipping meat, 71–72, 78
  and shipping produce, 125–27, 175
New York State Consumer Protection
    Board, 237
New York Times, The, 56, 68, 78, 113, 125,
    179, 219, 251
New Yorker, The, 91, 94, 252–53
New Zealand, 23, 69–70, 79–82, 99–100,
    175, 182, 186, 199
NewCold Wakefield, 202–5
Nkurunziza, Issa, 287
Nohi, Vered, 199
North Carolina State University, 18–20
North York Moors, 80–81
Numero, Joseph, 165–67, 170
nutrients
  depletion of, 269–70
  genetic engineering and, 252–53
  improved by canning, 39
  in meat/poultry, 33–35, 69, 71, 82
  in produce, 33–35, 138, 252–53, 269–70

and refrigeration, 82, 242, 269–70
and spray preservative, 305–6
nutrition, 288
chemists' knowledge of, 112–14
and climate change, 307
and early diets, 268–70
meat preservation and, 71
from perishables, 114, 265–66
from protein, 112–13
and refrigerated food, 40, 260, 269–70, 274–75
and vitamin-rich foods, 112–14
Nyce, Benjamin, 116

Odin, 321
Okazaki, Akira, 176–78
O'Neill, Eugene, 221
orange juice, 199
frozen concentrate, 146–56, 158, 172
industrialization of, 255
not-from-concentrate, 151–54
trading/speculating on, 154–56
Oscar Mayer, 191
OX trucks, 290–92
ozone layer, 295–97

Pabst, 61
Pabst, Fred, 244–45
Paleolithic ancestors, 40, 44, 46.
See also hunter-gatherers
pallets
and forklifts system, 196–97
freezing water tubes in, 292, 298
Panama Canal, 187, 199–200
pantry, 224, 238
Pasteur, Louis, 35
patents, 70, 74, 77, 168, 171, 207, 220–23, 228, 243, 297
Patra, Debasmita, 237
pawpaw, 158
Pennington, Polly, 90–95, 166, 187, 235, 311
Pennsylvania State University, 297
Pepsi, 151
perishability, hierarchy of, 111–12
Peru, 53, 178, 186, 188–90
pests, 131, 159, 307, 316–18
Peters, Toby, 283, 285, 293–94, 297–98, 311

Philadelphia, PA, 67–68, 90–91, 138, 200
Phyllis Wheatley Auxiliary, 168
pickling, 36, 171, 213, 273
pipeline refrigeration, 196
plant breeders, 122–23, 157, 253–55, 269, 307, 316–17
Pliny the Younger, 47
Poison Squad, 90, 270
political destabilization, 159–60
pollution, 61, 82, 189, 204, 263, 266, 297, 300
population growth, 67, 69, 205, 264, 274, 294, 299, 303
Port of Wilmington, DE, 146–47, 149–53, 199
Porta, Giambattista della, 47
poultry, 54, 84, 92–95, 106–7, 166, 169, 209, 269
Pound, Ezra, 47
Pratt, Barbara, 178–79, 183–91, 199
prerefrigeration-era food, 35–40, 225, 238–40, 249–50, 267–70
Preservaline, 88
preservation, of food, 54, 120, 307
cabinets for, 196
chemicals for, 88–90, 94
the future of, 316–21
plastic bags for, 5, 130–31, 301–2
in Rwanda, 286–87, 298
spray coating for, 301–8
traditional, 115–16, 239–41, 250, 273
and UK's Great Exhibition, 70–71
various methods of, 35–40, 238–41, 270, 301, 303, 311
See also drying food; pickling; salting food; sugaring food
Preston, Andrew, 140
Priest, Dave, 202–4
produce
atmospheric engineering of, 136, 146
available year round, 187–88, 268
cold storage of, 111–21, 136
decay of, 35, 114, 138–39, 157, 239, 284, 301–4, 307
as essential for diet, 113–14
exporting of, 298–99
precooling of, 283–84, 292, 298

produce *(cont.)*
  prerefrigeration and, 238–40
  preservation of, 111–21, 301–8
  in refrigerated warehouses, 198–99
  refrigeration's impact on, 249, 269–70
  shipping of, 139–41, 168–69, 181–83, 186, 307
  spray coating for, 301–8
  US consumption of, 188–89
Project Drawdown, 295
protein, 70–71, 94–95, 112–13, 265, 268, 274. *See also* meat; poultry
public health, 88–89, 260, 263, 265, 270, 310

Qianlong, Emperor, 220
Quartermaster Corps, 196–97
Quartermaster Subsistence Research and Development Laboratory, 147–48
Quechua people, 71

Ragnarök, 321
railways, 139, 200
  icing stations for, 75–76, 93, 125, 169
  invest in ice cooling, 169
  refrigerated railcars of, 170, 269, 310
  shipping livestock on, 71–75, 80
  See also iced railcars
Ralphs, 145–46, 200, 230–31, 237
Rathje, William, 233–34
Raymond, Joseph, 80–82, 87
reefers (refrigerated containers), 178–79, 182–91
Rees, Jonathan, 267
refrigerants, 29–31, 58, 60–61, 222, 294–96
refrigerated warehouses, 221
  annual rhythms in, 21–23
  and atmospheric engineering, 118
  automated, 202–5
  cold conditions in, 3–4, 9–17
  cooling machines used in, 29
  design of, 195–98, 204
  distribution-focused, 200–202
  earliest ones, 54, 62, 112, 195–98
  food products in, 3, 9–10, 12–14, 20–23
  and forklifts, 4, 8–9, 11–15, 96
  increasing number of, 6, 205

  location of, 195–202, 205
  underground, 191–95
  working in, 3–6, 9–23, 96
  See also Americold
refrigerators, 5–7, 254
  convenience of, 247–48
  design of, 8, 227–30, 234–35, 241
  downsides to, 238
  earliest ones, 55–62, 220–28, 234, 295–97
  electric, 221–25
  fear/suspicion of, 84–89, 249, 310
  of the future, 292, 294, 296–98
  how to build one, 27–32, 295
  judging people by, 218–20
  rearranges homes/cities, 220, 224
  science behind, 248–49
  size of, 6, 227, 232, 234, 241
  "smart," 234–35, 241–43
  See also specific companies
RefrigiWear, 17–21, 320
Reinke, Justin, 228, 232, 235, 241–42
ripening
  of meat, 98, 102–6, 173
  of produce, 115–17, 122, 124, 137–46, 158–59, 190, 252–55
Rogers, James, 301–8
Romans, 47–48, 68, 245
Roosevelt, Franklin, 225
Rosenblatt, Paul, 136–37, 141–44
"rot box" experiment, 32–35
Rowe, John Paul, 260–61
Royal Society (UK), 5–6, 8, 214
Royal Society of Arts (Society for the Encouragement of Arts, Manufactures and Commerce; UK), 69–71, 100
Royston Cave (UK), 43–44
Rubbish! (Rathje), 233–34
Rurban Revolution, 312
Rwanda, 279–93, 296–301, 310–11
Rwanda National Agricultural Export Development Board, 288, 299
Ryou, Jihyun, 238–41, 256

S. Liebmann's Sons (Brooklyn), 61
Sachse, Julius F., 68
SafePod, 120–21

"safety net" syndrome, 232–33
salad, 113, 125, 129–31, 255, 268
Salinas Valley (CA), 124–29, 131, 148, 303
salmonella, 241, 271
salting food, 36–38, 46–47, 67, 213, 270, 273
Samsung, 29, 228
Sanquan (China), 207–10
Schaefer, Jim, 121
Schwedes, Kyle, 4
Scott, Wentworth Lascelles, 69–71, 106
Sea-Land Service, Inc., 180, 184–87
seasonal foods, 21–22, 38–40, 66–67, 113,
    244, 255–56, 312
seasonality, 8, 40, 146, 236
seasonless foods, 22, 139, 152, 156, 172
second refrigerators, 231–32
seeds, 82, 115, 117, 135, 137, 225, 305
    seed banks, 316–20
Shanghai, China, 146, 176, 208–9, 211–2,
    271, 294
Shih-ching, 47
shipping containers, 5, 179–87, 199, 316.
    See also reefers
ships, refrigerated
    invention/evolution of, 75–79, 221, 268
    meat shipped in, 75–82, 174, 266
    produce shipped in, 140, 174, 181–83
Siebel, J. E., 59
Silberman, Ryan, 20–21
Silk soy milk, 3, 235–36
Simply Orange, 153
Sinclair, Upton, 65, 88
skeleton study, 260–63, 266
SmartFresh, 136
Smith, Adam, 128
Smithfield Foods, Inc., 21
Smithfield Market (London), 68, 80, 260
snow, 6, 40, 44–48, 107, 113, 126, 245, 250,
    316, 319, 321
"snowballing effect," 172
Soane, John, 47
social class, 49, 80, 138, 173, 221, 231–34,
    257–59, 262
"social storage," 303
solar energy/power, 77, 282, 291–92,
    294, 302
Solasz, Mark, 96, 98, 101–2, 105

Solasz, Max, 102–4
Solasz, Sam, 103
Sonnenburg, Justin, 272–74
South America, 69, 77–78, 87, 191.
    See also specific countries
South Bristol, ME, 40–43
Spiller, Henry, 134–35
Springfield Underground (MO), 191–95,
    198, 202
SS Frigorifique, 77
Stassopoulos, Tassos, 256–59
status symbol, 176, 233, 247–48
steam engines, 59–61
    steamships, 71–73, 139–40, 179–80
    See also specific ships
Steinbeck, John, 124, 126–27
Steinberg, Jon, 217–20, 233, 257
Stone Age, 46
submarines, 114, 119–20, 223
SubTropolis (Kansas City), 194–95
Sudan, Archer C., 133–34
sugaring food, 36, 38–39, 46, 67
supermarkets
    in China, 210, 212, 294
    cold room temperature and, 254
    and cooling technology, 297–98
    distribution centers and, 200–202
    in Europe/UK, 226–27, 311
    facilitated by refrigeration, 224
    food date labels of, 236–38
    industrialized produce of, 272
    paid for display, 235–36
    refrigerated tomatoes in, 251–52
    seasonless abundance of, 139, 187–88
    and spray preservative, 306
    stockpiling food and, 232, 235
    and Thermo King units, 169–71, 197
    See also specific names
supply chains, 195, 271
    and eggs, 241
    of food, 35, 89, 158, 171, 175, 200–202, 303
    global, 178–81
    and produce, 305–7
    refrigerated, 251, 255
sushi, 36, 176–78, 190, 211, 259
sustainability, 158, 293, 297–98, 311
Svalbard archipelago, 315–20

Svalbard Global Seed Vault, 316–20
Swanson Frozen Foods, 172
Swift, 82
    refrigerator car, 73–75, 264, 310
Swift, Gustavus, 65–67, 72–76, 79, 87, 90, 264
Swift, Louis, 65–66, 72–75, 79, 87

Tanatar, Selçuk, 289, 291
tankers, 149–50
taste, 111, 239, 251
    of cold-stored food, 243–49
    prerefrigeration, 36–37, 249–50
    receptors of, 246–47
    refrigeration's effects on, 5, 7, 14, 160, 247–49, 254–55
    and shipping method, 172
    and spray preservative, 306
Taylor, Ted, 129–30
technology
    for early refrigerators, 55, 58, 61, 221–22, 228, 309
    of future refrigerators, 291–92, 294, 296–98
    latest advances in, 242, 286
    public trust in, 95
Tellier, Charles, 54, 76–78, 87
Temple of Apollo (Delphi, Greece), 134–36
Thermal Protection Lab (North Carolina), 18–20
Thermal Technologies, 142–43, 146
Thermo King, 167–72, 174, 182, 197
thermodynamics, 27–31, 54
Theroux, Paul, 8–9
Thetcher, Thomas, 245–46
Thompson, Asa, 40, 53
Thompson Ice House (ME), 40–43, 51, 53, 62
tomatoes, 7, 39, 135, 145, 148, 152, 189, 239, 243–44, 248, 251–56, 268–69, 289, 305–6, 316
transportation, 181, 269, 308–9. See also specific types
Tropicana, 151–53, 255
trucks, refrigerated
    in China, 211, 284
    cold chain and, 201
    cooled by ice, 166, 168–69, 279–81, 284
    cooled mechanically, 5, 105, 166–70
    environmental impact of, 189, 295
    and food-production system, 170–71
    in Rwanda, 279–81
    shipping frozen foods in, 171–72
    shipping meat in, 170–73
    solar-powered, 291–92
    vast implications of, 174, 176
    See also OX trucks
True, Rodney, 133
Truman's brewery (London), 59
Trump, Donald, 176, 219
Tudor, Frederic, 49–55, 58, 60, 66, 73
Tudor, John Henry, 50
tuna, 10, 176–78, 181, 190, 211
TV dinners, 8, 22, 149, 172
Twitter, 219
Tyson, 173, 209

Umugenga, Jean de Dieu, 279–82
Umutoni, Louise, 290–91
UN Environment Programme, 285–87
UN Food and Agriculture Organization, 293, 317
Union Stock Yards (Chicago), 65–66, 79, 82, 88–89, 172–73
United Fruit Company, 140, 159–60, 181, 190
United Kingdom
    ACES program and, 285–86
    diet/health in, 268, 270
    "meat famine" in, 69–70, 78
    meat shipped to, 73, 76, 80–82, 266
    refrigeration in, 5–6, 8, 59
    See also Great Britain; London, England
University of Arizona, 233–34
University of California, 145, 230–32, 257, 302
University of Chicago, 133
University of Pennsylvania, 91
urbanization, 37–38, 60, 61, 67, 205, 209, 263, 264, 266, 268, 274, 300, 303
Uruguay, 77, 81
US Congress, 83–85, 89, 153

US Foods, 200
US International Trade Commission, 153
US military, 147–48, 167, 174–75,
    196–97, 297
US Thermo Control Company, 167
USDA, 151, 188, 225, 269, 284
Uwamahoro, Francine, 290

vaccines, 240–41, 263, 266, 298–99
vacuum
    cooling, 127–28
    packaging, 22, 105–6, 173
    technology, 29, 58, 116, 148, 226
Vacuum Cooling Corporation,
    127–28
Vacuum Foods Corporation, 148
vapor-compression refrigeration, 29–31,
    59, 297
vegetables
    ethylene and, 132–36
    harvesting of, 283–84
    Pratt's studies of, 186
    preservation of, 128, 156–57, 213
    refrigeration of, 242–43, 251
    See also iceberg lettuce; lettuce;
        produce; tomatoes
Velvet Underground, 222–23
Venus (fruit ship), 140
Verhoff, Clyde, 294
Vernon, CA, 21, 200
Veterans Administration, 170, 246
Veysey, Arthur, 175
Victoria, Queen, 76
vitamins, 39, 111–13, 126, 148–51, 242,
    268–70, 274

Walmart, 146, 195, 201, 210, 227,
    235, 250
Washington, DC, 83–84, 133, 167, 170
Washington, George, 48

Washington State, 118–19, 122–24, 130,
    189, 209, 239, 318
water availability, 189–90, 293, 296
Waters, Alice, 255–56
Wayside Inns on the Lancaster Roadside,
    The (Sachse), 68
Weightman, Gavin, 52
Werner, Harry, 166–67, 169
West, Constance Lane, 228
West, Cyril, 115–18, 120–23, 130, 132, 135,
    156, 158
Western, Gaynor, 261–64, 266
Whirlpool, 228–29, 242
    Tectrol technology, 119, 129–30
WhiteWave, 235–36
Whole Foods, 201–2
Wiley, Harvey Washington, 89–94, 270
Williams, Llewellyn, 140
Wolfe, Tom, 11
women, 40, 91–92, 221, 225–26, 247, 253,
    259, 262, 291, 300, 309, 316
Wondrich, David, 245
World Bank, 287–89
World Food Programme, 285
World Health Organization, 282
World War I, 114–15, 125, 164, 221
World War II, 103, 126, 147–48, 167–70, 172,
    174–76, 191–92, 196–97, 224–26, 261
Wulfraat, Marc, 200–202, 204
Wyeth, Nathaniel, 51–52
Wynder, Ernst L., 270
Wynter, Andrew, 67

Yankee of the Yards, The (Swift, Louis),
    65–66, 72–75
Yum China, 259
Yuma, AZ, 131

Zhengzhou, China, 206–8
Zhong, Gu, 296–97